Baking, Kneading and Mixing in a Trough with a Pestle Turned by an Ass.—From the Tomb of Eurysaces.

Grinding on Raised Saddle Stones, Pounding and Baking in Molds.—From an Egyptian Tomb of the Twelfth Dynasty.

Buying Grain, Grinding, Sifting, and Selling Flour.—From a Frieze on the Tomb of Eurysaces.

Reproduced by permission of the Northwestern Miller, Minneapolis, Minn.

THE CHEMISTRY OF WHEAT FLOUR

BY

C. H. BAILEY, Ph.D.

PROFESSOR OF AGRICULTURAL BIOCHEMISTRY
UNIVERSITY OF MINNESOTA

American Chemical Society
Monograph Series

BOOK DEPARTMENT
The CHEMICAL CATALOG COMPANY, *Inc.*
19 EAST 24TH STREET, NEW YORK, U. S. A.
1925

Copyright, 1925, by
The CHEMICAL CATALOG COMPANY, *Inc.*

All rights reserved

Printed in the United States of America by
J. J. LITTLE AND IVES COMPANY, NEW YORK

GENERAL INTRODUCTION

American Chemical Society Series of Scientific and Technologic Monographs

By arrangement with the Interallied Conference of Pure and Applied Chemistry, which met in London and Brussels in July, 1919, the American Chemical Society was to undertake the production and publication of Scientific and Technologic Monographs on chemical subjects. At the same time it was agreed that the National Research Council, in coöperation with the American Chemical Society and the American Physical Society, should undertake the production and publication of Critical Tables of Chemical and Physical Constants. The American Chemical Society and the National Research Council mutually agreed to care for these two fields of chemical development. The American Chemical Society named as Trustees, to make the necessary arrangements for the publication of the monographs, Charles L. Parsons, Secretary of the American Chemical Society, Washington, D. C.; John E. Teeple, Treasurer of the American Chemical Society, New York City; and Professor Gellert Alleman of Swarthmore College. The Trustees have arranged for the publication of the American Chemical Society series of (a) Scientific and (b) Technologic Monographs by the Chemical Catalog Company of New York City.

The Council, acting through the Committee on National Policy of the American Chemical Society, appointed the editors, named at the close of this introduction, to have charge of securing authors, and of considering critically the manuscripts prepared. The editors of each series will endeavor to select topics which are of current interest and authors who are recognized as authorities in their respective fields. The list of monographs thus far secured appears in the publisher's own announcement elsewhere in this volume.

GENERAL INTRODUCTION

The development of knowledge in all branches of science, and especially in chemistry, has been so rapid during the last fifty years and the fields covered by this development have been so varied that it is difficult for any individual to keep in touch with the progress in branches of science outside his own specialty. In spite of the facilities for the examination of the literature given by Chemical Abstracts and such compendia as Beilstein's Handbuch der Organischen Chemie, Richter's Lexikon, Ostwald's Lehrbuch der Allgemeinen Chemie, Abegg's and Gmelin-Kraut's Handbuch der Anorganischen Chemie and the English and French Dictionaries of Chemistry, it often takes a great deal of time to coördinate the knowledge available upon a single topic. Consequently when men who have spent years in the study of important subjects are willing to coördinate their knowledge and present it in concise, readable form, they perform a service of the highest value to their fellow chemists.

It was with a clear recognition of the usefulness of reviews of this character that a Committee of the American Chemical Society recommended the publication of the two series of monographs under the auspices of the Society.

Two rather distinct purposes are to be served by these monographs. The first purpose, whose fulfilment will probably render to chemists in general the most important service, is to present the knowledge available upon the chosen topic in a readable form, intelligible to those whose activities may be along a wholly different line. Many chemists fail to realize how closely their investigations may be connected with other work which on the surface appears far afield from their own. These monographs will enable such men to form closer contact with the work of chemists in other lines of research. The second purpose is to promote research in the branch of science covered by the monograph, by furnishing a well digested survey of the progress already made in that field and by pointing out directions in which investigation needs to be extended. To facilitate the attainment of this purpose, it is intended to include extended references to the literature, which will enable anyone interested to follow up the subject in more detail. If the literature is so voluminous that a complete bibliography is impracticable, a critical selection will be made of those papers which are most important.

GENERAL INTRODUCTION

The publication of these books marks a distinct departure in the policy of the American Chemical Society inasmuch as it is a serious attempt to found an American chemical literature without primary regard to commercial considerations. The success of the venture will depend in large part upon the measure of coöperation which can be secured in the preparation of books dealing adequately with topics of general interest; it is earnestly hoped, therefore, that every member of the various organizations in the chemical and allied industries will recognize the importance of the enterprise and take sufficient interest to justify it.

AMERICAN CHEMICAL SOCIETY

BOARD OF EDITORS

Scientific Series:—
  WILLIAM A. NOYES, *Editor*,
  GILBERT N. LEWIS,
  LAFAYETTE B. MENDEL,
  ARTHUR A. NOYES,
  JULIUS STIEGLITZ.

Technologic Series:—
  HARRISON E. HOWE, *Editor*,
  WALTER A. SCHMIDT,
  F. A. LIDBURY,
  ARTHUR D. LITTLE,
  FRED C. ZEISBERG,
  JOHN JOHNSTON,
  R. E. WILSON,
  E. R. WEIDLEIN,
  C. E. K. MEES,
  F. W. WILLARD.

American Chemical Society
MONOGRAPH SERIES
'PUBLISHED

The Chemistry of Enzyme Actions (Revised Edition).
By K. George Falk. Price $3.50.
The Chemical Effects of Alpha Particles and Electrons.
By Samuel C. Lind. 180 pages. Price $3.00.
Organic Compounds of Mercury.
By Frank C. Whitmore. 397 pages. Price $4.50.
Industrial Hydrogen.
By Hugh S. Taylor. Price $3.50.
Zirconium and Its Compounds.
By F. P. Venable. Price $2.50.
The Vitamins.
By H. C. Sherman and S. L. Smith. 273 pages. Price $4.00.
The Properties of Electrically Conducting Systems.
By Charles A. Kraus. Price $4.50.
The Origin of Spectra.
By Paul D. Foote and F. L. Mohler. Price $4.50.
Carotinoids and Related Pigments.
By Leroy S. Palmer. Price $4.50.
The Analysis of Rubber.
By John B. Tuttle. Price $2.50.
Glue and Gelatin.
By Jerome Alexander. Price $3.00.
The Chemistry of Leather Manufacture.
By John A. Wilson. Price $5.00.
Wood Distillation.
By L. F. Hawley. Price $3.00.
Valence, and the Structure of Atoms and Molecules.
By Gilbert N. Lewis. Price $3.00.
Organic Arsenical Compounds.
By George W. Raiziss and Jos. L. Gavron. Price $7.00.
Colloid Chemistry.
By The Svedberg. Price $3.00.
Solubility.
By Joel H. Hildebrand. Price $3.00.
Coal Carbonization.
By Horace C. Porter. Price $6.00.
The Structure of Crystals.
By Ralph W. G. Wyckoff. Price $6.00.
The Recovery of Gasoline from Natural Gas.
By George A. Burrell. Price $7.00.
The Chemical Aspects of Immunity.
By H. Gideon Wells. Price $4.00.
Molybdenum, Cerium and Related Alloy Steels.
By H. W. Gillett and E. L. Mack. Price $4.00.
The Animal as a Converter.
By H. P. Armsby and C. Robert Moulton. Price $3.00.
Organic Derivatives of Antimony.
By Walter G. Christiansen. Price $3.00.
Shale Oil.
By Ralph H. McKee. Price $4.50.

American Chemical Society
MONOGRAPH SERIES
IN PREPARATION

Thyroxin.
 By E. C. Kendall.
The Properties of Silica and Silicates.
 By Robert B. Sosman.
The Corrosion of Alloys.
 By C. G. Fink and R. J. McKay.
Piezo-Chemistry.
 By L. H. Adams.
Cyanamide.
 By Joseph M. Braham.
Liquid Ammonia as a Solvent.
 By E. C. Franklin.
Aluminothermic Reduction of Metals.
 By B. D. Saklatwalla.
Absorptive Carbon.
 By N. K. Chaney.
Refining of Petroleum.
 By George A. Burrell, *et al*.
Chemistry of Cellulose.
 By Harold Hibbert.
The Properties of Metallic Substances.
 By Charles A. Kraus.
Photosynthesis.
 By H. A. Spoehr.
Physical and Chemical Properties of Glass.
 By Geo. W. Morey.
The Chemistry of the Treatment of Water and Sewage.
 By A. M. Buswell.
The Rare Gases of the Atmosphere.
 By Richard B. Moore.
The Manufacture of Sulfuric Acid.
 By Andrew M. Fairlie.
Equilibrium in Aqueous Solutions of Soluble Salts.
 By Walter C. Blasdale.
The Biochemistry and the Biological Rôle of the Amino Acids.
 By H. H. Mitchell and T. S. Hamilton.
Protective Metallic Coatings.
 By Henry S. Rawdon.
Soluble Silicates in Industry.
 By James G. Vail.
The Industrial Development of Searles Lake Brines with Equilibrium Data.
 By John E. Teeple, *et al*.
The Chemistry of Wood.
 By L. F. Hawley and Louis E. Wise.
Diatomaceous Earth.
 By Robert Calvert.
Aromatic Coal Products.
 By Alexander Lowy.
Catalysis in Homogeneous Organic Systems.
 By F. O. Rice.
Nucleic Acids.
 By P. A. Levene.
Fixed Nitrogen.
 By Harry A. Curtis.
The Chemistry of Nitrocellulose Lacquers.
 By Bruce K. Brown.

INTRODUCTION

The data of cereal chemistry are widely scattered. They appear in thousands of published papers, not only in bulletins of agricultural experiment stations and government bureaus, but in numerous American and foreign scientific journals. During the several years that the author has been engaged in research in this field of specialization, an effort has been made to collect and organize these data. This monograph represents an attempt to present them in a condensed and systematic arrangement for the convenience of other workers in this field. Incidentally, an opportunity is thus afforded to collocate the results of the extensive researches along this line conducted at the Minnesota Agricultural Experiment Station during the past thirty years.

The general treatment accorded the subject is essentially similar to that which would be involved in discussing the properties of any like food material. Properties of flour must be considered in their relation to: (first) the raw material from which it is manufactured, or wheat; (second) the process of manufacture, or milling; (third) its adaptability to the principal use to which the flour is put, or baking.

It was not deemed appropriate to discuss the biochemistry of baking at any length in this volume. Such a discussion would inevitably become involved and extensive since a consideration of yeast nutrition, and the progressive changes which occur in a fermenting bread dough would be necessitated. The baking properties of flour are recognized as the most important criteria of its usefulness, and are referred to in the last two chapters of this monograph. It will be observed that an effort has been made to restrict this discussion of baking to the suggested correlations between composition, or physico-chemical properties of flour and baking qualities. A monograph devoted to the biochemistry of baking would constitute a valuable addition to our literature.

The scope of the monograph was accordingly limited to the three major considerations to which reference has been made. A further limitation was imposed as involved in the definition of the term "flour." For the purpose of this monograph, the definition included in the Federal standards of purity for food products is accepted in so far as the non-chemical portion of the definition is concerned. This definition states in part, "Flour is the fine, clean, sound product made by bolting wheat meal, etc." Such products as graham flour, gluten flour, self-rising flour, and mixed flours, made in whole or in part from wheat are excluded from consideration. Flours mixed with "flour improvers" are likewise excluded.

Critics of this volume may feel inclined to suggest that an undue amount of space has been used in developing the chemistry of the life cycle of the wheat plant, and the influence of environment on the composition of wheat. It was deemed advisable to consider these phases of the subject at some length because of the fact that roller milling is essentially a mechanical process and the flour miller can not incorporate anything in the flour (aside from bleaching agents and flour improvers) which is not found in the wheat. Hence, the influence of any factor affecting the properties of wheat becomes of immediate interest to the miller and baker since these influences may be reflected in the properties of flour milled from the wheat.

The author appreciates this opportunity of acknowledging the helpful criticisms offered by Dr. F. L. Dunlap, and Professors L. S. Palmer, H. K. Hayes and R. A. Gortner in the preparation of certain sections of the book. Particular mention should be made of the valuable aid given by Sophie A. Wilkins in checking portions of the bibliography.

St. Paul, Minnesota
  July 21, 1925.

Table of Contents

			PAGE
CHAPTER	1.	HISTORICAL	13
CHAPTER	2.	WHEAT IN ITS RELATION TO FLOUR COMPOSITION	16
CHAPTER	3.	THE GROWTH AND DEVELOPMENT OF THE WHEAT PLANT AND KERNEL	37
CHAPTER	4.	INFLUENCE OF ENVIRONMENT ON THE COMPOSITION OF WHEAT	56
CHAPTER	5.	DEFECTS OF, AND IMPURITIES IN COMMERCIAL WHEAT	86
CHAPTER	6.	STORAGE AND HANDLING OF WHEAT	108
CHAPTER	7.	CHEMISTRY OF ROLLER MILLING	121
CHAPTER	8.	CHANGES IN FLOUR INCIDENTAL TO AGING . .	177
CHAPTER	9.	THE COLOR OF FLOUR AND FLOUR BLEACHING .	194
CHAPTER	10.	FLOUR STRENGTH AND ENZYME PHENOMENA . .	228
CHAPTER	11.	FLOUR STRENGTH AS DETERMINED BY THE PROTEINS OF FLOUR AND COLLOIDAL BEHAVIOR OF DOUGH	242
		APPENDIX	291
		BIBLIOGRAPHY	293
		AUTHOR INDEX	315
		SUBJECT INDEX	321

Chapter 1.

Historical.

The history of the development of flour milling marches with the history of the development of civilization. Each major period in the improvement of creature comforts witnessed progress in the art of milling. Primitive man contented himself with the simplest of milling contrivances. Possibly one of the earliest of mechanical devices originated by man was the saddle stone and crushing cylinder employed in household grinding. During the New Stone Age such devices had replaced the older practice of munching parched grain. According to Manchester (1922) predynastic cemeteries of old Egypt have been found to contain concave sandstone and granite slabs or saddle stones, and semi-cylindrical upper stones. The latter were rocked in the concave hollow of the saddle stones, crushing the kernels of grain placed between them. The invention of pottery at about this time contributed to the ease of storing grain, and the handling and preservation of meal.

In the next stage of development pounding appliances replaced the original crude upper stone, and somewhat later a rotating upper stone was devised. This materially reduced the strain on the backs of the human operatives, enabling them to work in an upright posture. As early as the third Egyptian dynasty sieves had come into use, doubtless facilitating the production of a partially refined product. About 2000 B.C. leaven came to be used in bread making in Egypt, and there is evidence in the drawings on the tombs that professional millers and bakers were recognized as among the craftsmen of the community.

While the Romans may have borrowed much of their art, they made important contributions to the mechanical equipment of flour mills. Although mortar and pestle type mills were used in the early Roman period, the circular mill or quern was produced at least several centuries before the Christian era. This quern consisted of a lower conical stone, and a hollowed-out cap stone. The latter was provided with a vertical opening through the center, through which the grain could be poured. The upper stone was often turned by an ass, although slaves and the inmates of penal institutions sometimes provided the motive power. The Romans bolted their meal, and produced several grades of flour, including a relatively white flour. Soon after the development of the quern, public bakeries made their appearance in Rome. These bakeries and the mills constituted an invaluable military asset, and not infrequently the campaigns of Rome's enemies involved an effort to destroy the flour mills and granaries of the city on the Tiber.

During the first century B.C. water wheels were invented, and harnessed to the mills of Rome. Certain of these were apparently located on the open aqueducts which brought water into the city. Stress of life in this rapidly moving period of human history, and the constantly growing demands of the military organizations turned the attention of Roman engineers away from the arts of peace. Wars succeeded wars, Rome declined, and Latin and Greek culture waned. A new culture originating in the Arabian tribes made its appearance along the shores of the Mediterranean lake, and the Arabs contributed their quota to the development of this basic industry. They devised the wind mill, probably about the middle of the seventh century A.D., and its use spread over not only the domain of the Mohammedan, but through all Christendom as well. The two forms of motive power, the water wheel and the wind mill, continued in use until the invention of the steam engine, and the water wheel, in a modified form, is still employed in many mills.

An interesting description of the practices of these periods is given by Ashton (1904). Querns were largely used in feudal Europe, and in England were the occasion of numerous minor rebellions, owing to the granting of exclusive milling rights to the feudal lords and the clergy. While the Romans used flattened mill stones, it was not until recent times that the conical quern wholly gave place to flat mill stones. Once adopted, the latter continued in use until near the close of the last century. Often these flat mill stones were built up of segments, bound with a heavy hoop. Such stones were grooved and roughened in a manner which facilitated the movement of the material in grinding, from the eye or center to the periphery of the stone. Skillful millers found that by elevating the upper stone slightly "high grinding" could be effected, and the kernel thus ground gradually and by several successive treatments, instead of pulverizing it immediately on the first grinding.

These changes in the type of grinding appliance were accompanied by corresponding changes in the scale of milling operations. The primitive saddle stone process was a domestic or household occupation; the quern and the water wheel made community or manorial grinding a possibility; the mill stone, and the automatic elevating and air-cooling devices, which made their appearance late in the eighteenth century, made possible the construction and operation of merchant mills which served large areas.

During at least the first half of the nineteenth century American millers gave the greatest impetus to the development of milling practice. Kozmin (1917) states that they called attention to the excellent qualities of stones from certain French quarries, and substituted cylindrical reels for the old European sifting bags. An American, Oliver Evans, published several editions of his book "The Young Millwright and Millers' Guide," which was translated into French and German. This inspired the erection of completely automatic mills, hundreds of which were built in America and Europe.

In 1870 the purifier was introduced in Minneapolis, Minn. This facilitated the classification of the middlings particles, and their treatment with an air current effected the "purification" implied in the name of the machine. The principles of its operation will be discussed in a later section. In combination with high grinding, and a gradual reduction process, this machine, in the mills of Minneapolis, made that milling center famous. Its flours came to command premiums in domestic and foreign markets, and several large fortunes were founded with the profits of the period.

Edgar (1912) states that the inventor of the roller mill was Helfenberger, who constructed and experimented with the first roller mill at Rohrschach, Switzerland. It remained for Sulzberger, of Frauenfeld, Switzerland, to develop a successful roller milling system. This was in 1832-3; in 1839 a mill at Budapesth was equipped with Helfenberger rolls. In the early 80's of the last century American mills adopted the roller system. Apparently the first American rolls were cast in Connecticut about 1874, but it was several years later before they wholly replaced the old mill stones.

The next major development was the introduction of chemical bleaching of the finished flour, a practice which came into vogue in the first decade of the present century. The discussion of the roller milling process, and of flour bleaching, as these are now practiced constitute contemporary history, and will accordingly be reserved for a later section of this book. In tracing briefly this development of milling practise down through the centuries, it will be evident to the reader that the inventive miller has directed his efforts toward two major accomplishments, these being, first, the substitution of machines for the human operative, and rendering the process automatic, and, second, the production of a highly refined white product or flour. In both of these particulars remarkable success has attended his efforts until today the modern flour mill is one of the most completely automatic establishments of the food industries, and the whiteness of the flour which can be produced from the pigmented wheats with dark colored, vitreous kernels is remarkable.

Chapter 2.

Wheat in Its Relation to Flour Composition.

The Characteristics of Wheat Classes and Varieties.

The characteristics of any particular parcel of wheat, and of flour which can be milled therefrom, are determined not alone by the environment of the plants actually producing the grain in question, but also by the inheritance of those plants from their progenitors. Through uncounted years, moulded by nature, and consciously or unconsciously manipulated by the hand of man, our present varieties have been developed. These varieties present substantial differences in structure, and biochemical characteristics of the grain which they produce.

Hackel presented a classification of modern wheat varieties late in the last century, which classification was referred to in most treatises on this subject until about a decade ago. Hackel's grouping was based largely upon certain visible plant characters. More recently the genetic studies of Tschermak (1914) and others have resulted in a different view of the possible relations of our wheat varieties. Tschermak made species crosses, and placed the wheat species in three major groups as follows:

Group	Stem species	Cultivated covered	Cultivated naked
Einkorn	Triticum ægilopoides	T. monococcum	Unknown
Emmer	T. dicoccoides	T. dicoccum	T. durum T. turgidum T. polonicum
Spelt	T. spelta	T. spelta	T. vulgare T. compactum

Crosses between the naked wheats proved wholly fertile; likewise crosses between naked forms of the emmer group and the covered forms of the spelt group were fertile. Partial fertility to complete sterility resulted from other crosses that were attempted. The size of pollen grains, relative susceptibility to Puccinia triticina, and serological tests made by Zade (1912) tended to confirm Tschermak's grouping. Sakamura reported that the three groups differed in chromosome numbers, which was confirmed by Sax (1921). The haploid numbers found were:

7 chromosomes in T. monococcum
14   "   " T. dicoccum, T. durum, T. turgidum, T. polonicum
21   "   " T. vulgare, T. compactum.

These studies have not only been of value to geneticists and plant breeders, but attract the interest of chemists as well. They serve to indicate the possible explanation for certain substantial differences in the properties and practical usefulness of flours produced from wheats of the several groups. Thus it has been generally observed that durum, and polish (T. polonicum) wheat flours and semolinas possessed certain properties in common, and differed decidedly from like products milled from common or vulgare, and club (T. compactum) wheats. This may be due to their origin, derived, as they are, from different wild forms.

Since T. dicoccoides was produced synthetically by Love and Craig (1919) it has been argued that a loss of some hereditary condition would account for the production of cultivated forms. Thus by recombining the characters of certain cultivated varieties, all the characters of the wild form should be expected to be obtained occasionally in making a large number of crosses. The cereal chemist accordingly encounters qualities of practical value in cultivated varieties in consequence of the loss of certain characteristics of the primitive or wild parent stocks.

It is not surprising to find that in the experiments conducted by Le Clerc, Bailey and Wessling (1918), flour milled from T. dicoccum, T. spelta, and T. polonicum resembled that milled from T. vulgare more nearly than did flour milled from T. monococcum, and that none of them were equal to the vulgare (hard spring) wheat in baking qualities. Neither T. dicoccum nor T. spelta are given a high rating in terms of baking quality by most of the chemists who have tested them, and one commercial laboratory reported a loaf volume of 80 cubic inches in the case of bread baked from emmer flour, when good hard spring wheat flours baked into loaves displacing about 200 cubic inches when tested by the same method.

Cultivated wheat varieties of the United States have been studied and classified by Clark, Martin and Ball (1922). In their scheme of classification certain obvious or readily discernible plant characters have been employed, including presence or absence of awns, glabrous or pubescent glumes, glume color, kernel color, kernel length, kernel texture, and spring or winter habit. The species or subspecies are not grouped on a genetic basis, but solely on a basis of these characters. In addition to the key for use in determining the variety of an unclassified specimen, this bulletin contains much interesting data concerning the origin and distribution of the important varieties in the United States, to which the reader interested in this phase of the subject is referred.

Commercial wheats of the United States have been divided into five major classes in promulgating the Federal standards for wheat. These are (1) Hard red spring wheat; (2) Durum wheat; (3) Hard red winter wheat; (4) Soft red winter wheat; (5) White wheat. The hard red spring wheat of commerce is grown largely in the northern portion of the great plains area, in Minnesota, North Dakota, South Dakota, and Montana. Varieties of this class grown in quantity belong to the species T. vulgare. Durum wheat is grown in essentially the same region as the hard, red spring wheat, the area of greatest acreage being in the southeastern section of North Dakota, the northeastern section of South Dakota, and adjacent counties in Minnesota. As the name indicates, the varieties of this class belong to the species T. durum. They are not regarded as bread wheats, but are used extensively in the manufacture of macaroni and other edible pastes. Bread can be made from durum wheat, but not of the qualities and characteristics desired by the public.

Hard red winter wheats are grown in the southern portions of the great plains area, in adjacent regions to the eastward, and in scattering areas in the intermountain and Pacific slope regions. The chief varieties of this class, Turkey Red, Kharkof, Crimean, and Kanred, are very similar in their plant characteristics, and belong to the species T. vulgare. Soft red winter wheats are grown extensively in Missouri and the states to the eastward. Numerous varieties are to be found in this extensive area, chiefly of the species T. vulgare. In the Palouse and Walla Walla districts of the north Pacific region certain red club wheats are grown, which would be classified as of the species T. compactum, or hybrids of that species.

White wheat varieties are grown to a limited extent in the northeastern quarter of the United States, notably in Michigan and New York, in portions of the intermountain area, and in the Pacific districts. The varieties propagated in the eastern United States are chiefly of the species T. vulgare. In the intermountain area and Pacific districts limited quantities of white wheat are produced which would be classified as T. compactum. A still smaller quantity of poulard wheat, of the species T. turgidum, is produced in the same district, chiefly of the variety known as "Alaska."

The comparative milling and baking qualities and protein content of these five classes of American wheats are summarized in convenient form by Thomas (1917 b). These data will be presented at this point in Table 1, although the discussion of their significance will be reserved until later. The averages presented resulted from the tests of over 150 samples in each class except the white wheats. Owing to the limited amount of work done on the white wheats, only certain of the data are included.

From these data it appears that durum wheat, while averaging highest in test weight per bushel, did not mill as freely as the hard red winter wheat, which gave the highest average yield of flour. The flour

TABLE 1

Averages of the Results of Tests of the Five Commercial Classes of Wheat Recognized in the Federal Standards (Thomas, 1917 b)

	Hard Red Spring	Durum	Hard Red Winter	Soft Red Winter	White Wheat
Weight, per bushel, lbs........	60.5	62.8	62.1	61.4	
Yield of flour, per cent.......	70.2	70.3	72.0	69.7	
Loaf volume, cc...............	2421	2070	2219	1965	1907
Water absorption, per cent....	55.7	55.7	55.2	52.4	51.7
Color score	96.4	90.2	95.3	96.2	
Texture score	94.8	90.7	93.5	91.9	
Crude protein (N × 5.7)					
In wheat, per cent..........	12.9	14.3	12.1	10.6	
In flour, per cent...........	11.9		11.2	9.6	

produced from the durum wheat, although containing a relatively high percentage of crude protein, baked quite differently than did the hard wheat flours. The average loaf volume was smaller, and texture score lower than the average hard red spring wheat flour, even though the latter contained a smaller percentage of crude protein. In addition, the average color score of the durum wheat flour was the lowest re-

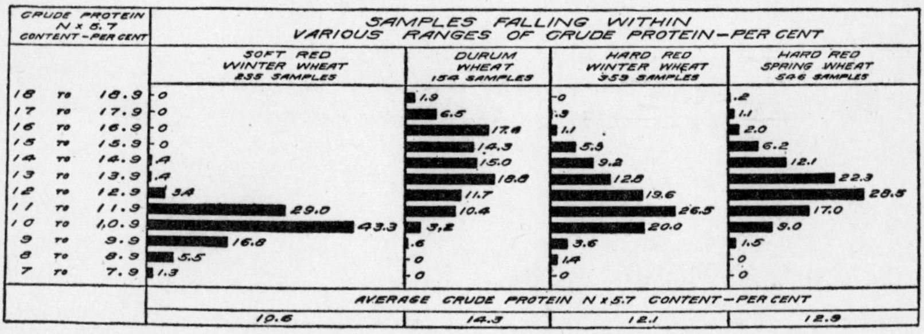

Fig. 1.—Diagram showing the variation in crude protein of four classes of wheat; crops of 1908 to 1913 inclusive. (Thomas, U. S. Dept. Agr. Bul. 557.)

ported, due to its yellow hue. Hard red spring wheat flours averaged highest in loaf volume and texture score, the commonly accepted criteria of baking strength. The color of the flour, rated on the visual appearance of the baked loaf, was also highest in case of the hard spring wheats. Hard red winter wheats rated second in average loaf volume and texture score, while the soft red winter wheat flour was lowest in loaf volume, and only slightly superior to the durum wheat flours in texture score. It should be again emphasized that durum wheat is not commonly used in the manufacture of flour for bread making, due to the peculiar properties of the flour made evident by the foregoing comparisons.

There are wide variations in protein content and baking strength among the individual lots of each class of wheat, however, and these are shown to advantage in the graphs presented in Thomas' paper. These graphs are reproduced here as Figures 1 and 2. Thus in the spring wheat group samples were analyzed which contained from less than 10 to more than 18 per cent. of crude protein. Corresponding variations in the percentage of this constituent were encountered in the other classes of wheat. One striking conclusion to be drawn from the data shown in Figure 1 is that there is a greater tendency toward uniformity in crude protein content in the soft red winter wheat class than

VOLUME OF LOAF—C.C.	SAMPLES FALLING WITHIN THE VARIOUS RANGES OF VOLUME OF LOAF — PER CENT				
	SOFT WHITE WHEAT 33 SAMPLES	SOFT RED WINTER WHEAT 244 SAMPLES	DURUM WHEAT 133 SAMPLES	HARD RED WINTER WHEAT 370 SAMPLES	HARD RED SPRING WHEAT 574 SAMPLES
3201 to 3300	0	0	0	0	.2
3101 to 3200	0	0	0	0	0
3001 to 3100	0	0	0	0	.3
2901 to 3000	0	0	0	0	.3
2801 to 2900	0	0	0	0	.5
2701 to 2800	0	0	0	1.1	3.0
2601 to 2700	0	0	.6	.8	6.9
2501 to 2600	0	0	1.3	2.4	18.3
2401 to 2500	0	.8	2.5	6.3	24.8
2301 to 2400	2.6	1.6	6.3	10.3	20.4
2201 to 2300	0	7.0	15.7	22.3	10.2
2101 to 2200	12.8	9.1	12.2	24.6	9.0
2001 to 2100	15.4	17.2	13.9	14.1	2.5
1901 to 2000	23.1	22.0	14.3	6.2	.7
1801 to 1900	12.8	18.9	10.1	2.7	.3
1701 to 1800	15.4	13.5	5.7	0	0
1601 to 1700	10.2	2.9	4.4	0	0
1501 to 1600	.5/	1.6	1.3	0	0
1401 to 1500	2.6	.8	1.6	0	0
	COMPARISON OF MAXIMUM, MINIMUM AND AVERAGE VOLUME OF LOAF—C.C.				
MAXIMUM	2320	2480	2550	2755	3260
MINIMUM	1435	1470	1450	1610	1675
AVERAGE	1907	1965	2070	2213	2421

Fig. 2.—Diagram showing the variations in loaf volume of test loaves baked from samples of the five classes of American wheat. (Thomas, U. S. Dept. Agr. Bul. 557.)

in the other three classes reported. Thus 72.3 per cent of the soft red winter wheat samples studied fell into two groups with a range of two per cent of crude protein, which is far from being the case in the hard wheat and durum wheat classes.

Similar variations in the loaf volume of the several samples in the group of flours milled from the five classes of wheat may be observed from the data expressed graphically in Figure 2. In the matter of both loaf volume and crude protein content it is evident that there is a decided overlapping in these properties when individual lots of hard spring and hard winter wheats are compared. Thus many samples of hard winter wheat flour exceed the average of the hard spring wheat flours in strength and protein content.

Shollenberger and Clark (1924) examined a large number of samples of the principal varieties of each class of wheat, and presented summaries of the data resulting from the tests and analyses which they made. The average of the tests of each market class are shown in Table 2. The several classes occupy about the same relations to each other as in the summary of Thomas' (1917 b) data. Hard red winter wheat averaged highest in yield of total flour, and ranked next to hard spring wheat in loaf volume. In the latter particular less difference was found in these studies than was observed by Thomas. Durum wheat had the highest weight per bushel and content of crude protein, but ranked next to the lowest in loaf volume, and lowest in color score of the loaf. A calculation of the averages of the data resulting from the tests of the common white, and the white club wheats has been made by the author, and are included in the tabulation. The common white

TABLE 2

Summary of Milling and Baking Data on the Classes of Wheat Grown During the Seven Years from 1915 to 1921, Inclusive, as Reported by Shollenberger and Clark (1924)

Class of Wheat	Wt. per Bushel, Dockage-Free, Lbs.	Crude Protein of Wheat, Per Cent	Yield of Straight Flour, Per Cent	Water Absorption, Per Cent	Baking Results		
					Volume of Loaf, cc.	Texture Score	Color Score
Hard red spring..	56.9	13.6	69.3	59.4	2,142	89.5	89.3
Durum	59.3	14.9	70.6	62.0	1,945	89.7	88.1
Hard red winter..	58.8	12.6	72.0	60.0	2,121	90.3	90.4
Soft red winter ..	58.6	11.3	71.1	55.9	2,001	88.9	89.1
White	58.5	12.0	70.7	56.8	1,872	87.2	90.2
Common white[1]		12.4	70.5	57.7	1,908	87.3	90.6
White club[1] ...		11.3	71.3	54.9	1,736	85.3	87.2

[1] Calculated by the author from Shollenberger and Clark's data, Table 60, p. 62.

wheats were, on the average, superior to the white club wheats in every particular except yield of total flour, when considered as bread-producing wheats.

Numerous investigations of the comparative qualities of commercial wheat varieties have been conducted in different parts of the world. Not all of these can be cited in this connection. Nor are all of the data submitted in terms of chemical composition. In many instances the most useful comparisons are those of bread-making quality, or, in case of the durum wheats, of macaroni-making properties. The differences observed in these properties must be due to variations which, while not always definitely correlated with the constituents commonly determined and reported by analysts, are nevertheless associated with physico-chemical characteristics, or biochemical phenomena involved in enzyme activity. Hence, in a biochemical treatment of this subject we should be privileged to utilize at this juncture the comparisons afforded by these baking tests, and endeavor to develop the biochemical basis for the differences observed.

Hays and Boss (1899) at the Minnesota Experiment Station, employed gluten tests and bakers' sponge tests in the selection and breeding of bread wheats. These early breeding efforts resulted in the selection and propagation of two excellent spring wheats, Minn. 163 (Glyndon Fife), and Minn. 169 (Haynes' Bluestem). The latter was subsequently grown extensively in the spring wheat area of North America.

In the closing years of the last century William Farrer was engaged in endeavoring to produce hybrid wheats in Australia which would possess to an increased degree that strength so desired in wheats of that country. These efforts are referred to by Guthrie (1899, 1900), Gurney and Norris (1901), and Guthrie and Norris (1907). Thus Guthrie (1900) states: "It appears not too much to expect that we shall soon be in the possession of a grain that is at the same time prolific and a good milling wheat, yielding readily a good quantity of flour of high strength and color and gluten content"; while Gurney and Norris (1901) comment that "The flours possessing the colour most appreciated by the public were, as a rule, those deficient in gluten. It is due to Mr. Farrer's very successful work that upon examining the tables it will be found that this fault no longer exists." As the result of an examination of samples collected in several Australian states Guthrie and Norris (1907) found that Farrer's varieties known as Bobs, Comeback, Jonathan, and Federation behaved in a gratifying manner in the several states, showing a power of retaining their high flour strength in districts where the tendency is toward a gradual deterioration in this important particular.

Fulcaster wheat was found to contain a higher percentage of protein than other varieties grown at the Kentucky and the Pennsylvania Experiment Stations in the studies reported by Gardner (1910). In baking qualities the Fulcaster was likewise superior.

Of the Ohio wheats, Gladden, Trumbull, Poole, Ohio 127, and Ohio 9920 milled into flour which was characterized by Corbould (1921) as having the best bread-baking quality; that from Velvet Chaff and Mediterranean was intermediate in this particular, while the flour milled from Red Wave and Dawson's Golden Chaff was of "medium" quality. Early Ripe and Mealy yielded flour inferior for all purposes. These varieties and several others tested by Corbould are soft winter wheats. Hard winter wheat varieties grown in Ohio, of the Turkey Red and Kharkof varieties, milled into strong baking flours.

The results of tests and analyses of Michigan wheats were presented by Spragg (1912). The 1911 crop winter wheats contained from 10.94 to 14.53 per cent of crude protein. In baking quality Turkey, Berkeley, and Shepherd's Perfection rated highest. Spragg and Clark (1916) later indicated that Red Rock, a selection from the variety known as

Plymouth Rock, has proven superior in baking quality (as well as yield per acre) to the other common varieties in Michigan.

The protein content and baking quality of the varieties of wheat grown at Guelph, Ontario (Canada), were found by Harcourt (1901) to vary appreciably. In order of comparative superiority Michigan headed the list of Guelph wheats, followed by Genesee Giant, Dawson's Golden Chaff, and Early Red Clawson in the order named. Turkey Red grown at Waterloo was superior to any of the wheats grown at Guelph, as was Fife grown at Bowmanville, the latter being the only wheat yielding bread in this series of tests with a quality rated as 100. Harcourt (1909) found Crimean Red and Buda Pesth of the 1907 crop to be superior in baking strength to the other varieties tested. Early Genesee Giant, Michigan Amber, and other varieties were intermediate in this particular, while Red Wave, Forty Fold, Dawson's Golden Chaff, and several others, were inferior to the standard used as a basis of comparison. A large number of varieties were included in this series of tests, which was continued with the crop of 1908 (Harcourt, 1910), with essentially similar results. Harcourt (1911) also tested a number of spring wheats grown in Ontario in 1909, and compared the flour milled from them with a Manitoba flour. Early Java was about equal to the standard in loaf volume, with White Fife, Red Fife, and Preston rating high in baking strength.

Red Fife and White Fife of the 1910 crop continued to hold a high rating among the Ontario spring wheats in point of baking quality in Harcourt's (1912) experiments, although Preston, Gatineau and Hungarian yielded loaves of larger volume and about equal in quality. The comparisons of the baking qualities of Ontario winter wheats, begun with the crop of 1907, was continued through the crop season of 1911. A winter wheat called Banatka appeared to head the list in the three crops of 1909, 1910, and 1911. Alberta Red, Crimean Red and Buda Pesth remained in the strong flour group, with Early Genesee Giant following closely, and Dawson's Golden Chaff, American Banner, and others still appearing in the inferior group.

A few of the winter wheats grown on the plots at Guelph, Ontario, were contrasted by Harcourt and Purdy (1910). Early Genesee Giant gave the largest and best loaf, while Dawson's Golden Chaff and Early Red Clawson gave the smallest loaves of the poorest quality.

Wisconsin Pedigree No. 2 (Turkey Red winter wheat) was found equal to the Marquis (hard spring) wheat grown at Madison, Wisconsin, in milling and baking quality, and considerably superior in yield per acre, in a six-year test conducted by Leith (1919). The percentage of yellow or soft starchy berries in hard wheat was found to vary both with the season and with the variety. Pure lines of hard winter wheat may, in Leith's opinion, be almost identical in appearance, but have widely different baking qualities. Two selections from Beloglina, a hard winter wheat variety, numbered 70 and 71, were quite different in baking quality, the former proving equal to the best hard winter wheat, while the latter ranked with the semi-hard winter wheats, which gave an inferior loaf.

Composition and the results of baking tests of flour milled from Kansas winter wheats were reported by Willard and Swanson (1911). The series of tests is hardly sufficient to afford an accurate comparison between the several varieties, but it is interesting to note the wide variation in the protein content of the varieties grown at Manhattan, Kansas, in 1906. Thus the variety designated as Minnesota contained only 7.92 per cent of crude protein, while Weissenberg contained 13.81 per cent.

Kanred (hard winter) wheat was shown by Clark and Salmon (1921) to compare favorably in composition and milling and baking qualities with Turkey and Kharkof wheats grown at Manhattan, Kansas, from 1912 to 1919. Swanson (1924) also reported that during a period of nine seasons Kanred wheat contained an average of 16.17 per cent of crude protein, while Turkey Red wheat contained 15.83 per cent.

The composition of the progeny of Minnesota spring wheats grown in Maine, in comparison with the progeny of the same parents grown in Minnesota, were contrasted by Woods and Merrill (1903). The Maine and Minnesota crops contained practically the same percentage of protein; such advantage as was found in 1902 was in favor of the Maine crop. The latter was likewise heavier, plumper grain, as evidenced by a greater weight per 1000 kernels.

Zinn (1920) concluded that in Maine "The Red Fife strains yielded the strongest flour. . . . The Preston strains are good yielders, but only a few excel in quality, although a number of them showed a higher protein content. . . . Under Aroostook conditions the Marquis strains did not make a good showing. They all yielded flour with a short, stiff gluten of only fair quality." This experience with Marquis is quite different from the author's in Minnesota, to which reference will later be made.

A series of comparisons of common spring and durum wheat was presented by Ladd and Bailey (1911). Flour milled from the common spring wheats baked into loaves with a larger average loaf volume and higher color score than durum wheat flours. The durum flours contained the higher percentage of ash, the average in the patent flours during three seasons being 0.69 per cent, as compared with 0.49 per cent in the spring wheat flours during the same period. While the average protein content of all durum wheat samples examined was higher than the common spring wheat samples, a careful comparison of samples grown on the same farms during a period of four seasons did not reveal any material difference in the percentage of this component when the two classes of wheat were contrasted. (Note page 246.) Ladd (1912) later reported that at five stations in North Dakota during the crop season of 1911 the average protein content of durum wheat (N × 5.7) was 17.48 per cent, and of spring wheat 16.55 per cent.

The results of milling and baking tests of several spring and durum wheat varieties grown at Fargo and at Dickinson, North Dakota, were summarized by Stoa (1921). At Fargo, Marquis exhibited the greatest flour strength, as indicated by loaf volume, during a four-year period. During the last two of the four seasons Kota compared favorably in these particulars. At Dickinson the same general relations were observed. Durum varieties baked into smaller loaves of darker color.

Kota wheat flour baked into loaves of slightly smaller size or volume, with inferior crumb texture than comparable Marquis samples in the experiments conducted by Waldron, Stoa and Mangels (1922). Tests of Kota wheat reported by Clark and Waldron (1923), covering five seasons and including forty-eight comparable samples, indicated that the wheat contained a higher percentage of crude protein and yielded a higher percentage of flour than Marquis wheat. In the instance of baking tests conducted by the U. S. Department of Agriculture the volume of loaves baked from the Kota was 5 per cent smaller and the color score was two points lower than the loaves baked from Marquis wheat flour. In similar experiments conducted at the North Dakota Agricultural Experiment Station the advantage in these particulars was slightly in favor of the flour milled from the Kota wheat. The latter is usually yellower, indicating a higher carotin content.

Marquis spring wheat was found by Bailey (1914 a) to average somewhat higher in protein content, and to be superior in baking qualities to Bluestem wheat grown under the same conditions. He later (1914 b) reported that Minnesota grown winter wheats proved inferior in milling and baking qualities to hard spring wheat grown in the same sections. They were likewise substantially lower in gluten content. Attention was also called to the uniformly inferior baking characteristics of Humpback spring wheat to the bluestem grown in the state. This opinion of Humpback was confirmed by Thomas (1916).

While several of the common spring wheats grown at eleven experiment stations on the northern great plains were observed by Clark, Martin and Smith (1920) to contain higher percentages of nitrogen than did Marquis, loaves baked from flours milled from the other varieties were smaller than the Marquis loaves with one exception. In three comparisons with Galgalos, grown under extremely dry conditions, which were unfavorable to the quality of Marquis, the loaves baked from the Galgalos wheat flour were slightly larger than the Marquis wheat loaves. The difference found is not regarded as conclusive proof of the superiority of Galgalos. Two early-maturing common wheat varieties, Prelude and Pioneer, had a higher protein content than, and yielded loaves practically equal to, Marquis in these trials.

Decided differences in the properties and composition of the several classes of wheat grown in Montana were encountered by Thomas (1917 a). Excluding the durum wheats, certain of the average data resulting from the tests are given in Table 3. The flour milled from the hard spring wheat samples was superior in baking strength, the hard winter wheat flours were intermediate in this respect, while the western red and western white wheat flours were decidedly inferior.

TABLE 3

COMPARATIVE MILLING AND BAKING TESTS OF SAMPLES OF THE SEVERAL CLASSES OF MONTANA WHEATS AS REPORTED BY THOMAS (1917 a)

	Yield of Straight Flour, Per Cent	Volume of Loaf, cc.	Bread Texture, Score	Color of Bread, Score	Crude Protein in Flour ($N \times 5.7$), Per Cent
Hard spring	71.1	2,342	96	98	11.98
Hard winter	71.8	2,142	94	97	11.73
Western red	68.5	1,787	84	98	10.38
Western white	66.7	1,756	85	96	9.16

While these are not comparisons of varieties grown under identical conditions, they afford an interesting comparison of the market classes of the region. The small difference in the average crude protein content of the hard spring and hard winter wheat classes is not sufficient to account for the substantial difference in their flour strength as indicated by loaf volume and texture. Reference will be made in the discussion of flour strength to the probable explanation of this peculiarity of Montana hard winter wheat.

Montana wheats studied by Whitcomb, Day and Blish (1921) were rated on the basis of baking qualities with the common spring wheats as superior, followed by winter wheats and durum wheats in the order named. Of the spring wheat group, the varieties Humpback and Quality were outstandingly inferior to the Marquis with which they were compared. The variety called Quality was low in gluten, but the Humpback sample was the highest in the group in this constituent.

The wheats grown in Idaho were subjected to study by Jones, Fishburn and Colver (1911). In North Idaho the Turkey Red and Bluestem varieties contained a higher average percentage of crude protein than did Little Club and Red Russian. Much the same relations were observed in samples from South Idaho. The authors state that if flours milled from Little Club, Red Russian and like inferior varieties were replaced by flour milled from Turkey Red, fewer complaints would be registered against Idaho wheat flours.

In a later publication the composition of Turkey Red winter wheat and fife and bluestem wheat grown in Idaho were compared by Jones and Colver (1918). The spring wheat contained a higher percentage of crude protein than did the Turkey Red.

Alaska wheat was found to be inferior to the hard spring and durum wheats with which it was compared by Ball and Leighty (1916) in the matter of milling and baking properties. The loaves baked from the Alaska wheat flour were of very unsatisfactory quality.

Hard Federation wheat was found by Clark, Stephens and Florell (1920) to be superior to Early Baart and Pacific Bluestem in the important milling and baking factors. These are all of Australian origin. At Moro, Oregon, and Chico, California, the average percentage of crude protein in winter wheats of these varieties, crops of 1918 and 1919, were 12.6, 11.5 and 12.8 respectively.

Five of the wheat varieties extensively grown in California were compared by Shaw and Gaumnitz (1911). The average protein content and loaf volume of samples of these varieties are given in Table 4.

TABLE 4

Protein Content of Wheat and Loaf Volume of Bread Baked from California Wheats (Shaw and Gaumnitz, 1911)

Variety	Average Protein, Per Cent	Average Loaf Volume, Cu. In.
Propo	10.64	84
Bluestem	10.18	85
White Australian	9.89	83
Sonora	9.71	77
Little Club	9.35	76

It thus appears that the Propo, Bluestem and White Australian varieties were distinctly superior in these important characteristics to the Sonora and Club varieties. The last two varieties were evidently similar in their properties. Californian wheats appear, from the data in this paper, to be abnormally low in protein content. Thus (see p. 382) a flour milled from their sample No. 253 Little Club wheat, contained only 4.71 per cent of total protein ($N \times 5.68$), which is the lowest reported percentage of protein in an American wheat that has come to the attention of the author. The average protein content of all the California wheat samples which Shaw and Gaumnitz analyzed was only 9.95 per cent, calculated to the dry basis. Recalculated to a 13 per cent moisture basis this is equivalent to about 8.7 per cent of crude protein.

After several years of study of Washington wheats, Thatcher, Olson and Hadlock (1911) concluded that (eliminating durum varieties from consideration) Bluestem and Red Allen stood in a class by themselves as superior in milling qualities to all other varieties commonly grown in that state. A second group includes Turkey Red and Jones' winter Fife, while Little Club and Forty Fold form a third group, with Red Russian the lowest in milling quality.

This opinion was confirmed in general by Bailey (1917 c), but Early Baart was rated with the Bluestem variety. Marquis wheat grown at Pullman, Washington, was found to be equal or superior to Bluestem from the same station.

The high gluten content of Utah wheats was established by Stewart and Greaves (1908), and it was indicated that the spring varieties contained higher percentages of this constituent than did the winter varieties. Gold Coin was the lowest in protein content of the varieties grown on the arid farms. Stewart and Hirst (1913) report higher percentages of protein in the hard spring varieties, such as Fife, when grown in Utah, than in the semi-hard winter varieties, such as Ghirka and Galgalos. No substantial difference in the average protein content of the hard and semi-hard winter wheat varieties was observed in the 1907 and 1908-9 crops, while the soft winter wheats were somewhat lower in protein content. The average for the latter class, 15.62 per cent in 1908-9, is substantially greater than the average reported by Thomas (1917 b) for this class of American wheats. In the baking tests the flours milled from the semi-hard spring wheat varieties gave the largest average loaves, with a volume of 1853 cc., while the average for the hard winter varieties was 1771 cc., for the semi-hard winter 1600 cc., and for the soft red winter 1556 cc.

An extensive study of American wheat varieties was conducted by Shollenberger and Clark (1924) during the seven seasons from 1915 to 1921 inclusive. In the hard spring wheat class it appears that while half of the 15 varieties tested contained higher percentage of crude protein than the Marquis used as the basis of comparison, none exceeded Marquis in flour strength, as indicated by loaf volume and texture. Only two varieties, Pioneer and Ruby, scored higher in color of loaf (crumb). In the hard winter wheat class Kharkof was used as the basis of comparison, and 12 varieties were contrasted with it. The crude protein content of Kharkof averaged higher than all the other varieties except two, Montana No. 36, and Nebraska No. 6, which are selections of Kharkof and Turkey respectively. In loaf volume and texture Minturki proved superior. This is a winter-hardy hybrid of Turkey and Odessa, propagated at the Minnesota Experiment Station. Its kernels generally appear softer than Turkey, and it averaged somewhat lower in protein content than did Kharkof. Several other varieties or strains appeared somewhat superior to Kharkof in flour strength, while Kanred was almost identical with it in strength and color. Red Rock appeared to yield the strongest flour of the soft red winter wheats, although not averaging the highest in protein content. Of the soft white wheats Pacific Bluestem no longer occupies the leading position in the Pacific coast states, so far as flour strength is concerned, being inferior to Baart, Bobs, Bunyip, Federation, Hard Federation, and White Federation. Sonora, Hybrid 63 (club type) and Hybrid 128 (club type) were, on the other hand, distinctly inferior to Pacific Bluestem in this respect.

The results of baking tests of spring and winter wheats of the 1906 crop grown at Ottawa, Canada, were reported by Saunders (1907). Fife type wheats, including Marquis, as well as Haynes Bluestem, were given high ratings, while Dawson's Golden Chaff, Riga, Ladoga and Grant were given a low rating. Preston was intermediate in this respect. Shutt (1907) reported the chemical composition of flours milled from the same series of wheats. The durum varieties were found to contain the highest percentage of protein (of the Ottawa 1906 crop), followed closely by Red Fife, with Dawson's Golden Chaff the lowest in this constituent. The variation was from 14.71 per cent to 7.75 per cent of protein.

In England the Home Grown Wheat Committee of the National Association of British and Irish millers endeavored to improve the strength of English wheat. From their reports (1911, 1913) it appears that Red Fife wheat, introduced from Canada, evidenced "ability to maintain continuously without any appreciable diminution its relative superiority in quality of endosperm." Professor R. H. Biffin was a member of this committee and reference will be made later to his work in selecting and hybridizing wheats to improve strength and baking qualities.

Several samples of wheat grown in India were submitted by Howard and Howard (1908) to A. E. Humphries, of England, who subjected them to milling and baking tests. Substantial differences, attributed to varietal characteristics, were noted in these tests, Pusa 6 being preeminent in its capacity for making large, shapely loaves. They later (1911) called attention to the superior baking qualities of an Indian wheat, Pusa 4, which they had selected and propagated, when compared with Muzaffarnager White. Howard, Leake and Howard (1914) found that Pusa 12, a large-grained white wheat, maintained its high baking qualities when grown under different conditions in India. Its baking strength appeared to be quite superior to other varieties grown in that country. A. E. Humphries coöperated in these studies, in conducting the milling and baking tests.

Wheats of France were studied by Arpin and Pecaud (1923 a), who gave particular attention to the percentage and characteristics of the gluten and the qualities of loaves baked from a 60 per cent flour. On the basis of these qualities the flours were divided into three groups. The total range in percentage of gluten in the flour was from 4.53 per cent in the case of a wheat grown at Colmar, and known as Alsace 22 Colmar, to 15.84 per cent in the instance of "Manitoba" wheat from Rochetaillee-Almont. Twenty of the 45 samples contained between 8.00 and 10.00 per cent of gluten, while only 5 contained in excess of 10 per cent. All but one of the 45 scored lower in total points than did the Australian, Plata and Hard Winter wheats with which they were compared.

Data resulting from the baking tests of durum wheat are of doubtful value in appraising the macaroni-making qualities of such wheat. Thus a distinct yellowness is desirable in macaroni, but might result in a lower color score in the bread. Again, there are properties of flour, such as diastatic activity, which function in determining loaf volume, but may be relatively insignificant in their influence on macaroni properties. Mindum wheat rated low in Shollenberger and Clark's (1924) baking tests, but they state that it has proven satisfactory in macaroni manufacture. Pentad, likewise rated low in the same series of baking tests, has not been regarded favorably by the manufacturers of macaroni.

An extended series of durum wheat studies were reported by Shepard (1902, 1903, 1905, 1906). Certain varieties, notably Kubanka, Arnautka, and Gharnovka, made superior macaroni, while Black Don, Velvet Don, Saragolla, Iumillo, and others were notably inferior in this particular. The flour and semolina milled from durum wheat was found to possess more of the yellow and orange tints than similar products milled from Fife wheat grown in North Dakota. In baking qualities the durum flour gave smaller loaves with a deeper color.

Kubanka, Arnautka and Mindum durum varieties were found by McLaren (1923) to be the most satisfactory for the manufacture of macaroni. Semolinas made from wheat of these varieties were amber in hue, and otherwise of good quality. Acme and Monad varieties of durum wheat yielded semolinas of dull or grayish hue and otherwise of inferior quality. Red Durum proved undesirable for the manufacture of macaroni.

The significance of the gasoline soluble pigments in semolina milled from durum wheats used in macaroni production was stressed by Clark (1924). According to Clark, yellow semolina, having a high gasoline color value, is preferred by the macaroni industry. When a number of durum wheat varieties were examined by this method, the results were as shown in Table 5. Kahla appeared somewhat superior to the other varieties, but only nine samples were tested, which is too small a number to permit of drawing final conclusions. There was no material difference between the average of the tests of Kubanka, Nodak, and Mindum, all rating high in this particular. Arnautka and Peliss were intermediate in this particular, while Pentad, Acme, and Monad were distinctly inferior. The author doubts if the content of carotinoid pigments, measured roughly by the gasoline color method, is the sole factor determining the visual appearance of macaroni. It seems probable that the physical structure of the semolina particles may also serve to determine the apparent "color" of the finished sticks of macaroni, much as the visual appearance of the endosperm of the wheat berry is determined in large part by its relative density.

Macaroni produced from the Kubanka, Arnautka and Mindum varieties was found by Mangels (1923) to be of good quality, or at least superior to that made from Monad, Acme and D-5 (red durum) varieties. The last three named have been grown by farmers chiefly because of their greater resistance to black stem rust, which resulted in higher yields of grain per unit of area.

Red durum (Pentad) wheat was not a good bread producer, either in point of volume or quality, in the opinion of Ladd (1915). Sanderson (1920) confirmed this opinion as the result of a study extending through five seasons.

TABLE 5

Gasoline Color Value of Nine Varieties of Durum Wheat (Clark, 1924)

Variety	Average Gasoline Color Value
Kahla	1.75
Kubanka	1.56
Nodak	1.51
Mindum	1.51
Arnautka	1.33
Peliss	1.32
Pentad	1.14
Acme	1.06
Monad	1.04

This review of papers dealing with the relative composition and quality of wheat classes is not exhaustive, and numerous citations to papers on this phase of the subject have been purposely omitted. Among the omissions are several papers by European investigators, who worked with varieties unfamiliar to the author. In these cases it is difficult, if not impossible, to determine from the published work which American wheat varieties these Continental wheats most nearly resemble. Hence the data of such examinations cannot be compared directly with data resulting from American studies. It is evident, moreover, that the most comprehensive studies along this line have been conducted in America.

From the data here presented it appears that substantial differences may be found in the composition and properties of the several varieties of each wheat class when grown under similar conditions. Of the varieties in the hard spring wheat class, Marquis, and related members of the Fife wheat group, are apparently superior in bread-making qualities to the other varieties of this class. Kota, a common bearded hard spring wheat, may have plumper kernels than the Marquis wheat when the two varieties are grown under rust conditions. This results in greater flour yields from the Kota wheat in such cases. Humpback is generally conceded to be of very poor quality. Of the hard winter wheats Turkey Red and the related varieties, such as Crimean and Kharkof, have generally proven superior to the other varieties of their class so far as baking strength is concerned. Certain selections of these varieties appear even better than the original mixed stocks. Kanred is apparently very similar to the Turkey Red, so far as can be determined from published data.

Of the soft red winter wheats grown in the eastern half of the United States, Red Rock seems to be recognized as one of the most glutinous and strongest wheats for bread making. In general, there is less variation in the composition and properties of the soft red winter wheats than is encountered in the hard spring and hard winter wheats.

In the Pacific wheat districts Baart and Pacific Bluestem or White Australian were formerly regarded as superior wheats, when contrasted with the Club varieties, club hybrids, Sonora and Red Russian. More recently some of the later Australian importations, such as Federation and Hard Federation, appear to be crowding Pacific Bluestem from the position which it occupied as the choice milling wheat of the Pacific districts. Alaska wheat, on the other hand, has received a very low rating in baking strength.

Durum wheats cannot be accurately rated on the basis of baking value, since they are not extensively used in bread production. The data available indicate that Kubanka, Arnautka, and a selection propagated at the Minnesota Experiment Station known as Mindum, can be converted into choice macaroni. Iumillo, a rust resistant variety used as one of the parents in hybridizing for rust resistance, has received a low rating as a macaroni-producing wheat. The pigmented or red durum variety, Pentad, has been regarded with marked disfavor by durum wheat millers because of the undesirable appearance of macaroni manufactured from it.

The common wheat classes (thus omitting club and durum) may be rated in the following order so far as gluten content and baking strength are concerned: (1) Hard red spring; (2) Hard red winter; (3) Soft red winter; (4) White wheat. Compactum or club wheats rate lower than the common wheat varieties of the white wheat class. There is substantial overlapping in so far as these properties of the individual parcels of wheat of the several classes are concerned. Thus certain lots of any class may be superior to the average of the next higher class. Soft wheats, rating low in baking strength, may be superior for other purposes such as cracker, biscuit, and pastry production.

Breeding Wheat for Strength of Flour.

The differences observed in the baking properties of the several wheat varieties have attracted the attention of certain plant breeders, who have been desirous of selecting and propagating superior strains of wheat, or of combining the baking strength of superior varieties with certain characteristics of other varieties. Efforts in the direction of line selection for composition and quality as might be anticipated have been accompanied by indifferent success. Hays and Boss (1899) evidently selected several good varieties of hard spring wheat, but they probably started with mixed races, rather than pure lines, and from these mixtures the best individuals were isolated and propagated. Similar results have attended like efforts elsewhere.

Seed wheats, separated into starchy and vitreous portions by Feilitzen (1904) were propagated separately, but with only slight effect upon the starchiness of the resultant crops. Humphries and Biffin (1907) likewise failed to effect any improvement in the strength of the stocks with which they worked, after four years of line selection.

Parent plants were selected on the basis of their protein content by Lyon (1905) and their progeny were compared on the same basis. There was less difference observed between the extreme classes in the offspring than in the parent plants. In other words, the progeny tended toward similarity even though the parents differed markedly. While Lyon believed that the protein content of wheat might be increased by continuous reproduction from plants of high protein content, Thatcher (1913 b) states that a continuation of Lyon's work by Montgomery led to diametrically opposite conclusions.

A number of hard winter wheat parent stocks were classified into groups on the basis of their percentage of yellow-berry by Roberts and Freeman (1908). The progeny were examined and it was found that the percentage increase of yellow-berry in the parents follows a mean percentage increase of yellow-berry in the offspring. Since increasing percentages of yellow-berry generally accompany decreasing percentages of protein, it follows that Roberts and Freeman were attempting a selection leading in the direction of increasing the protein content.

Roberts (1919) later concluded that "the operation of common causes for the production of yellow-berry overshadowed any differences that may have been due to hereditary tendencies, and precludes a definite statement regarding the relation of hereditary tendencies in hard winter wheats towards the production of yellow-berry. That some isolated pure strains of wheat are freer from yellow-berry than others growing in the same field and under identical conditions of soil and climate is, however, possible."

After one year's work on line-selection for protein content, Thatcher (1908) was convinced that a definite and positive effect was to be observed. He (1913 b) later advanced the opinion, based on four year's observations, that further attempts to improve the chemical composition of Washington wheats by line-selection breeding would be absolutely useless.

It is a matter of common knowledge among plant breeders who have had considerable experience in wheat breeding that pure lines (each the progeny of a single plant) of wheat of the same variety, which cannot be differentiated on the basis of morphological plant characters, sometimes differ in texture of grain. Thus some pure lines of the Crimean group of winter wheats when grown at University Farm, St. Paul, Minnesota, have a rather consistently higher protein content than others when the lines compared are grown under similar conditions. Environmental conditions greatly modify the percentage of protein produced in the grain and in order to study hereditary differences it is necessary that the progeny lines which are to be compared be grown on the same land and be similarly handled. That pure lines of the same variety may differ consistently has been proved by a comparison made by Hayes (1923) of several lines of Kota wheat. One pure line for several years produced seed with a softer texture than others. This line, however, could not be differentiated from others on the basis of morphological characters or rust resistance.

Since line-selection in general proved futile in improving the baking strength, vitreousness, or protein content of desirable types of wheat, plant breeders turned their attention to the problem of crossing or hybridizing unlike wheats, using one parent which possessed superior baking strength and another parent which had other desirable characters. Biffin (1905) presents evidence to show that hardness and softness of endosperm constitute Mendelian units. When polish, a wheat with hard endosperm, was crossed with Rivet, having soft endosperm, an examination of 200 heads of the resulting F_3 generation of hybrids gave 152 hard to 48 soft endospermed forms, or a ratio of 3 to 1. Hard endosperm was recessive.

Freeman (1918) crossed vitreous durum wheats with Sonora (T. vulgare), a soft-seeded wheat. The F_1 plants produced hard, intermediate, and soft kernels. The hard or vitreous kernels of the F_1 tended to give more hard-kerneled plants in F_2, and the soft kernels tended likewise to give more plants yielding soft kernels. This tendency continued into two succeeding generations. These results were explained by Freeman as due to two factors governing the proportion of gluten and starch, or degree of softness. Each of these two factors was believed to be inherited independently. Assuming that the endosperm results from double fertilization, or the fusion of two polar nuclei of the egg with one of the male generative nuclei, there may be a range of from 0 to 6 factors for starchiness of the endosperm. This assumption explained the experimental findings in a satisfactory manner.

Humphries and Biffin (1907) comment "that it would be premature to state without any reservation that strength and 'weakness' form a pair of Mendelian characteristics, but the assumption that they are so has proved a valuable one in building up desirable varieties." Of the hybrids which they produced with Fife (a strong wheat) as one of the parents, many, while diverse in habit, retained the strength of the Fife parent.

Strength (baking) did not appear to Saunders (1907) to be a Mendelian character. In the case of four cross-bred wheats, their baking qualities were intermediate between those of the two parents. Biffin (1908) then presented further data supporting his contention that strength was inherited as a Mendelian unit, and contended that Saunders did not isolate hybrids which were homozygous with respect to strength, and also that actually in some of these hybrids strength was inherited in its entirety. Saunders (1909) then stressed the care which had been taken in the work at Ottawa, Canada, to secure pure strains for hybridization work. He suggested that the failure to agree as to the manner of inheritance of strength might be due to the fact that strength is itself a composite rather than a simple quality. Biffin (1909) replied briefly, concluding this polemic with the suggestion that definitions of strength and weakness as variety characteristics are terms too elastic to permit of scientific conclusions on the points in controversy.

A number of wheat hybrids were subjected to milling and baking tests in comparison with the parents by Hayes, in collaboration with the author, at the Minnesota Agricultural Experiment Station. The resulting data, as yet unpublished, lead to essentially the same conclusion as that reached by Saunders. Thus when Marquis, a superior bread wheat, was crossed with Iumillo, a durum wheat, of inferior baking qualities, the baking strength of the hybrids ranged all the way between the extremes represented by the two parents. Incidentally it may be remarked that the relative resistance to certain biological races of stem rust possessed by Iumillo was inherited by many of the hybrids, which simultaneously inherited satisfactory baking qualities from the other parent.

Most economic plant characters have been shown by plant breeders to be a result of the interaction of many inherited factors plus environment. Thus from the same cross numerous pure lines can be isolated which range in value from one parent to the other. In some crosses some new varieties can be isolated which are superior to either parent. Results of this nature are due to the interaction of cumulative factors, some obtained from one parent and some from the other. While in some crosses strength and weakness of flour may be dependent on a single Mendelizing pair of factors as a rule many inherited factors are without doubt involved.

It should be emphasized that tracing strength through several generations of progeny of one of these crosses is difficult when baking data alone are employed. The volume and texture of test loaves are determined by several variables, certain of which might be inheritable, while others would be conditioned almost wholly by environment. Such studies, as in the case of flour strength researches in general, should be resolved into a separate consideration of the several variables, so far as that is feasible. Until this is done, one is left in doubt as to whether bread-making qualities observed are a matter of an inherited character, or the influence of environment upon the particular plants involved. This is especially true in so far as varying diastatic activity is concerned: a quality, the degree of which is undoubtedly induced by the environment of the plant and the ripened grain, but which, as will be shown in the later discussion of flour strength, is of importance in determining the quality of the resulting loaves.

Chapter 3.

The Growth and Development of the Wheat Plant and Kernel.

There are, of course, any number of stages in the life cycle of the wheat plant at which this discussion could begin. It seems most convenient and logical to start with the germinating kernel of wheat, however, since at this state metabolic activity is resumed after a more or less prolonged resting period. When the after-ripening period is concluded, on bringing the sound, living wheat kernel into a favorable environment, including temperature, the presence of moisture, and air, its enzymes become active and germination ensues. The first significant changes in the kernel incident to germination may be observed in the scutellum, that portion of the germ or embryo which lies next to the endosperm. The epithelial cells of the scutellum become filled with granular material, such as is characteristic of glandular tissues, and Choate (1921) observed that these cells elongate, more than doubling in length during 7 days of germination. Mann and Harlan (1915) state, as a result of their observations on germinating barley: "The scutellum as a feeding organ is endowed with all the functions of digestion, being able to utilize all foods occurring in its natural storehouse, the endosperm." This implies that the scutellum constitutes a glandular organ of the developing embryo, secreting digestive enzymes, and thus enabling the embryo to feed on the reserves of the endosperm. The latter, as shown by Mann and Harlan, is progressively digested, the area of liquefaction moving rapidly through the portion adjacent to the aleurone layer, and even more rapidly along the furrow. The path of disintegration at the end of a few days appears in the form of a crescent, the horns of which are advanced beneath the aleurone layer. "The aleurone has a protective function in preventing the inroads of moulds and bacteria, and also in furnishing nitrogenous food to the seedling after its green tissues are capable of photosynthesis."

The starch granules of the endosperm are not gradually dissolved, but appear rather to be attacked by powerful enzymes which cause them to become irregularly pitted and rapidly dissolved. The reserve tissues first attacked by these enzymes become entirely liquefied as the conversion proceeds. Aleurone cells are apparently not absorbed until the starchy endosperm has been nearly exhausted. While translocation diastase in the starchy endosperm cells, and that secreted by the aleurone layer may be responsible for a part of the endosperm disintegration, Mann and Harlan believe that in the case of germinating barley the scutellum secretes the diastase and other enzymes responsible for the liquefaction of the endosperm reserves.

The products of this process of digestion of the reserves in the endosperm are transported in solution to the embryo, where vigorous growth occurs. The formation of new tissue takes place first in the radical, which soon breaks through the protecting layer and emerges from the grain. The secondary rootlets follow, while the plumule develops more slowly.

The translocation of plant food from wheat kernel to seedling was followed by Le Clerc and Breazeale (1911), who found that at the end of 12 days of germination only 4 per cent of the potash, 17 per cent of the nitrogen, and 20 per cent of the phosphoric acid remained in the seed. The seedlings when grown in a nutrient solution took up potash very rapidly, and at the end of 4 days contained about as much as was present in the seed. At the end of 12 days the seedlings contained half again as much potash as was present in the original seed. The seed residues contained about two-thirds of the fat originally present, although fat was found in the seedlings to the extent of 70-100 per cent of that present in the seeds. Fiber in the seed decreased only about 6 per cent in 15 days, while at the end of that period only 40-50 per cent of the pentosans remained. In the seedlings the pentosans were apparently converted in part into fiber after the ninth day. As a result of the hydrolysis of endosperm starch, the quantity of sugars in the seedlings increased rapidly up to the ninth day, after which there was a decrease. The dry weight of the seed decreased from 3.50 grams to 0.36 grams per 100 kernels after 15 days of germination; the weight of seedlings and seed residues from 100 kernels was then 1.84 grams, and the seedling alone 1.48 grams.

A negative test for proteins in the germinating wheat kernel after 7 days was obtained by Choate (1921). Asparagin could be identified by microchemical methods in the coleoptile after the fourth day of germination, and slightly later in the root. The amino nitrogen content (on dry basis) of the ungerminated wheat was 0.0275 per cent. After 3 days' germination this increased to 0.29 per cent, and after 6 days to 0.63 per cent. Catalase activity of germinating wheat kernels increased steadily to the seventh day. The rate, in terms of oxygen in cubic centimeters released from hydrogen peroxide by each gram of dry weight of grain, increased from 84.5 cc. in the ungerminated wheat to 1,726 cc. after 7 days of germination.

The presence of amino acids in the sap of corn sprouts during germination was detected by Pettibone and Kennedy (1916), and they concluded that the protein reserves of the grain were transported to the seedling in the form of amino acids, and, possibly, in the form of soluble protein or protein hydrolytic products of the peptide type. The presence of the latter was suggested by the increase in amino-nitrogen found on acid hydrolysis of this sap.

That the temperature determined the mineral requirement for good germination and initial growth in wheat was the conclusion reached by Gericke (1921), and he suggested that possibly other climatic conditions might also exert an effect. He further suggested that discordant results secured in various fertilizer experiments might perhaps be due to varying climatic influences.

When the seedling has absorbed most of the reserve material stored in the endosperm of the grain, it has usually reached a condition where it is capable of maintaining an independent existence if in contact with a suitable substratum. As observed by Le Clerc and Breazeale (1911), it has probably assimilated some mineral matter, particularly potash, from the nutrient solution in the substratum during the seedling stage. The young plant is then concerned with the production of a structure capable of bearing fruit. Its energies are centered for a time on absorbing and working over such nutrient substances as are essential to this end.

Wheat at different stages of growth was sampled by Lawes and Gilbert (1884), and it was found that in 5 weeks following June 21 there was little increase in the quantity of nitrogen in the plants occupying a unit area, although more than half of the total carbon of the crop was accumulated during that period.

Liebscher (1887) states that the absorption of plant food proceeds more rapidly in proportion to dry matter formation during the early stages of growth than during the period just preceding maturity of the grain. The proportion of nitrogen to dry matter formed decreased rapidly during the first three stages of growth in the wheat plants which he had under observation.

During the first period of growth (50 days) the wheat plant took up 86 per cent of the nitrogen, 75 per cent of the potassium, 80 per cent of the phosphorus, and 70 per cent of the silica which it finally contained, in the study conducted by Snyder (1893). The starch was formed mainly during the second and third periods of 15 and 16 days respectively. In this instance the plants headed in 65 days, and were in the kernel milk stage in 81 days. Adorjan's (1902) findings agree substantially with those of Snyder. He states that the greater portion of the plant food is taken up in the earlier stages of growth, and later drawn upon in the growth and functioning of the various plant structures. At the time of blossoming the phosphorus assimilation exceeded that of nitrogen and is at its maximum. After blossoming the phosphorus assimilation ceases while that of nitrogen is reduced to the needs of the plant for the formation of grain. Henry (1903) found the absorption of nitrogen varied at different stages in the life of the wheat plant. The maximum rate was reached between May 7 and 17, and at the time of the formation of the kernel.

The maximum quantities of the various substances (per unit of area) were found by Wilfarth, Römer and Wimmer (1905) to have been present in the developing wheat plants at the following stages of growth: starch at ripeness; dry matter P_2O_5, and nitrogen at the third period; potassium and sodium at the second, or period of bloom. During the development of the plant the loss of dry matter and of phosphorus is small, while that of the potassium is 38 per cent, and of the nitrogen 23 per cent. These materials were assumed to have been returned to the soil by the downward movement of the sap. Le Clerc and Breazeale (1908) contended, however, that the loss of mineral matter from plants was due to its removal from leaves and stems by rain or dew. Thus four rainfalls on ripe wheat removed 27 to 32 per cent of the nitrogen, 20 to 22 per cent of the P_2O_5, 63 to 66 per cent of the potassium, 46 to 65 per cent of the sodium, and 90 per cent of the chlorine.

Nitrogen and mineral matter was taken up largely in the younger stages, and at decreasing rates as growth proceeded, in the experiment conducted by Haigh (1912). The maximum dry matter was in the stalks at the time of blossoming, while the fiber had all formed when the blossom had fallen.

The sodium salts commonly encountered in alkali soils, when in solutions in concentrations up to 1000 parts per million, did not affect the absorption of nitrogen from culture solutions by young wheat plants, according to Breazeale (1916). Sodium sulfate and sodium carbonate in concentrations of 1000 p.p.m. depressed the absorption of potash and phosphoric acid by these plants.

Wheat plants were analyzed by Maschaupt (1922) at weekly intervals from May to August. An analysis of the ash showed ripe wheat to contain more of the common ash constituents than the unripe, with the exception of K_2O, CaO, Cl, and SO_3.

The proportion of mineral nutrients for best growth of wheat are practically the same during the first two growth periods, in the opinion of McCall and Richards (1918), being 3.4 parts of mono-potassium phosphate, 4.7 parts calcium nitrate, and 1.9 parts magnesium sulfate. For the third and final growth period the best results were secured with somewhat different ratios of salts, the proportion of potassium phosphate being reduced, and the calcium nitrate and magnesium sulfate being increased.

Lipman and Taylor (1922, 1924) claim to have proven that growing wheat plants can fix nitrogen from the air. Wheat plants grown in nutrient solutions gained in nitrogen from the air during a six weeks' growth period from 13 to 21 per cent of the total amount of nitrogen found in the plants.

Azzi (1921, 1922) concluded that there is a critical period as regards moisture during the 15 days previous to spike formation by the wheat plant. If insufficient moisture is available during this period the yields will be reduced regardless of conditions later in the life of the plant. On the other hand, if the plant has an adequate supply of moisture available at this period, it can subsequently withstand less favorable conditions and still produce good yields of grain.

The carbohydrates of the green parts of wheat plants were studied by Colin and Belval (1922), who found that the soluble carbohydrates of the wheat leaves were sucrose and its products of hydrolysis. The stalk receives carbohydrates from the leaves, the ratio of reducing sugars to sucrose being greater in the stalk than in the leaves. As the head was formed a change in the sugars of the stalk occurred. While the sucrose increased the rotation of polarized light changed from dextro to lævo, and hydrolysis by acids yielded increasing amounts of levulose. Where the ratio of dextrose to levulose was 1.48 in May, it fell to 0.28 by July 1. The sugar content of 100 grams of fresh tissues on two dates is shown in Table 6. Levulosanes appeared in the stem toward the latter part of this period.

TABLE 6

Sugars in Wheat Leaves at Two Stages of Growth (Colin and Belval, 1922)

	Total Sugars		Reducing Sugars		
	May 18	July 1	May 18	May 29	July 1
Fresh leaves	1.9	3.35	0.49	...	0.73
Upper half of stalks	1.57	8.1	0.79	2.33	0.81
Lower half of stalks	3.4	8.74	1.52	2.31	1.20

These researches indicate that mineral matter is absorbed from the substratum by the developing wheat plant at a rapid rate in the early stages of growth. At the time of blossoming practically enough has been accumulated in the tissues of the plant to enable it to function normally during the remainder of the growth period. These mineral substances are translocated from the dying to the living parts as they are required. Nitrogen continues to be absorbed during kernel development, however, although at a slower rate. Investigations reported further along in this chapter indicate that increasing the concentration of nitrates in the substratum late in the life of the plant may result in an increased absorption of nitrogen and a higher percentage of protein in the grain.

After the wheat florets blossom and fertilization of the ovule has occurred, a development of the kernel or fruit ensues. It is with this stage of the development of the plant that we are especially concerned, since the kernel is the source of the flour which is presently to be discussed. The morphology of the development of the wheat kernel has been traced by a number of investigators, including Brenchley (1909), Jensen (1918), and Gordon (1922). Brenchley found the peripheral layer of the endosperm to be marked off about 2 weeks after fertilization, and this developed into the aleurone layer. The deposition of starch began in the middle of the flanks of the endosperm, at the lower or germ end, and proceeded upwards and outwards. The reserve nitrogenous materials entered at the same time as the starch. Disorganization of endosperm nuclei resulted from increasing pressure during the process of ripening.

The first cells of the endosperm are formed from the secondary nucleus of the embryo sac, according to Gordon (1922). These form a lining having the characteristics of a cambium, which divides only on its inner surface, thus giving rise to the successive layers or tiers of starchy endosperm cells. After division of the cells in the lining cease, these become filled with aleurone granules and form the aleurone layer. It is suggested that the presence of vitamins in this layer, and its greater respiratory activity are a consequence of its cambium characteristics.

The developing wheat kernel was subjected to microchemical study by Eckerson (1917), who found that no storage protein was formed in the endosperm until desiccation began. Gliadin and glutenin were formed when drying of the grain caused the amino-acids in the endosperm to condense into proteins. Thus grain which, when brought into the laboratory gave no protein reaction but reacted positively to tests for asparagine, arginine, histidine, and leucine, on drying for 12 hours gave positive protein reactions and contained gluten.

Lucanus (1862) collected samples of the grain of rye at five stages of development, and showed that while the weight of nitrogenous and non-nitrogenous substances increased regularly from the green condition to ripeness, the percentage of these components did not change materially from the milk stage to the ripe condition (freely shelling). The percentage of nitrogenous material was somewhat higher on June 28 than was the case on July 3, when the grains were in the milk stage.

Analyses of wheat kernels at several stages of development were published by Heinrich (1867, 1871). The percentage of ash and protein was found to decrease regularly until an over-ripe stage was reached, while the percentage of starch increased.

The composition of the wheat grain at different stages of development was determined by Kedzie (1882). This work was repeated, and in a later publication Kedzie (1893) showed that the percentage of starch increased very rapidly for a time, and later at a slower rate until the grain was ripe. The percentage of protein decreased regularly with the approach of ripeness. When the grain was allowed to become over-ripe the percentage of starch decreased, and the protein increased somewhat, which suggested that possibly some of the starch served as a source of energy to the plant after it had normally ripened.

The percentage of amids, ash, fat, fiber, dextrin, and pentosans was found by Teller (1898, 1912) to decrease in the grain up to ripeness, while the starch rapidly increased. The percentage of total protein decreased up to a week before ripeness, and then increased.

The rise of nitrogenous material from the lower to the upper leaves, and finally to the grain, was traced by Deherain and Dupont (1902). In addition to the leaves, the upper portion of the stem elaborates starch for the grain, while the glumes were believed to be incapable of photosynthesis. The significance of the awns in elaborating material for storage in the grain is supported by the work of Harlan and Anthony (1920) with barley. They clipped the awns from barley spikes or heads, and found that the kernels from these heads had a smaller volume and a lower weight of dry matter at maturity than kernels from normal, bearded heads. This could not be attributed to the injury or shock of removing the awns, since the rate of development of kernels for several days after clipping was normal. Daily deposit of nitrogen and ash was more nearly equal in the awned and clipped heads than was the deposit of starch. Rachises of clipped spikes contained about one-fourth more ash than rachises of normal spikes, and it was suggested that this might account for the increased tendency of the former to shatter.

The development of the wheat kernel was divided by Brenchley and Hall (1909) into three stages, termed (1) pericarp formation, (2) endosperm filling, and (3) desiccation. In the filling of the endosperm there is moved into it the same ratio of nitrogenous to non-nitrogenous materials. The character of the "mould," that is, the actual ratio of these materials, is determined by variety, soil, season, and other factors.

An elaborate study of the chemistry of wheat kernel development was reported by Thatcher (1913 a, 1915), which tended to confirm the findings of Brenchley and Hall. In the studies of the first season, a winter wheat (Turkey Red) and a spring wheat (Bluestem) were employed, while in the second season's work three varieties of spring wheat were included. The data from the analyses of the Bluestem samples are given in Tables 7-10 as being typical of the results. The wheat heads used in this case blossomed on July 7. It appears that the dry weight of the kernels increased rapidly during the last few days of the ripening period. Thus in the 12 days from July 18 to August 2 the Bluestem kernels more than doubled in weight. Any condition which interferes with the process of translocation of material to the kernel during this period, such as hot dry winds and stem rust infections, may operate to substantially reduce the yield and the grade of the wheat crop, in consequence of the shriveled, light-weight grain which will result. In all of the five series of wheats studied there was the same tendency toward a reduction in the percentage of nitrogen in the grain during the first half of the kernel-development period, followed by an increase in the percentage of this constituent during the latter half of the period. The percentage of ether extract and of fiber diminished regularly during the entire period, due, no doubt, to the fact that the pericarp and embryo, which are high in their content of fiber and of fat respectively, constitute progressively a smaller proportion of the kernel as it approaches maturity. The percentage of sugars decreased and of starch increased as ripeness approached. The sharp decrease in sugar content occurred some days before the grain was ripe, however. Gain in milligrams of protein per kernel each day was less variable than the gain in carbohydrates, the latter decreasing toward the end of the ripening period. In consequence the ratio of carbohydrates to protein moved into the kernels diminished appreciably, and the percentage of protein in the kernels accordingly increased as the grain approached ripeness.

TABLE 7

YIELD AND PHYSICAL PROPERTIES OF KERNELS OF BLUESTEM WHEAT CUT AT SUCCESSIVE STAGES OF DEVELOPMENT

Date of Cutting	Average No. of Kernels per Spike	Weight of 1,000 Kernels, Grams	Dry Matter per 1,000 Kernels, Grams	Specific Gravity of Kernels	Volume of 1,000 Kernels, cc.
July 15	24.6	6.98	6.47	1.4130	4.94
July 18	24.4	10.94	10.17	1.4367	7.61
July 21	25.2	15.85	14.65	1.4202	11.22
July 24	23.4	21.54	19.85	1.4107	15.27
July 27	26.9	23.29	21.57	1.4102	16.50
July 30	25.5	26.02	23.90	1.4088	18.47
Aug. 2	24.7	27.95	25.23	1.4045	19.91

TABLE 8

ANALYSES OF SAMPLE KERNELS OF BLUESTEM WHEAT CUT AT SUCCESSIVE STAGES OF DEVELOPMENT

Date of Cutting	Percentage Composition of the Dry Matter						
	Ash	Protein (N × 5.7)	Ether Extract	Fiber	Total Sugars[1]	Starch	Undetermined
July 15	2.94	14.20	5.39	5.26	9.08	53.93	9.20
July 18	2.70	13.00	4.04	4.33	7.67	54.84	13.42
July 21	2.42	13.01	3.28	4.08	3.13	61.00	13.08
July 24	2.30	13.91	3.16	4.29	1.56	63.42	11.36
July 27	2.05	14.23	3.16	3.51	1.38	65.11	10.56
July 30	2.01	14.77	3.02	3.83	1.27	64.82	10.28
Aug. 2	2.24	15.62	2.91	3.40	1.08	65.50	9.25

[1] Calculated as dextrose.

The writer made observations on Thatcher's samples, not heretofore published, in an effort to determine the changes in the sizes of the kernels, and the percentage of endosperm by weight. These data are

TABLE 9

Weight in Milligrams of Material per Kernel of Bluestem Wheat, Cut at Successive Stages of Development

Date of Cutting	Weight per Kernel in Milligrams							
	Ash	Protein (N × 5.7)	Ether Extract	Fiber	Total Sugars	Starch	Undetermined	Dry Matter
July 15	.190	.919	.349	.340	.587	3.489	.595	6.47
July 18	.275	1.322	.411	.440	.780	5.577	1.365	10.17
July 21	.354	1.906	.480	.598	.458	8.937	1.916	14.65
July 24	.426	2.761	.627	.851	.310	12.589	2.285	19.85
July 27	.442	3.069	.681	.757	.298	14.044	2.278	21.57
July 30	.480	3.530	.722	.915	.303	15.492	2.457	23.90
Aug. 2	.565	3.941	.734	.858	.272	16.525	2.334	25.23

TABLE 10

Gain in Milligrams of Protein and Carbohydrates in Each Kernel of Bluestem Wheat per Day, with the Ratio of the Two Materials for Each Period

Period	Gain in Milligrams per Day		Ratio of Carbohydrates to Protein
	Protein	Carbohydrates	
7/15-7/18	0.134	1.050	7.82
7/18-7/21	0.195	1.249	6.44
7/21-7/24	0.285	1.365	4.82
7/24-7/27	0.103	0.457	4.35
7/27-7/30	0.157	0.597	3.88
7/30-8/2	0.137	0.274	2.00

given in Tables 11 and 12 and show that the kernels increased in size up to the last three days, when the shrinking incident to desiccation became evident. While the fibrous envelope, consisting of ovary wall and testa, steadily diminished in thickness, its total weight increased for a time because of its increased size and density. About 9 or 10 days before the kernels were ripe, this fibrous coat began to decrease

TABLE 11

Dimensions of Kernels of Bluestem Wheat Cut at Successive Stages of Development. Season of 1914

Date Sampled	Width of Kernel on Cross-section	Length of Kernel on Long-section	Thickness of		
			Ovary Wall	Testa	Ovary Wall and Testa
	mm.	mm.	mm.	mm.	mm.
July 15	3.1	5.7	.1071	.0381	.1452
July 18	3.3	5.9	.0690	.0262	.0952
July 21	3.3	5.9	....	....	.0881
July 24	4.6	6.6	....	....	.0809
July 27	4.5	6.9	....	....	.0690
July 30	5.2	7.3	....	....	.0571
Aug. 2	4.5	7.1	....	....	.0452

in weight, indicating a translocation of material to the endosperm. The latter increased in weight until the kernel was fully ripe, when it constituted nearly 82 per cent of the weight of the grain.

TABLE 12

WEIGHT OF THE SEVERAL KERNEL STRUCTURES OF BLUESTEM WHEAT CUT AT SUCCESSIVE STAGES OF DEVELOPMENT. SEASON OF 1915

Date Sampled	Weight per 1000 Kernels				Volume per 1000 Kernels
	Total	Seed-coat and Germ	Endosperm (by Difference)	Endosperm	
	Grams	Grams	Grams	Per Cent	cc.
July 28	12.702	4.752	7.950	62.6	9.897
July 31	17.171	5.728	11.443	66.6	12.317
Aug. 3	22.483	6.801	15.682	69.8	16.114
Aug. 6	25.062	7.236	17.826	71.2	17.871
Aug. 9	27.666	6.391	21.275	76.9	19.417
Aug. 12	30.432	5.826	24.606	81.7	21.372

The size of the developing barley kernel following flowering was measured by Harlan (1920). The length of the kernel increased rapidly during the first 7 days, after which there was little change in this dimension. Diameter of the kernel measured laterally, or dorso-laterally, increased quite steadily during the first 15 or 16 days, after which there was little change in the lateral diameter, and slight increase in the dorso-ventral diameter almost to maturity. Weight of dry matter per kernel increased quite regularly during 24 days following flowering, as did also the weight of nitrogen and ash. At the end of this period the kernels contained about 42 per cent of moisture. For the first 10 days after flowering the day and night gains in weight appeared to be fairly equal, but thereafter until maturity the day gain was greater. Starch was observed in the kernels on the fourth day. Up to the sixth day the starch granules seemed to be of a lower density than normal barley starch. They did not stain readily and were indefinite in outline. Rapid infiltration of starch was observed after the increase in length had practically ceased.

The percentage of protein in the developing wheat kernel was found to increase as the kernel approached ripeness in the study reported by Woodman and Engledow (1924). The non-protein nitrogen decreased from 0.81 per cent on the 33rd day after the emergence of the ear (25 days after fertilization of the ovule) to 0.20 per cent on the 65th day. During the same period the amino-acid nitrogen decreased from 0.413 per cent to 0.021 per cent, and the crude fiber from 4.26 per cent to 2.32 per cent. The crude fat remained fairly stationary during the last 20 days of this period. This is shown by the data in Table 13. It is noticeable that from the 50th to the 54th day when the moisture content of the grain decreased from 45.01 to 40.44 per cent, the per-

TABLE 13

CHEMICAL DEVELOPMENT OF THE WHEAT KERNEL AS REPORTED BY WOODMAN AND ENGLEDOW (1924)

Analysis of Samples of Grain (Calculated on Dry Matter Basis)

Number of sample	1	2	3	4	5	6	7	8	9
Days from ear-emergence	33	40	43	47	50	54	57	62	65
	%	%	%	%	%	%	%	%	%
Dry matter	27.54	37.77	44.80	50.62	54.99	59.56	64.79	77.18	83.54
Total nitrogen	2.56	2.48	2.44	2.59	2.67	2.70	2.69	2.80	2.78
Protein nitrogen	1.75	2.10	2.11	2.04	2.25	2.50	2.44	2.56	2.58
Non-protein nitrogen	0.81	0.38	0.33	0.55	0.42	0.20	0.25	0.24	0.20
Nitrogen soluble in NaCl solution	1.77	1.00	0.87	0.76	0.68	0.62	0.53	0.55	0.52
Gliadine nitrogen	0.15	0.42	0.78	0.80	0.79	0.91	0.90	1.02	1.02
Glutenine nitrogen	0.15	0.26	0.36	0.44	0.58	0.54	0.58	0.54	0.56
Amino acid nitrogen	0.413	0.109	0.076	0.071	0.064	0.039	0.038	0.020	0.021
Ammonia nitrogen	0.508	0.091	0.059	0.163	0.137	0.139	0.102	0.111	0.111
Nitrate nitrogen	...	...	...	...	...	...	...	...	...
Crude fat	2.12	2.65	2.69	2.83	2.45	2.29	2.29	2.27	2.21
Crude fiber	4.26	3.59	2.93	2.36	2.41	2.36	2.34	2.32	2.32
Ash	3.06	2.31	2.17	1.93	1.69	1.70	1.65	1.68	1.67
Carbohydrates (by difference)	75.97	77.31	78.30	78.12	78.23	78.26	78.39	77.77	77.95
Acid values (mg. KOH per gm. fat)	90.0	69.1	49.5	47.3	33.3	20.9	18.4	19.0	18.5
Acidity of acqueous extract (cc. N)	12.7	4.0	5.1	4.1	4.1	4.0	2.8	1.6	2.3

Amounts of Constituents Calculated on Basis of Grains in 100 Ears

Number of sample	1	2	3	4	5	6	7	8	9
	Gm.	Gm.	Gm.	Gm.	Gm.	Gm.	Gm.	Gm.	Gm.
Fresh weight	85.2	198.0	242.4	267.0	281.0	255.0	242.4	201.3	185.0
Dry matter	23.5	74.8	108.6	135.2	154.5	151.9	157.1	155.4	154.6
Water	61.7	123.2	133.8	131.8	126.5	103.1	85.3	45.9	30.4
Total nitrogen	0.60	1.86	2.65	3.50	4.13	4.10	4.23	4.35	4.30
Protein nitrogen	0.41	1.57	2.29	2.76	3.48	3.80	3.83	3.98	3.99
Non-protein nitrogen	0.19	0.29	0.36	0.74	0.65	0.30	0.40	0.37	0.31
Nitrogen soluble in NaCl solution	0.42	0.75	0.94	1.03	1.05	0.94	0.83	0.85	0.80
Gliadine nitrogen	0.04	0.31	0.85	1.08	1.22	1.38	1.41	1.59	1.58
Glutenine nitrogen	0.04	0.19	0.39	0.60	0.90	0.82	0.91	0.84	0.87
Amino acid nitrogen	0.10	0.08	0.08	0.10	0.10	0.06	0.06	0.03	0.03
Ammonia nitrogen	0.12	0.07	0.06	0.22	0.21	0.21	0.16	0.17	0.17
Crude fat	0.50	1.98	2.92	3.83	3.78	3.48	3.59	3.53	3.42
Crude fiber	1.00	2.69	3.18	3.19	3.72	3.59	3.67	3.61	3.59
Ash	0.72	1.73	2.36	2.61	2.61	2.58	2.59	2.61	2.58
Carbohydrates (by difference)	17.85	57.82	85.03	105.62	120.87	118.88	123.15	120.85	120.51

centage of amino-acid nitrogen decreased from 0.064 to 0.039 per cent. Another sharp decrease in amino-acid nitrogen was registered between the 57th and the 63rd day, when the moisture content decreased from 35.21 to 22.82 per cent.

In either a very dry or a very wet soil the period of growth was increased over the period in a soil of medium moisture, according to Harris (1914). Dry weight of the tops of wheat plants increased between blooming and maturity, as shown by the data in Table 14, while the weight of roots diminished. With a soil of high moisture content (30 per cent), or the use of a complete or of a high nitrogen fertilizer, the dry weight of tops at maturity was increased. The percentage of nitrogen in tops and roots diminished during this period. With similar fertilizer treatments the tops of wheat grown on the dry soil (15 per cent moisture) contained the highest percentage of nitrogen, although the actual weight of nitrogen in the plants is less, because of the smaller weight of tops.

TABLE 14

Effect of Moisture and of Fertilizers on the Dry Weight and the Nitrogen Content of Tops and Roots of Wheat at Various Stages of Growth (Harris, 1914)

Fertilizer	Soil Moisture	Dry Weight of Tops			Dry Weight of Roots		
		Root Stage	Bloom Stage	Maturity	Root Stage	Bloom Stage	Maturity
None	30	25.31	50.54	74.95	15.80	10.73	9.31
None	15	15.21	19.38	29.28	10.02	8.82	9.59
Complete	30	45.48	76.24	102.02	22.87	14.36	11.71
Complete	15	27.79	39.96	52.32	20.76	12.46	13.19
High nitrogen	30	39.69	65.97	112.12	14.13	12.65	11.45
High nitrogen	15	26.50	39.47	37.78	18.29	11.40	13.19
		Percentage of Nitrogen in Tops			Percentage of Nitrogen in Roots		
None	30	1.23	0.94	0.83	1.36	1.23	0.96
None	15	1.62	1.32	1.25	1.68	1.59	1.36
Complete	30	1.31	0.88	0.77	1.20	1.15	0.98
Complete	15	1.62	1.17	0.96	1.50	1.35	1.30
High nitrogen	30	2.30	1.30	0.95	1.98	1.26	1.03
High nitrogen	15	1.85	1.52	1.20	1.62	1.58	1.16

The carbohydrates of the wheat grain from time of pollination to maturity was studied by Colin and Belval (1923). In the newly formed kernels the warm alcoholic extract contained about 6 per cent of levulosine (Tanret), 1.0 per cent of sucrose, and 1.0 per cent of reducing sugars, while at maturity the percentages were about 0.6, 0.5 and 0.15 respectively. They contend that the soluble carbohydrate reserve of the ripened grain is chiefly a levulosan the same as exists in the stems.

Development of the two varieties of wheat was followed by Passerini (1894), who determined the weight and the percentage of organic matter, moisture and ash from June 9 until July 8. Increases in weight occurred until June 22, when the kernels assumed a color indicating ripeness. The percentage of moisture in the grain decreased sharply after a rise of several degrees in atmospheric temperature.

Acidity and formol-titrable nitrogen in yellow-ripe and fully ripe wheat were studied by Lüers (1920), and it was found that these decreased as the grain approached complete ripeness. A further but slower decrease was observed in these values on prolonged storage subsequent to harvest. The decrease in acidity was attributed to the transformation of the acids into organic compounds, while the formol-titrable substances were condensed into more highly complex compounds.

Nitrogen and P_2O_5 content of wheat kernels during development and maturation were determined by Rousseaux and Sirot (1920). The nitrogen content of the grain increased as the grain developed, reaching the maximum before the grain became yellow. There was then a decrease in nitrogen content, followed by a second slight increase as the grain ripened. The percentage of soluble nitrogen decreased steadily and at a fairly uniform rate throughout the entire period. The grain contained 1.03 to 1.06 per cent of soluble nitrogen on June 23-25, and 0.31 per cent on July 26. The phosphoric acid content tended to parallel the nitrogen content during the development of the grain.

The development of jet barley kernels was traced by Harlan and Pope (1923). It appeared that translocation of dry matter to the kernel ceased when the moisture content fell below 42 per cent. The suggestion is made that the endosperm cells mature abruptly, the protein content reaching a density "beyond which it cannot function." In the case of this pigmented barley variety the appearance of the black pigment signalized the stage when the pericarp tissues ceased to be active. The first appearance of the color was on the dorsal surface near the tip, which, thus, is probably the first to mature. Artificial shading, protection of the head by the leaf sheaf, or cool weather can postpone the date of final maturation, and increase the size of the barley kernels.

Moisture content of the ripening grain was also observed by Olson (1917 a) to bear an important relation to the deposition and transformation of materials in the endosperm. The process of desiccation was found to accelerate gluten formation, probably through the concentration of the solution of materials from which it is formed. Wheat kernels apparently ceased to fill after their moisture content had dropped to 40 per cent. The movement of the nitrogenous material in the wheat plant was also traced by determining its distribution in the various parts of the plant. Thus the weight of nitrogen in the entire plant remained fairly constant during the first 29 days of July, while the weight in the kernels steadily increased, and that in other parts of the plant decreased. At the time of the filling of the kernel the highest percentage of nitrogen was found above the top node and in the chaff. The nitrogenous material in the chaff entered the kernel first, followed by a draft on the nitrogen in the straw above the top node. Then, if conditions remained favorable, nitrogenous material moved upward into the kernel from the lower parts of the plant. Five harvestings of wheat at weekly intervals, beginning July 1, showed a regular increase in weight of the kernel, and the percentage of nitrogen decreased from the first to the second week, and thereafter remained fairly constant. Olson further concluded that no translocation to wheat kernels occurred after the moisture content was reduced below 40 per cent. A fairly uniform rate of increase in weight of kernels was observed while the moisture content of the kernels progressively diminished until it reached 40 per cent, after which the weight of the kernels remained fairly constant. Nitrogen and phosphorus entered the grain simultaneously except in early stages of development.

Sharp (1925) traced the relation of moisture content to amino-nitrogen content and found a direct correlation between these constituents during the middle period in the development of the wheat kernel. Apparently this middle period was limited at the beginning of kernel development by a period when such a relation did not maintain. Again, at the end of the period of development there was no correlation, since the moisture content continued to decrease while the amino-nitrogen content did not decrease. While desiccation facilitated a synthesis of protein from amino-nitrogen, thus causing a decrease in the latter constituent, it was observed that the amino-nitrogen fraction decreased slowly when there was no desiccation. When immature wheat kernels were dried slowly in the unheated atmosphere, the amino-nitrogen content diminished to less than half the original quantity. Sharp accordingly contended that a decrease in moisture content of the wheat kernel is not necessary for the stage in protein synthesis which involves a decrease in amino-nitrogen content.

Since the earlier investigations tended to indicate that the protein content of the wheat kernel decreased as it approached ripeness, numerous trials were made to ascertain if grain cut on the green side would contain more of this important constituent. The experiments conducted by Bedford (1893, 1894), Briggs (1895), McDowell (1895), and others, showed that reduced yields of wheat were secured when the grain was cut before it ripened.

Wheat harvested in the soft dough stage by Failyer and Willard (1891) was found to contain 16.98 per cent of total protein, while at ripeness 7 days later it contained 15.94 per cent. Humphries and Biffin (1907) found there was no appreciable difference in the baking strength of flours milled from grain cut green, normally ripe, and dead ripe. Guthrie and Norris (1912 a) concluded that the milling quality of wheat is not greatly affected by the stage at which it is harvested. Baking tests of flours milled from wheats cut at the soft-dough, hard-

GROWTH AND DEVELOPMENT OF THE WHEAT

dough, and dead-ripe stages showed that cutting at the hard-dough stage did not affect the size of the loaf unfavorably. Shaw (1913) conducted numerous trials of early and late harvesting at two points in California, and found no marked difference in the percentage of protein in kernels of the lots classified on the basis of time of harvest. Of 61 trials, the average protein content of the early cut wheat was 11.80 per cent and of the late cut 11.51 per cent.

Marquis wheat was harvested at different dates by Saunders (1921), and the progressive changes in weight and composition of the grain were determined. While the weight of total protein in the kernels increased steadily from July 21 to Aug. 15, the percentage of protein decreased somewhat during the first week of this period, but remained fairly constant during the latter half, as shown by the data in Table 15.

TABLE 15
Analyses of Marquis Wheat at Different Stages of Growth, Reported by Saunders (1921)

Date Harvested	Protein in Wheat Kernels	
	Per Cent	Grams per 100 Kernels
July 21	20.02	0.884
July 24	16.42	1.337
July 27	14.09	1.925
July 30	14.62	2.783
Aug. 2	15.09	3.814
Aug. 4	15.41	4.287
Aug. 6	15.57	4.610
Aug. 8	15.87	4.890
Aug. 11	15.89	4.940
Aug. 13	16.03	5.015
Aug. 15	15.92	5.104

It appears from these various studies that no advantage is to be gained, in so far as gluten content of the grain is concerned, by harvesting the wheat early, and reduced yields per acre may result from cutting the grain before it is ripe.

Changes in weight of wheat kernels, which resulted from cutting the plants at different stages of maturity and subjecting them to different treatments, were determined by Briggs (1895). It appeared that the grain derived material from the straw after cutting, the greatest gains in this respect being made by those kernels of the plants cut before the plants were ripe. In these cases the more of the plant that was left attached to the head, the greater the increase in weight of the kernels. These data are shown in Table 16.

The proportionate weight of head to weight of straw increased from 14.4:85.6 at time of flowering, to 58.4:41.6 at time of dead ripeness in the plants examined by Shutt (1911 a). The percentage of nitrogen in the dry matter of the straw decreased during the same interval from 1.72 per cent to 0.46 per cent.

TABLE 16

WHEAT CUT AT SUCCESSIVE STAGES OF MATURITY AND SUBJECTED TO DIFFERENT TREATMENTS (BRIGGS, 1895)

(The figures indicate the weights, in milligrams, of 100 kernels.)

Time of Harvesting	Shelled and Weighed Immediately upon Cutting	Ripened in the Head	Ripened in the Straw	Ripened with the Roots on	Average Weights at Successive Stages of Maturity
June 19, milky stage	1057	1112	1521	1905	1400
June 30, dough stage	3868	3852	4051	4083	3962
July 13, yellow ripe	4850	4955	5035	5190	5008
July 24, dead ripe	4835	4917	5030	4665	4862
Average weights for the different treatments	3652	3709	3909	3960	....

Data resulting from a continuation of the work previously reported by Saunders (1921) were presented by Shutt (1922). Three lots of samples were taken on each date of cutting, (1) with kernels removed from heads at once, (2) heads with 6 inches of straw attached, which were allowed to dry in the laboratory, and (3) heads with full length of straw, allowed to dry under "stook" (shock) conditions in the field. From these data, given in Table 17, it may be concluded that in the case of immature wheat, when cut with straw attached to heads, there is an appreciable increase in the weight of the grains, probably in conse-

TABLE 17

PRELUDE WHEAT: PROTEIN (ON DRY BASIS) AND WEIGHT OF KERNELS (SHUTT, 1922)

Date of Cutting	Kernels Removed at Once			Kernels with 6-in. Straw			Kernels with Full Length Straw		
	Weight of 1000 Kernels, Grams	Per Cent	Grams per 1000 Kernels	Weight of 1000 Kernels, Grams	Per Cent	Grams per 1000 Kernels	Weight of 1000 Kernels, Grams	Per Cent	Grams per 1000 Kernels
June 27, 1921	11.24	18.64	1.93	11.27	18.70	1.93	13.17	19.73	2.37
June 29, 1921	15.26	17.24	2.41	15.20	18.43	2.57	18.31	19.48	3.25
July 1, 1921	18.73	17.26	2.97	18.71	17.31	2.94	19.91	18.25	3.29
July 4, 1921	24.29	18.07	3.99	25.14	18.40	4.18	25.17	18.88	4.33
July 6, 1921	27.60	18.64	4.65	28.29	18.80	4.83	29.16	18.80	4.97
July 8, 1921	30.21	19.28	5.25	30.37	19.07	5.24	29.18	19.21	5.09
July 11, 1921	30.55	19.58	5.29	30.35	19.00	5.22	29.62	19.38	5.24
July 13, 1921	30.92	19.51	5.47	30.79	19.53	5.44	30.59	19.70	5.41

quence of a continued translocation of material from straw to grain. Weight of the kernels increased steadily up to ripeness, but the protein content was fairly constant during the latter part of the ripening period.

No mention has been made in the foregoing paragraphs of the changes incident to the wintering of fall sown wheats. These were discussed by Schulze (1904), who found 100 wheat plants weighed 6½ times as much on April 22 as they did at the end of the preceding October. The amids present in the roots were moved into the upper parts of the plant when development commenced in the spring. Wheat took up most of its nitrogen during the period from the end of April until kernels began to form, differing from rye, which took up one-half of its nitrogen before the close of winter. Wheat continued to take up nitrogen from the soil while the heads were forming. Potash and phosphorus were assimilated largely from the beginning of spring growth until blossoming. The greatest use of potash was simultaneous with the active formation of carbohydrates and cellular tissue. The needs for lime and magnesia by the young plants were small, but increased as the plants developed, and Schulze believes this to be connected with the hardening of the tissue.

No constant relation between depression of the freezing point, specific conductivity, or hydrogen-ion concentration of the cell sap, and relative frost hardiness of winter wheats, was found by Newton (1922). Neither was the relation between dry matter content and hardiness constant. Hardened wheat plant tissue retained its water content with great force, no appreciable amount of sap being expressed by 400 atmospheres pressure. In a later paper Newton (1923) stated that the imbibition pressure of the fresh leaves of wheat plants in the winter hardened condition was found in most cases to be directly related to hardiness. The volume of press juice from a unit quantity of hardened leaves afforded an index of hardiness, being inversely proportional to the latter character. When plants in an unhardened condition were examined by similar methods, no such correlations could be detected, indicating that exposure to diminishing temperatures sets up some mechanism in the hard varieties which progressively modifies the colloidal condition of the protoplasm of exposed structures.

A considerable loss of material, presumably carbohydrates, in consequence of respiration during the ripening of wheat, was observed by McGinnis and Taylor (1923). The greatest loss, calculated in terms of carbohydrates oxidized per 24 hours, and amounting to from 1.0 to about 1.5 grams per 100 grams of dry matter daily, occurred while the plants contained in excess of 40 per cent of moisture. After the moisture content dropped below 20 per cent the daily loss was very small, ranging around 0.2 grams of carbohydrates. These observations suggest that conditions determining the relative rate of respiration of the ripening plants may affect, in a measure, the relative protein and carbohydrate content of the grain that is produced.

From the available data it appears that germination of the wheat kernel is attended by a change in the physiological activity of the scutellum, an organ of the germ or embryo. This secretes active enzymes, which digest, and render soluble and diffusible the reserve proteins, carbohydrates and other nutrients stored in the endosperm. The resulting materials are translocated to the embryo, and serve as nutrients for the seedling. The latter develops roots, which draw upon the substratum for inorganic nutrients. Mineral nutrients, including nitrates, are absorbed by the young plant, and serve in the growth of a structure which proceeds to fruit. Most of the mineral matter is taken up by the plant before blossoming, used in plant metabolism, and later translocated in part to the head, and partially lost by washing back into the soil. Nitrogenous constituents in particular are progressively translocated to the developing kernel. If nitrates are applied to the soil about the time the plant flowers, they will be absorbed, and an increase in the nitrogen content of the grain will result. In the usual course of field culture of wheat such increases in nitrate content of the soil do not occur at midseason. Nitrates are not adsorbed by the soil particles, and since nitrification is generally proceeding in the soil at the highest rate early in the season, and nitrates not absorbed by the plants are leached out of the surface soil, it follows that as blossoming time is approached the soil nitrate content is progressively diminishing rather than increasing. Thus the nitrogen metabolism of the wheat plant during the first half or two-thirds of the season is of more significance than is this particular metabolism after kernel development begins. For it is during the pre-flowering period that, under ordinary field conditions, the bulk of the nitrogen enters the wheat plant.

After flowering the carbohydrate metabolism becomes of prime importance. Data presented in the next section justify the conclusion that the rate of starch deposition in the kernel may be determined in large measure by the moisture relations of the plant. If the atmospheric humidity is fairly high, and the substratum moist, the flow of carbohydrates into the kernel is rapid, and a plump, "starchy" kernel will result. When an abundance of nitrogen has previously been absorbed by the plant, and carbohydrate synthesis does not rise above an average rate, the kernel should be plump, and high in gluten. Under drought conditions late in the growing period, with a restricted carbohydrate synthesis, but with the flow of nitrogen to the kernel not seriously interfered with, a small-berried kernel rich in gluten will result.

Nitrogenous material is not built up into gluten as fast as it is moved into the wheat kernel. Gluten synthesis is evidently conditioned by the moisture content of the grain. When the moisture content falls below 40 per cent, translocation of nitrogenous material to the kernel practically ceases, and gluten synthesis ensues. The latter may be effected in part by an increased concentration of the "bausteine" or amino-compounds resulting from the loss of moisture, although a periodicity in the cycle of changes may also contribute to the rate of protein synthesis.

When the wheat plant is cut from the roots while still somewhat green, translocation of material from the straw and leaves to the kernel may continue for a time, with a slight increase in the (dry) weight of the latter.

Chapter 4.

Influence of Environment on the Composition of Wheat.

It has long been known that the environment of the growing wheat plant exerted a profound effect upon the composition of the ripened grain. Environment in this case is determined by so many variables that it becomes difficult to ascertain the exact influence of each factor. While soil and climate may be assumed in a superficial way to constitute the totality of environment, both of these are exceedingly complex, and neither is the least stable during an extended period. Climate may be considered and recorded in such terms as temperature, precipitation, humidity, wind velocities, proportion of sunshine, cloudiness and darkness, barometric pressures, and other variables, all of which fluctuate widely. Hence climate presents an ever-changing kaleidoscopic picture, with corresponding variations in the effect on living things in the region. Records accumulated at one geographical location are exact for a limited area only, and substantially different conditions may be encountered but a short distance from the recording station.

Soil, likewise, is not stable, and as the substratum for the developing wheat plant must be regarded as dynamic in its potentialities. The soil bacterial flora is responsible for a succession of changes on the one hand, while leaching and other physical treatments may operate in an opposite direction in translocating the products of bacterial changes. These soil changes are tied up with climatic conditions and are responsible for both chemical and physical modifications which may influence the life in the soil. Because of these interrelations of soil and climate it may be contended that certain properties of the ripened wheat grain which have been correlated with climate are affected by the climate only because the latter in turn has induced certain changes in the substratum or soil. It thus becomes apparent that the discussion of the effect of climate cannot always be resolved into a precise correlation of individual climatic (and soil) factors with the composition of the wheat plant and grain.

A survey of the literature in this field reveals wide variations in the composition of wheat grown in various parts of the world. König (1903) reports variations in the protein content of wheat ranging from 7.07 per cent, in the instance of a sample from Scotland, to 24.16 per cent in a sample from the Russian Caucasus. Like variations in the percentage of protein have been reported in the case of wheats grown within the confines of the United States. Thus Shollenberger and Clark (1924) have reported a range of from 7.5 to 19.6 per cent of crude protein in samples of a single hard spring wheat variety (Marquis) grown in the western United States. The sample lowest in protein was grown at Chico, California, in 1919, the highest at Havre, Montana, in 1920, and the samples reported in this connection were collected during only five seasons. The variations in the percentage of other constituents of the wheat kernel, while marked, are generally not so great as are encountered in the percentage of protein. The percentage of starch generally varies inversely as the percentage of protein; the percentage of fiber and of ash is often higher than normal when the kernels are shrunken and shriveled. Since commercial value in any particular class of wheat is associated with the percentage of protein (or gluten), the variations in this constituent become a matter of importance and will receive the most attention in this treatment of the subject. In many of the researches reported, in fact, the protein percentages are the only analytical data recorded.

Climate and the Composition of Wheat.

About the middle of the last century Lawes and Gilbert (1857) had under way a series of experiments which led to their observation that a long ripening period after heading resulted in plump wheat kernels with a low percentage of nitrogen, whereas when the ripening period was relatively short the reverse was true. Schindler (1893) confirmed this opinion and attributed the low percentage of nitrogen found in wheat produced during certain crop seasons to a long period between flowering and ripeness. Jensen (1899) showed that as the climate became more continental in character the size of the wheat kernel decreased and the percentage of nitrogen increased. This is indicated by the fact that cold and dry winters, rains in the late spring and early summer, and high temperatures just previous to and during harvest, as in Hungary, Russia, Manitoba, and the northwestern part of the United States, result in the production of wheat containing a high percentage of nitrogen, and hard and flinty in texture. Insular climates, on the other hand, result in wheats low in gluten, high in starch, but plump and soft in texture.

The percentage of nitrogenous substances was found by Melikov (1900) to be highest (21.2 per cent) in wheat of southern Russia in dry years, while in years of good crops it fell to 14 per cent. The Ploti (Russia) experiment station (1901) reported that those conditions which favor an increase in the yield of grain result in a decrease in the percentage of nitrogen. Excessive humidity favored the deposition of carbohydrates, while drought hastened maturity and produced high-protein grain.

The nitrogen content of wheat varies widely, and depends more upon the conditions under which the wheat is grown than upon the variety, according to Schribaux (1901). Circumstances which shortened the period of growth were found to favor the formation of nitrogenous substances in the grain. Schribaux further observed that both starchy and glutinous seed wheat would, under similar conditions of growth, produce crops having the same composition.

The inferiority of the yield and the higher nitrogen content of the wheat crop of 1881 at Grignon, France, when compared with the crop of 1880, was attributed by Deherain and Meyer (1882) to the unfavorable summer weather of 1881. Deherain and Dupont (1902) report that in 1888 the harvest at Grignon was late and the process of ripening slow, the wheat containing 12.60 per cent of gluten and 77.2 per cent of starch. The following season was hot and dry, resulting in a rapid ripening of the grain, which contained 15.3 per cent of gluten and 61.9 per cent of starch, although the yield of grain was smaller.

The influence of various factors on the composition and characteristics of wheat was summarized by Hall (1905). Soil was asserted to have a comparatively small effect on the composition of the grain, while the effect of climate was large, though less than anticipated.

A negative correlation of rainfall with the percentage of nitrogen in wheat, as well as an increased nitrogen content of the grain in hot summers, was observed by Kharchenko (1903). Gurney and Norris (1901) report an increased gluten content in consequence of hot, dry weather while the grain is ripening. Wohltmann (1906) concluded that cloudy, rainy summers decreased the percentage of nitrogen in the grain, while dry, sunny, warm summers produced the opposite effect.

A variety of Hungarian wheat was grown in California, Maryland, Kentucky and Missouri, and on analyzing the crops produced during the same season, Wiley (1901) found the percentage of protein in the wheat ranged from 9.68 per cent in the Maryland sample to 17.86 per cent in that which was raised in California. It was accordingly concluded that the wide variation is due almost entirely to meteorological conditions, which influence the rate of growth and rapidity of ripening.

Attention was called by Soule and Vanatter (1903) to the influence of the season on the protein content of wheat. As an example, the variety known as Beech Woods Hybrid contained 21.92 per cent of crude protein in 1900, 13.77 per cent in 1901, 15.74 per cent in 1902, and 15 per cent in 1903. During the crop seasons of 1900 and 1902, when the protein content of the wheat was highest, the rainfall was somewhat deficient during the ripening period. This resulted in a short ripening period, retarding the elaboration and transformation of starch from the leaves and stems to the grains.

A tendency in the direction of a higher protein content in wheat grown in dry soils when contrasted with that grown on soils with a higher moisture content, was observed by Hare (1904). Feilitzen (1904) likewise concluded that weather conditions induced a marked effect upon the protein content and vitreousness of wheat.

The results of investigations in this field were reviewed by Lyon (1910), and it was pointed out that the production of hard, glutinous wheat resulted from (a) an abundant supply of nitrogen in the soil, (b) hot weather early in the growing season to promote nitrification in the soil, and a large absorption by the plant, (c) light rainfall and large evaporation, which concentrates the nitrates in the upper portion of the soil where they may be readily obtained by the plant, and (d) a short ripening period, preventing the transference of starch to the kernel. The short ripening period is usually due to high atmospheric temperatures.

A large number of durum wheat samples from different parts of the United States, which were grown under arid, semi-arid and humid conditions, were examined by Le Clerc (1906). The results of the analyses of these samples are given in Table 18.

TABLE 18

Composition and Physical Characteristics of Durum Wheat Grown Under Arid and Humid Conditions (Le Clerc, 1906)

	Nitrogen in Wheat, Per Cent	Ash, Per Cent	Weight per 1000 Kernels, Grams	Flinty Kernels, Per Cent
Average of wheat grown in arid and semi-arid regions	2.79	2.13	30.3	98
Average of wheat grown in humid and irrigated regions	2.22	2.15	33.5	69

The effect of the length of the growing season upon the protein content of wheat was also noted by Le Clerc. Wheat grown in Colorado and South Dakota, when the length of the growing period averaged 114 days, contained 3.13 per cent of nitrogen, and when the length of the growing period in the same regions averaged 137 days the nitrogen content was 2.33 per cent. Wheat grown in different localities contained an average of 2.46 per cent of nitrogen when the period of growth averaged 162 days, and 2.80 per cent when 90 days, the average rainfall being nearly the same in both cases.

Crimean winter wheat from 1905 Kansas seed was grown in California, Kansas and Texas, under supervision of Le Clerc and Leavitt (1909, 1910). The 1906 wheat crop had the composition and characteristics shown in Table 19.

TABLE 19

COMPOSITION AND CHARACTERISTICS OF CRIMEAN WHEAT GROWN IN THREE LOCALITIES BY LE CLERC AND LEAVITT

	Crude Protein, Per Cent	Ash, Per Cent	Weight per 1000 Kernels, Grams	Weight per Bushel, Pounds	Flinty Kernels, Per Cent
Original seed	16.22	2.41	21.0	56.5	98
Kansas, 1906 crop	19.13	2.05	22.7	58.8	100
California crop	10.38	1.81	34.0	59.4	36
Texas crop	12.18	1.80	30.8	58.9	..

Seed from each point was grown at each of the other points the following season, the analysis of the crop being shown in Table 20. This shows that low protein seed produced wheat high in protein under certain climatic conditions, and that wheat high in protein, such as the 1906 Kansas seed, when grown under such conditions as prevailed in California during 1907 produced a crop containing as low percentages of this constituent as locally grown seed. In other words, the protein content of the wheats in any section could not be increased by planting high protein seed wheat of the same variety grown in other sections. Attention is also called by Le Clerc and Leavitt to the fact that the absolute quantity of nitrogen per kernel was not the same in all cases, as has sometimes been assumed. One thousand kernels of the Kansas grown seed contained 4.6 grams of protein, while the same number of California grown kernels contained but 3.7 grams of protein.

TABLE 20

COMPOSITION OF WHEAT GROWN IN THE TRI-LOCAL EXPERIMENTS OF LE CLERC AND LEAVITT (1909, 1910)

	Grown in Kansas			Grown in California			Grown in Texas		
	Kans. seed	Calif. seed	Tex. seed	Kans. seed	Calif. seed	Tex. seed	Kans. seed	Calif. seed	Tex. seed
Protein, per cent	22.23	22.23	22.81	11.00	11.33	11.37	16.97	18.22	18.21
Ash, per cent	2.22	2.28	2.21	2.21	2.21	2.26	2.46	2.45	2.46
Wt. per 1000 kernels, grms.	20.5	21.3	20.4	33.8	33.3	33.0	23.6	22.7	23.6
Wt. per bu. lbs.	51.3	51.3	50.7	61.3	61.8	62.3	58.5	57.3	58.6
Flinty kernels, per cent	100	100	100	50	60	50	98	100	95

As the result of several years' work with Washington wheats Thatcher (1911) concluded that the ripening process was simply one of desiccation, and that climatic conditions favoring a gradual but complete maturity resulted in the production of grain of a lower protein content than did rapid ripening. It was further shown that there was a definite relationship between the amount of rainfall and the composition of wheat grown in Washington. In Table 21 the percentage of crude protein, where the total rainfall from 1905 to 1909 was 45 inches, was shown to be 12.82 per cent, while in the case of wheat grown where the rainfall for the same period was 116 inches, the crude protein content was 9.03 per cent.

TABLE 21

PROTEIN CONTENT OF WASHINGTON WHEATS FROM DIFFERENT COUNTIES IN COMPARISON WITH THE RAINFALL (THATCHER, 1911)

County	Average Protein, Per Cent	Rainfall at	Rainfall, Inches 1905 to 1909
Adams and Franklin	12.82	Hatton	45
S. W. Whitman	12.53		
Okanogan	12.37	Omak	50
Douglas	12.25	Waterville	58
N. W. Whitman	11.65		
Walla Walla	11.56	Eureka	69
Lincoln	11.16	Wilbur	73
Garfield and Asotin	10.96	Pomeroy	82
N. E. Whitman	10.75	Rosalia	97
S. E. Whitman	10.63	Pullman	103
Klickitat	9.03	Goldendale	116

A general negative correlation between the total rainfall from April to August, inclusive, at different points in Minnesota, and the protein content of spring wheat grown in the several localities, was found by Bailey (1913 a). Bailey (1913 b) later stressed the significance of the distribution of rainfall during this period, showing that moderate precipitation at frequent intervals resulted in a lower protein content than occasional heavy rains, even though the total precipitation might be the same during the four months of the growing period.

A greater change in the protein content of wheat on increasing the moisture content of the soil from 17.6 per cent to 42.8 per cent than resulted from the application of fertilizers, was observed by Morgan (1911). Certain of his data are presented in the section of this chapter in which the influence of soil composition and of fertilizers is discussed.

When the same wheat variety was sown both in the fall and in the spring in the same field, Thatcher (1910) found that the spring-sown grain, ripening later and more rapidly, invariably contained a higher percentage of protein than the fall-sown wheat. Wheat grown in the hotter, drier sections of Washington, where the ripening period was shorter, was invariably higher in percentage of protein than the same varieties grown in sections where the rainfall was heavier, the harvest weather cooler, and the ripening slower.

When wheat plants were shaded with duck and burlap, Thatcher (1913 b) found that in most instances the protein content of the grain was lower and the starch content higher than was true of the grain from the unshaded plants. When the length of the period of kernel development was studied as a variable, it was found that in the case of Washington wheats it varied inversely with the percentage of nitrogen in the grain.

The effect of several factors on the protein content of wheat was discussed by Shaw (1913), who notes wide seasonal variations. Glutinous seed commonly gives progeny in California of low gluten content. Rainfall induced substantial variations, and the supply of moisture in the soil during the latter part of the growing period seemed to be especially significant. Late seeded grain possessed a higher protein content than that which was seeded early. A long growing period tended to reduce the gluten content of the grain.

The milling, baking and analytical data resulting from the examination of wheat samples grown in regions (in North Dakota) where the rainfall varied during the months of May to August, inclusive, were presented by Stockham (1912 a). No exact correlations between the protein content and the precipitation are shown, although in a general way it appeared that the percentage of protein in the wheat was highest in the regions of highest rainfall. Stockham (1912 b) found, however, that the lower the moisture content of the soil at the time of the maturity of the wheat plants, the higher the protein content of the wheat kernels. In 1910 and 1911 the relation between average soil moisture and average protein content of the hard spring wheat was summarized, with the result shown in Table 22.

TABLE 22

Soil Moisture at Time of Maturity and Protein Content of Wheat as Shown by Stockham (1912 b)

Moisture in Soil at Maturity of Wheat Plants	Protein in Wheat ($N \times 5.7$)	
	1910 Per Cent	1911 Per Cent
Below 10 per cent	15.33	16.24
Between 10 and 15 per cent	14.59	14.11
Above 15 per cent	12.86	13.27

It was noted by Stockham (1912 c) that when atmospheric temperatures alone are considered as a factor, there must be marked deviations from the normal when considering the mean of an extended period, to establish any correlation with the composition of wheat. It appears that the day to day variations are more important than the mean temperature for extended periods.

The data collected over a period of five crop seasons relating to certain correlations of factors involved in environment and the composition and quality of wheat were then summarized by Stockham (1914). In addition to the factors discussed in the foregoing paragraphs, his conclusion that the general baking quality of wheat is better in seasons of low yields per acre than when abundant yields are secured justifies special emphasis. Moisture conditions apparently induced greater variations than did the other factors studied.

In Roumania the quality of the wheat crop was controlled by the light and temperature relations maintained during the formation of the kernels, in the opinion of Dobrescu (1921). Dry, warm weather from blooming to maturity favored the production of high quality grain. Assimilation of plant nutrients, and hence the relative yield of grain appear dependent on the water content of the soil.

The length of the growing period of the head-bearing stalks was observed by Gericke (1922 a) to be decidedly shorter in those cultures that produced high protein wheat than in those which produced wheat low in protein. Thus, when the length of the growing period of head-bearing stalks was 193 days in one instance, and 125 days in the second instance, the protein content of the wheats produced were 8.6 per cent and 15.2 per cent respectively, with regular inverse gradations between the two extremes. When the variations in length of growing period were conditioned by the environment of the plant, a greater effect on the protein and carbohydrate content of the grain was apt to be encountered, in Gericke's opinion, than when occasioned by genetic characteristics of the variety. In the experiments of Gericke, nitrogen nutrition of the wheat plants determined, in large measure, the length of this period.

That climate, rather than modified soil conditions resulting from cropping to wheat, is responsible for fluctuating gluten content of wheat, is suggested by the data presented by Balland (1914), and Briggs (1923). Balland summarized the data resulting from the examination of French wheat crops from 1869 to 1914 inclusive. The range and average of the gluten content of the three periods in which these data were grouped did not indicate any degeneration in the wheat during the 46 seasons. This is shown by the data in Table 23. Balland pointed out that variations in climate induced the lower average gluten content in the third or last period; also that more highly refined flours were

TABLE 23

Gluten Content of Flours Milled from French Wheat Crops of 1869-1914, Arranged in Three Groups, as Reported by Balland (1914)

Period	Wet Crude Gluten, Per Cent		
	Average	Minimum	Maximum
1869-1880	29.1	27.5	30.3
1881-1895	25.1	23.4	26.8
1896-1914	24.7	23.1	26.8

produced during the later years, which flours contained less of the gluten in the wheats from which they were milled.

The average percentage of gluten in 16 successive crops of northwestern spring wheat, as determined by Briggs (1923) and shown in Table 24, justifies the conclusion that recent crops are as glutinous, and presumably as satisfactory in baking qualities, as the crops of 15 and 16 years earlier.

TABLE 24

Average Gluten Content of Northwestern Spring Wheat Flour by Crops, as Reported by Briggs (1923)

Season	Gluten in Flour, Per Cent	Season	Gluten in Flour, Per Cent
1907	11.70	1915	10.93
1908	11.50	1916	11.51
1909	12.64	1917	12.99
1910	12.42	1918	12.18
1911	13.80	1919	11.51
1912	11.67	1920	12.35
1913	11.56	1921	12.37
1914	11.29	1922	12.02

Effect of Irrigation on the Composition of Wheat.

In the discussion of the effect of climate on wheat it was made evident that the moisture relations of the growing wheat plant, as induced by precipitation and other conditions, were responsible in large part for the composition of the ripened plants. If increasing rainfall tends to produce diminishing protein content of the grain, it would be reasonable to expect that increasing applications of irrigation water would have a like effect.

Increasing the percentage of moisture in the soil was found by Prianishinkov (1900) to decrease the percentage of nitrogen in the grain of wheat. Traphagen (1902) found 8.81 per cent of protein in wheat grown in irrigated soil, and 14.44 per cent in wheat from plots which were not irrigated. The percentage of moisture, ash, crude fiber and ether extract did not differ materially in the two lots of wheat.

An elaborate investigation of the effect of varying quantities of irrigation water on the yield and composition of wheat was conducted by Widtsoe (1902). Twelve plots, receiving progressively increasing applications of irrigation water, were employed. Certain of the data resulting from this study are included in Table 25, and show a regularly diminishing percentage of crude protein with increasing amounts of irrigation water, until the latter exceeded 30 inches. Probably any additional water over 30 inches rendered the soil too moist for normal plant growth, as evidenced by the decreased yield of grain on the plot receiving 40 inches. It is interesting to note that the total weight of nitrogen in the grain from an acre increased with increasing applications of irrigation water up to 30 inches of water.

TABLE 25

RELATION BETWEEN IRRIGATION WATER APPLIED, COMPOSITION OF WHEAT, AND YIELD OF NITROGEN AND ASH PER ACRE, AS REPORTED BY WIDTSOE (1901)

Water Applied, Inches	Yield Per Acre, Bushels	Crude Protein, Per Cent	Ash, Per Cent	Yield of Nitrogen Per Acre, Pounds	Yield of Ash Per Acre, Pounds
4.63	4.50	24.8	2.50	10.7	6.75
5.14	3.83	23.2	3.07	8.5	7.05
8.73	10.33	19.9	2.54	19.7	15.74
8.89	11.33	19.4	2.93	21.1	19.72
10.30	14.66	18.4	2.34	25.9	20.24
12.09	11.16	21.3	3.25	22.8	21.44
12.18	11.66	23.1	2.88	25.8	20.30
12.80	13.00	17.1	2.52	21.3	21.50
17.50	15.33	17.2	2.57	25.3	23.64
21.11	17.33	15.9	2.34	26.4	24.33
30.00	26.66	14.0	4.14	35.8	26.20
40.00	14.50	17.1	2.52	23.8	21.92

The moisture content of soil in pots in which spring wheat was grown was varied by von Seelhorst (1900). In lot 1 the soil moisture was maintained at 47.4 per cent of saturation; in lot 2 at 47.4 per cent until the plants began to shoot and then at 84.1 per cent; in lot 3 at 84.1 per cent until the plants began to shoot and then at 47.4 per cent; while in lot 4 the soil moisture was maintained at 84.1 per cent of saturation throughout the experiment. The protein content of the grain grown in these pots is recorded in Table 26. The highest percentage of protein was found in the grain grown on the dryest soil, and the next in order of protein content was the grain raised on the soil which had a low content of moisture at the outset, with a high water content later in the life of the plant. That grown on the soil which was moist throughout the life of the plant contained the least protein.

TABLE 26

EFFECT OF MOISTURE IN SOIL ON NITROGEN CONTENT OF WHEAT
(VON SEELHORST, 1900)

Group	Moisture in Soil in Terms of Saturation		Nitrogen Content of the Air-Dry Kernels, Per Cent
	First Stage	Second Stage	
1	47.4	47.4	2.769
2	47.4	84.1	2.501
3	84.1	47.4	2.391
4	84.1	84.1	2.354

Durum wheat grown with and without irrigation was analyzed by Le Clerc (1906), and the chemical and physical characteristics of these two lots of wheat are shown in Table 27. A reduction in the percentage of protein and of flinty kernels apparently resulted in consequence of the application of irrigation water, the plumpness of the kernels, as shown by the weight per 1000 kernels, being the same in both cases.

TABLE 27

CHARACTERISTICS OF DURUM WHEAT GROWN WITH AND WITHOUT IRRIGATION (LE CLERC, 1906)

	Protein, Per Cent	Flinty Grains, Per Cent	Weight Per 1000 Kernels, Grams
Irrigated wheat	11.1	20	29.4
Non-irrigated wheat	17.7	100	29.2

Preul (1908) varied the applications of water to soils rich and poor in fertility, and found that the nitrogen content of wheat grown on these soils was high when the application of water was constantly low.

Application of irrigation water to black soil by Evans (1909) resulted in softer, less vitreous durum and hard wheats, and therefore probably reduced their strength. Similar applications of water to sandy, silt soil had much less effect upon the hardness of the grain.

The effect of irrigation upon the moisture content of soils, and the composition of wheat grown on the soils at a station in Alberta, Canada, was determined by Shutt (1909 a). The irrigated plot received one application of water on July 15, 1908, after the soil sample had been taken. Table 28 shows the moisture content of soil samples taken from the irrigated and non-irrigated plots at three times during the season. The irrigation on July 15 evidently maintained the moisture content of the irrigated plot substantially higher than that of the non-irrigated plot during the last month of the growing period.

TABLE 28

MOISTURE IN THE SOIL ON WHICH WHEAT WAS GROWN BY SHUTT (1909 a)

Date Sampled	Moisture in the Soil	
	Irrigated Plot Per Cent	Non-irrigated Plot Per Cent
May 14, 1908	16.56	15.61
July 15, 1908	8.78	8.11
Aug. 17, 1908	10.37	6.38

Red Fife spring wheat grown on the irrigated plot contained 13.70 per cent of crude protein, while wheat of the same variety grown on the non-irrigated plot contained 16.37 per cent of crude protein.

These studies of the effect of irrigation were continued by Shutt (1911 a), who presented the results of analysis of Red Fife wheat grown in Alberta during 1909. The irrigated plot in this instance received one application of irrigation water on July 10, 1909. On Aug. 25, 1909, the moisture content of the irrigated and non-irrigated soils was 8.16 per cent and 5.99 per cent respectively, while the protein in the wheat from the two plots was 11.74 per cent in the case of the irrigated plot, and 16.13 per cent in the wheat from the non-irrigated plot.

Wheats grown on irrigated soil were compared with dry-farmed wheats of the same region by Stewart and Hirst (1913). While the differences in the composition of their samples were not as marked as those noted by certain other investigators, it appears from their data that the dry-farmed wheats were characterized by a higher protein content than were the irrigated wheats. Baking tests resulted in larger loaves from the dry-farmed than from the irrigated wheats. Their data are found in Table 29.

TABLE 29

COMPOSITION AND BAKING QUALITY OF WHEATS GROWN UNDER IRRIGATION BY STEWART AND HIRST (1913)

	Weight per 100 Kernels, Grams	Crude Protein Wheat, Per Cent	Crude Protein Flour, Per Cent	Volume of Bread, cc.
Irrigated with 25 in. water....	4.008	14.00	12.63	1605
Irrigated with 15 in. water....	4.065	14.35	12.92	1630
No irrigation	3.569	15.45	13.62	1655
Winter dry-farmed wheat.....	3.004	15.76	14.64	1681
Spring dry-farmed wheat.....	3.106	16.85	15.74	1841

Lowering the nitrogen content of the soil in the third stage of plant development was found by Harris (1914) to tend to reduce the length of (a) the period from heading to maturity, and (b) the period from planting to maturity. Table 30 shows the percentage of nitrogen in

TABLE 30

MOISTURE IN SOIL AND NITROGEN CONTENT OF WHEAT, AS REPORTED BY HARRIS (1914)

Percentage of Soil Moisture			Percentage of Nitrogen in Wheat	
First Stage	Second Stage	Third Stage	1908-9	1909-10
30	30	30	2.62	2.05
30	30	15	2.82	2.19
30	15	15	2.98	2.31
15	15	15	3.24	2.42
15	15	30	3.24	2.52
15	30	15	2.96	2.21

the grain when the moisture content of the soil was varied during the three stages of plant growth. The highest average percentage of nitrogen resulted when the soil was comparatively dry throughout all three stages, or through the first of the two stages of growth.

In a series of plots in which the soil moisture was maintained uniformly at a predetermined percentage, the nitrogen content of the grain diminished regularly with increasing soil moisture content up to 37.5 per cent, as shown by Harris in Table 31.

TABLE 31

Soil Moisture Throughout Growing Period and Nitrogen Content of Wheat Plants (Harris, 1914)

Soil Moisture, Per Cent	Nitrogen, Per Cent		
	Grain	Straw	Grain and Straw
11	3.24	.69	1.38
13	3.13	.49	1.28
20	2.51	.42	1.04
25	2.30	.34	.94
37½	2.26	.40	.93
45	2.39	.40	.90

The slight effect on the protein content of wheat resulting from progressively increasing the amount of irrigation water applied to wheat plots at Boise, Idaho, was stressed by Headden (1916 b). Headden (1918 a, 1918 b) emphasizes his conclusion that in the case of Colorado wheats the amount of water used in irrigating did not affect either hardness or nitrogen content of the grains of wheat. No differences were observed in the baking qualities of flours milled from these wheats. Olson's (1923) experience in studies of irrigation at Grandview, Washington, was similar to that of Headden.

Larger yields of flour were obtained in milling irrigated wheat than in milling dry-farmed wheat of the same variety grown in Idaho by Jones and Colver (1916). The characteristics and composition of 51 dry-farmed and 48 irrigated wheat samples that were compared are shown in Table 32. These data show that the dry-farmed wheat, and flour milled therefrom contained higher percentages of crude protein than the irrigated wheat. Baking tests of the flours were conducted, the average volume of loaves baked with 340 grams of flour being 1369 cc. in case of the dry-farmed, and 1376 cc. in case of the irrigated samples. The total scores of the two lots of bread were nearly the same. From this it appears that there was little difference in the baking qualities of the two lots of flour.

Data and graphs indicating a negative correlation between amount of irrigation water applied and percentage of protein in wheat were presented by Jones, Colver and Fishburn (1918). Emphasis is placed on their conclusion that irrigation does not of necessity imply that the wheat

TABLE 32
Composition of Dry-farmed and Irrigated Wheat and Flour Milled Therefrom, Analyzed by Jones and Colver (1916)

Wheat	Dry-farmed	Irrigated
Weight per bushel, pounds	59.01	58.82
Weight per 1000 kernels, grams	35.72	35.36
Moisture, per cent	10.43	10.46
Ash, per cent	1.76	1.76
Ether extract, per cent	2.09	2.19
Crude protein (N × 5.7), per cent	12.01	11.08
Flour		
Moisture, per cent	12.99	13.40
Ash, per cent	.47	.50
Ether extract, per cent	1.14	1.19
Crude protein (N × 5.7), per cent	9.43	8.73
Gluten, wet, per cent	26.18	27.70
Gluten, dry, per cent	10.43	9.32
Gliadin, per cent	4.96	4.77

crop will be soft, nor that all soft wheat is grown on irrigated fields. Available nitrogen in the soils appears to be a conditioning factor, and this in turn may be influenced by irrigation treatment. Thus (p. 31) it appeared that irrigation may translocate nitrates downward in the soil and thus move such nutrients beyond the reach of the roots. Suitable cropping methods, as the inclusion of alfalfa or red clover in the rotation, may aid in maintaining a suitable nitrate concentration in the surface soil, and thus stimulate the elaboration of protein in the grain of wheat subsequently grown on such soil.

Analyses of Idaho wheats, summarized by Jones and Colver (1918), show that in the case of spring fife and bluestem and of Turkey Red winter wheat grown on dry-farmed and irrigated fields at Aberdeen, Idaho, the grain from the dry-farmed fields was substantially higher in its content of crude protein than that from the irrigated fields.

A gradual decrease in the percentage of protein in wheat kernels as the quantity of irrigation water applied was increased, was reported by Widtsoe and Stewart (1920). Thus with 5, 15, 25, and 35 inches of irrigation, the protein content of the grain was 18.05, 16.45, 16.22, and 15.89 per cent respectively. Similar results were secured by Greaves and Carter (1923). The difference was pronounced with small applications of water, but when large applications were made the "nitrogen reaches what may be termed an irreducible minimum." The percentage of total ash, potassium, phosphorus, calcium and magnesium increased quite regularly with increasing applications of irrigation water within the limits studied.

Influence of Soil and of Fertilizers on the Composition of Wheat.

There is probably a greater diversity of opinion concerning the influence of soil and of added fertilizers on the composition and quality of wheat than is expressed concerning any other factor. So frequently it is contended that wheat quality is being reduced in consequence of a continual cropping to wheat, and the resulting loss of fertility. It accordingly becomes necessary to examine carefully the significant data, since this may lead to important conclusions relative to the future of the milling and baking industries. No extended reference will be made in this treatment of the subject to the relation of soil to yields of crop per unit of area.

Three general types of experiments were involved in the studies reported in the following paragraphs of this section. In the first, fertilizers singly or in combination were superimposed on the soil, or were used in the nutrient substratum on which the wheat plants were grown. In the second, nutrients were applied at different stages in the life of the wheat plant; while in the third, soils were transferred from one geographical location to another, in an effort to determine whether the soil type or the physical environment played the most important rôle in affecting the composition of the crop.

One of the earliest systematic studies of cropping methods and fertilizer treatments was conducted at Rothamsted, England. Lawes and Gilbert (1884) review the results of the first 20 years of experimentation, and in Table 33 an outline of the manurial treatment followed, and the average results obtained from each of the 9 plots are presented. These data appear to indicate that the use of nitrogenous fertilizers, either as ammonia salts or barnyard manure, increased somewhat the average percentage of nitrogen in the grain, and that "mineral" fertilizers failed to effect any marked difference when compared with the control or unmanured plot. When certain favorable seasons were contrasted with unfavorable seasons, it appeared that the effect of the barnyard manure was somewhat different. In the case of the former group of seasons the percentage of nitrogen in grain from plots receiving barnyard manure was lower than in the case of the untreated plots.

TABLE 33

MANURIAL TREATMENT AND COMPOSITION OF WHEAT GROWN ON ROTHAMSTED PLOTS

Manurial treatment followed in the Rothamsted experiments.

Plot 2. Farmyard manure.
" 3. Unmanured.
" 5b. Mixed mineral fertilizer.
" 7b. Mixed mineral and ammonia fertilizer.
" 10b. Ammonia salts alone.
" 11b. Ammonia and superphosphate.
" 12b. Ammonia, soda and superphosphate.
" 13b. Ammonia, potash and superphosphate.
" 14b. Ammonia, magnesia and superphosphate.

Average Composition of the Twenty Wheat Crops from the Rothamsted Plots

Plot	Weight per Bushel, Lbs.	Yield per Acre, Bus.	Nitrogen in Grain, Per Cent	Total Ash, Per Cent	Phosphoric Anhydrid, Per Cent	Potash (Oxid), Per Cent	Nitrogen per Acre, Lbs.
2........	60.6	37.8	1.95	1.99	1.044	.635	36.9
3........	57.6	15.2	1.84	2.04	1.022	.686	14.1
5b.......	58.9	18.0	1.83	2.07	1.061	.673	16.8
7b.......	59.3	37.3	1.97	1.94	.981	.643	37.1
10b......	58.0	27.4	1.95	1.82	.870	.619	27.0
11b......	57.4	29.8	1.96	1.88	.944	.599	29.6
12b......	59.2	35.6	1.92	1.91	.963	.630	34.6
13b......	59.3	36.7	1.87	1.91	.968	.633	34.6
14b......	59.3	35.9	1.90	1.92	.977	.629	34.5

The studies of Tod and Wels, Siegert, Hartstein, Sopp and Töpler, Werner and Stutzer, Zöller, Märcker and others during this period advanced the knowledge of the effect of certain fertilizers. In many of these experiments the combinations used were such as to preclude conclusions as to the effect of the individual elements in the mixture. Ritthäusen and Pott (1873) conducted an interesting experiment to determine the effect of fertilizers rich in nitrogen and phosphoric acid.

TABLE 34

Fertilizer Treatment in the Experiments of Ritthäusen and Pott (1873), and Composition of the Wheat

Plot	Fertilizer Treatment	Moisture, Per Cent	Protein, Per Cent	Ash, Per Cent
1	Not fertilized	13.42	15.04	2.56
7	" "	13.59	14.26	2.35
12	" "	13.91	12.86	2.34
4	4 kg. superphosphate	13.16	17.04	2.36
8	4 " "	13.80	14.92	2.11
11	6 " "	13.53	13.84	2.18
2	2.5 kg. ammonium sulfate...............	13.94	18.50	1.97
5	3.0 " sodium nitrate	13.71	18.28	1.57
9	1.25 " ammonium sulfate, and 2.0 kg. sodium nitrate	13.45	18.82	2.16
3	2.5 kg. ammonium sulfate and 4.0 kg. superphosphate	13.45	20.66	2.39
6	3.0 kg. sodium nitrate and 4.0 kg. superphosphate	13.71	18.12	2.36
10	1.25 kg. ammonium sulfate, 2.0 kg. sodium nitrate, and 4.0 kg. superphosphate......	13.65	19.85	2.56
	Average of unfertilized plots	13.64	14.03	2.42
	Average of plots with phosphoric fertilizers	13.50	15.17	2.08
	Average of plots with nitrogenous fertilizers ...	13.70	18.55	2.11
	Average of plots with nitrogenous and phosphoric fertilizers	13.60	19.54	2.44

The seed wheat used was hard, glassy and dark in color, but the resulting crop from the unfertilized plots was half-mealy, or between a dark and a light shade in color, large and plump, with a smooth and shining seed coat. The wheat kernels raised on the plots fertilized with phosphoric acid were similar in appearance, but varied in size, and there were about as many small, glassy kernels as large, plump kernels. The grain from the plots fertilized with nitrogen lodged, due to a heavy rain after the straw was of considerable height. The kernels of wheat from these plots were small, but hard, glassy and dark. The mixed fertilizer had, in the main, the same effect as nitrogen, although many kernels were shriveled. The analysis of these several lots of wheat is given in Table 34, and it appears that the use of nitrogenous fertilizers increased the percentage of protein in the grain appreciably.

Superimposing increasing amounts of nitrogenous fertilizer on a mixture containing phosphorous and potassium resulted in progressively increasing the percentage of nitrogen in the resulting crop in Jordan's experiments (described by Richardson, 1883) recorded in Table 35. Stable manure had little effect.

TABLE 35

FERTILIZER TREATMENT EMPLOYED BY JORDAN (1882) AND THE COMPOSITION OF THE WHEAT CROP

Plot 1. Not fertilized.
" 2. Phosphoric acid and potassium.
" 3. Phosphoric acid and potassium, and 1 portion of nitrogen.
" 4. "   "   "   "   " 2 "   "   "
" 5. "   "   "   "   " 3 "   "   "
" 6. Stable manure.

COMPOSITION OF THE WHEATS

Plot	Moisture, Per Cent	Protein, Per Cent	Fat, Per Cent	Nitrogen Free Extract, Per Cent	Fiber, Per Cent	Ash, Per Cent
1	13.33	10.86	1.99	69.02	2.76	2.04
2	13.04	10.50	1.97	69.85	2.65	1.99
3	13.16	11.16	1.90	69.24	2.51	2.03
4	13.06	11.69	1.90	67.90	2.47	2.98
5	12.59	11.70	1.92	69.43	2.53	1.83
6	12.41	11.04	1.89	70.10	2.37	2.09

A study of alkali soils conducted by Bogdan (1900) indicated that an increase in the percentage of soluble salts in the soil produced an increased percentage of nitrogen and ash in wheat, although the weight of the kernels decreased. This is regarded as the reason for the high protein content of wheat from eastern and southeastern Russia.

The percentage of nitrogen in the wheat grown on three plots to which Vignon and Conturier (1901) applied phosphates decreased regularly with increasing applications of phosphate fertilizer. Thus in the case of Goldendrop wheat the grain from the plots which received 75 kg. of phosphoric acid per hectare contained 1.83 per cent of nitrogen, while that from the plot which received 225 kg. per hectare contained but 1.54 per cent of nitrogen.

In a few experiments no beneficial results followed the application of nitrogenous fertilizer, either in terms of protein content of the wheat or in baking strength of the flour. Guthrie and Norris (1902) did not increase the gluten content of wheat grown at Bathhurst and Wagga, N. S. W., Australia, by the application of ammonium sulfate to the soil on which they were grown.

No definite relation existed between the composition of ash of plants and the soil in which they were grown in the opinion of Tollens (1902). The important factors in influencing the composition of plant ash were stated to be the stage of growth, thickness of the stand, available moisture, the soil, and the fertilizers.

Relatively large effects were induced by climate on the composition of wheat in the opinion of Soule and Vanatter (1903), but a rich soil or the use of fertilizers did not effect an appreciable increase in the nitrogen content of wheat.

Increasing the available nitrogen, phosphate, or potash content of soils at Davis, California, had only a slight effect on the percentage of protein in wheat grown on these soils, in Shaw's (1913) opinion.

The results of analyses of wheats grown at 12 locations in Minnesota on plots receiving varying fertilizer treatment were reported by Snyder (1907, 1908 b). The average composition of the wheat grown at these several points is given in Table 36, from which it appears that a slight increase in the nitrogen content of the grain resulted from the application of nitrogen fertilizers to the soil on which it was grown. Potash fertilizer had no effect on either nitrogen or ash content of the grain crop, while phosphate fertilizers decreased the nitrogen (or crude protein) content and increased the percentage of ash in the grain somewhat.

TABLE 36

Nitrogen and Ash Content of Wheat Grown by Snyder (1907, 1908b) on Plots Receiving Different Fertilizer Treatments

Fertilizer	Number of Samples	Crude Protein (N × 6.25), Per Cent	Ash, Per Cent
None	12	13.04	1.64
Phosphoric acid	12	12.65	1.73
Potash	12	13.02	1.62
Nitrogen	12	13.63	1.58
Complete (N, K, and P)	12	13.17	1.69

Applications of manure slightly increased the protein content of the wheat crop, while manure and superphosphate decreased the percentage of protein in the grain, in the experiments reported by Raynaud, Brunerie and Paturel (1910).

The effect of various fertilizers on the composition of wheat grown at State College, Pennsylvania, was studied by MacIntire (1911). Nitrogen in the form of dried blood, with bone black and potassium chloride, resulted in a slight decrease in the percentage of crude protein in the wheat, when compared with the control or unfertilized sample. Increasing the quantity of dried blood stepped up the protein content of the grain in proportion to the quantity applied. The other fertilizers used in this series, including manure and lime, had comparatively little effect on the percentage of crude protein in the wheat crop.

The composition of wheats grown by Ames (1910) on two unfertilized soils parallel the composition of the soils. The addition of potassium, phosphorus and nitrogen to the soil increased the percentage of these elements in the ripened wheat. The percentage of nitrogen in the grain decreased when phosphates were applied to the soil, either alone or mixed with other fertilizers, while the effect of potassium in this respect was less marked. Nitrogenous fertilizers (inorganic), when used singly or mixed with potassium, resulted in an increase in the percentage of nitrogen in the grain. The composition of wheat harvested from the plots used by Ames, at Wooster, Ohio, is shown in Table 37.

TABLE 37

AVERAGE COMPOSITION OF WHEAT OF THE CROPS OF 1907 AND 1908, GROWN IN A 5-YEAR ROTATION AT WOOSTER, OHIO (AMES, 1910)

Fertilizers per Acre			Composition of Grain		
Phosphorus, Pounds	Potassium, Pounds	Nitrogen, Pounds	Phosphorus, Per Cent	Potassium, Per Cent	Nitrogen, Per Cent
0	0	0	.3143	.3591	1.841
0	108	0	.3300	.3269	1.832
0	0	76	.2985	.3367	2.103
0	108	76	.2955	.3438	2.084
10	41	25	.3016	.3429	1.782
15	75	51	.3192	.3489	1.811
20	0	76	.3326	.3371	1.831
20	108	76	.3323	.3366	1.823
20	108	76	.3265	.3598	1.741
20	108	76	.3362	.3468	1.714
20	108	76	.3419	.3519	1.742
20	108	114	.3241	.3547	2.004
30	108	38	.3388	.3652	1.661
30	108	38	.3658	.3709	1.613
30	108	38	.3650	.3681	1.581
30	108	38	.3562	.3633	1.634
30	108	38	.3794	.3685	1.651
24	56	72	.3350	.3413	1.692
48	112	144	.3678	.3681	1.751
20	0	0	.3692	.3621	1.660
20	108	0	.3687	.3528	1.691

ENVIRONMENT ON THE COMPOSITION OF WHEAT

The work instituted by Ames was continued by Ames, Boltz and Stenius (1912), who reported the effect of various fertilizers upon the yield, quality and composition of wheat of the 1910 crop. In Table 38 is shown the quantity of fertilizer applied per acre, the percentage of plump and shriveled wheat kernels in the resulting crop, and the yield of grain per acre. The plumpest grain was secured through the application of mixed fertilizers, or a combination of phosphorus and potassium. Barnyard manure and the mixed treatment given to plot A resulted in the largest yields per acre.

TABLE 38

FERTILIZER APPLICATION, AND RELATIVE PLUMPNESS AND YIELD OF WHEAT GIVEN BY AMES, BOLTZ AND STENIUS (1912)

Plot No.	Acid Phosphate, Pounds	Potassium Chlorid, Pounds	Dried Blood, Pounds	Nitrate of Soda, Pounds	Plump Kernels, Per Cent	Shriveled Kernels, Per Cent	Yield per Acre, Bushels
0	...	...	...	...	51	49	8.45
3	...	260	...	...	66	34	10.33
5	...	...	50	440	40	60	10.75
9	...	260	50	440	51	49	13.00
2	320	...	...	...	88	12	20.33
6	320	...	50	440	79	21	25.25
8	320	260	...	...	88	12	19.75
11	320	260	50	440	81	19	27.00
12	320	260	50	680	84	16	28.33
30[1]	...	260	...	...	87	13	29.25
18[2]	...	...	...	...	87	13	33.42
A	...	...	...	...	94	6	34.15

[1] Phosphorus and potassium supplied by tankage containing 6 per cent of nitrogen and 6 per cent phosphorus.
[2] The three elements supplied by manure, 8 tons on wheat and 8 tons on corn.
A—Elements supplied by 10 tons manure, with the addition of 400 pounds of untreated rock phosphate for the corn crop, and a fertilizer on wheat consisting of 100 pounds steamed bone meal, 130 pounds acid phosphate, 20 pounds potassium chlorid, and 60 pounds sodium nitrate. The sodium nitrate was applied broadcast over the wheat in April.

Table 39 gives the quantity of fertilizing elements thus supplied per acre in the several treatments, and the composition of the grain raised on the several plots in 1910. These investigators conclude that an increase in available nitrogen increased the percentage of protein, the addition of phosphorus resulted in a decrease in the protein content of the grain, while potassium had little effect in that particular.

Yellow-berry in hard wheat, that soft condition which usually denotes a low gluten content, can be prevented by the application of a sufficient quantity of available nitrogen to the soil, in the opinion of Headden (1915 a), and intensified or increased by the application of available potassium. The ratio of potassium to available nitrogen was deemed to be the cause of yellow-berry.

TABLE 39

Fertilizing Elements Added Per Acre to Wheat Plots, and Composition of Grain, as Reported by Ames, Boltz, and Stenius (1912)

Plot No.	Phosphorus, Pounds	Potassium, Pounds	Nitrogen, Pounds	Starch and Soluble Carbohydrates, Per Cent	Protein, Per Cent	Phosphorus, Per Cent
0	...	...	...	60.04	14.50	.3222
3	...	108	...	60.39	14.75	.3399
5	...	...	76	58.37	15.81	.3217
9	...	108	76	58.23	15.88	.3112
2	20	...	...	64.10	12.56	.3777
6	20	...	76	61.99	13.94	.3318
8	20	108	...	64.25	12.00	.3954
11	20	108	76	61.16	13.88	.3376
12	20	108	114	60.80	14.69	.3262
30	30	108	38	64.46	12.88	.3993
18	.48	112	144	64.04	11.44	.4253
A	112	81	103	64.56	12.32	.4358

An extended study of the relation of plant nutrients in the soil to the hardness and composition of wheat was detailed in a series of bulletins by Headden (1915 b, 1916 a, 1916 b, 1918 a, 1918 b). It was concluded that available nitrates determined these properties in a large measure. Increasing the nitrate content of the soil by additions of sodium nitrate resulted in harder wheat of a higher protein content. A survey of Headden's data (note 1918 b, pp. 12-18), resulting from the analysis of 1913, 1914, 1915 and 1917 crops of Defiance, Red Fife, and Kubanka (durum) wheats indicates that the addition of potassium in fertilizers effected no significant change in the average nitrogen content of flours milled from the wheats. Phosphorus applied in fertilizers apparently tended to reduce the percentage of nitrogen in the flour somewhat. These data, as summarized and averaged by the writer from Headden's data, are given in Table 40.

TABLE 40

Summary of Certain of Headden's Data Relating to Fertilizer Treatment and Nitrogen Content of Flour Milled from Wheat Crop

Treatment of Plot with Fertilizer	Average Percentage of Nitrogen in Flour, Crops of 1913, 1914, 1915 and 1917			Average of Three Varieties
	Defiance	Red Fife	Kubanka	
Check	1.60	1.72	1.81	1.71
Nitrogen	1.92	1.89	1.93	1.91
Phosphorus	1.57	1.69	1.69	1.65
Potassium	1.58	1.75	1.80	1.71

Headden, however, stresses the significance of the nitrogen-to-potassium ratio in the soil, contending (1916 b, p. 47) that at least so far as mealiness or relative softness of the grain is concerned, this is caused by too high a ratio of potassium to available nitrogen.

Wheat was grown by Neidig and Snyder (1922) in pots in a greenhouse on soil low in nitrogen, to which was added varying amounts of readily available nitrogen. The percentages of protein in the wheats from three such series are given in Table 41. Three sources of nitrogen were represented in the applications made: (a) sodium nitrate, (b) ammonium sulfate, and (c) hydrolysed wheat extract, the dosages of which were graduated so that the same quantities of nitrogen were added in comparable plots of the three series. Much the same effect was observed from the addition of nitrogen in either of the three forms, the tendency being in the direction of a higher protein content in the grain in consequence of increased additions of available nitrogen to the soil.

TABLE 41

Protein Content of Wheat Grown by Neidig and Snyder (1922) in Pots Receiving Different Forms of Nitrogen

Addition to Soil	Crude Protein in Wheat, Per Cent		
	Sodium Nitrate Series	Ammonium Sulfate Series	Hydrolysed Wheat Extract Series
None	15.25	15.25	15.25
P + K	19.24	19.24	19.24
.3708 grams N, + P + K	21.63	20.69	22.00
.7514 " N, + P + K	20.56	21.00	22.00
1.4830 " N, + P + K	21.94	23.00	23.56

The variations in baking strength of flours milled from wheats grown on plots to which Saunders, Nichols, and Cowan (1921) accorded varying treatment with fertilizers was small. The range in strength was from 96, in case of the wheat grown on the plot to which barnyard manure was applied, to 100, in case of the wheat from the plot which received muriate of potash. These variations fall within the limits of probable error that might be anticipated in testing different samples from the same lot of wheat. The effect of fertilizing the wheat land on the baking quality of flour milled from the resulting wheat crop was small, in the opinion of Neumann (1924).

An extensive study of the influence of soil moisture and various fertilizer treatments on the yield and composition of wheat was conducted by Morgan (1911). The fertilizers used were applied to pots containing 10.5 kilograms of soil, at different stages of growth of the wheat, the three stages enumerated being: (first stage) from time of planting to five well developed leaves; (second stage) from end of first stage to indications of spike in culms; (third stage) from end of second stage to beginning of blooming. The composition of the grain from the several pots is shown in Table 42, from which it appears that increasing the soil moisture had more effect upon both yield and composition of the grain than did the use of either a complete, or a highly nitrogenous fertilizer. In the soil containing 17.6 per cent of moisture the effect of either the complete or the high-nitrogen fertilizer was much less than was the case when the same fertilizer treatment was applied to the soil containing 42.8 per cent of moisture.

TABLE 42

INFLUENCE OF SOIL MOISTURE AND FERTILIZER APPLICATIONS ON THE PERCENTAGE OF NITROGEN IN WHEAT (MORGAN, 1911)

	Yield of Grain		Per Cent of Nitrogen in Grain	
Fertilizer Treatment	17.6 Per Cent Moisture in Soil	42.8 Per Cent Moisture in Soil	17.6 Per Cent Moisture in Soil	42.8 Per Cent Moisture in Soil
No fertilizer	9.93	23.74	2.74	2.36
Complete fertilizer, 1st stage	12.79	26.49	2.86	2.10
" " 2nd "	15.63	28.58	2.80	2.04
" " 3rd "	14.65	29.03	2.80	2.13
High-nitrogen fertilizer, 1st stage	16.84	30.18	2.84	2.50
" " " 2nd "	15.40	31.66	2.86	2.30
" " " 3rd "	13.46	33.68	2.87	2.16

Increasing the nitrate content of the soil, by the addition of sodium nitrate, at the time of heading of the wheat plants was found by Davidson and Le Clerc (1917) to result in a better quality of grain with reference to color and protein content. Similar additions of sodium nitrate at the milk stage of the grain had no such effect. Additions of potassium chloride did not apparently affect the composition of the grain, but it did seem to increase the amount of yellow-berry when used alone. Davidson and Le Clerc (1918) observed an increased protein content of the straw as the result of applying nitrate to wheat plots at the time the plants headed. Potassium chloride, similarly applied, depressed the protein content of the straw. Davidson (1922), working with soft red winter wheat at College Park, Maryland, found that the effectiveness of nitrates in increasing yields decreased consistently as the time of their application in the spring approached the stage of heading. The effectiveness of these nitrate applications in increasing the protein content was in inverse ratio to their effectiveness in increasing yields. Davidson and Le Clerc (1923) applied various forms of nitrogen, as nitrates and ammonium salts, to plots of wheat at three stages of growth (1st) when the crop was 2 inches high, (2nd) when the wheat was heading, and (3rd) when the wheat kernels were in the milk stage. These wheat plots were near Beatrice, Nebraska. In general, the application of nitrogen at the second stage resulted in a higher percentage of crude protein in the grain than did similar applications at either of the other two stages.

The stage of growth of spring wheat plants at which nitrate was applied to the soil in which they were growing was found by Gericke (1920) to have a pronounced effect on the protein content of the grain. From the data in Table 43 it appears that delaying the application of nitrates effected a progressive increase in the protein content of the wheat crop. In this series the grain was planted November 14, 1919, and the application of $NaNO_3$ was at the rate of 100 pounds of nitrogen per acre.

TABLE 43
Relation of Date of Nitrate Application to Yield, Grade and Protein Content of the Wheat (Gericke, 1920)

Date of Nitrogen Application	Days After Planting When Nitrogen Was Applied	Yield of Grain, Grams	Commercial Grade	Crude Protein, Per Cent
Nov. 14	..	9.4	No. 2 soft white	8.6
Dec. 1	17	10.6	No. 2 " "	9.3
Dec. 16	33	21.0	No. 2 " "	10.4
Jan. 1	48	19.9	No. 2 hard "	11.8
Jan. 24	72	21.9	No. 1 " "	13.2
Mar. 2	110	13.1	No. 1 " "	15.2

In another series of experiments with Turkey Red winter wheat Gericke (1922 b) applied sodium nitrate at time of planting (Nov. 18, 1919), and at variable intervals of time after planting, and found that, unlike the spring wheat, there was no consistent response to a delay in the application of the fertilizer. This difference in behavior of the two wheat types, the spring wheat reported in the foregoing paragraph and this winter wheat, was attributed to the relative dormancy of growth early in the life of the winter wheat. The result was that while the nitrates were applied about the same number of days after planting in both series of studies, the two types of wheat were not in exactly the same growth phase.

Gericke (1922 b) indicates that the nutrition of wheat seedlings influences the development of the root system. When grown in tap water alone for 4 weeks the roots were 62 cm. long and weighed 0.18 grams, while in a nutrient solution the roots were 12 cm. long and weighed only 0.015 grams. The former averaged 5.4 tillers per plant, the latter only 1.2 tillers. The addition of nitrogen to cultures grown in a nitrogen-poor soil late in the life of the plant might possibly result in a larger absorption and more immediate utilization of this nutrient in building proteins in the kernel, owing to the larger root development, than would be the case when nitrogen was added earlier in the life of the plant.

Sufficient soil from Hays, Kansas, was secured by Shaw and Walters (1911) to fill a plot at Davis, California, 6 feet by 3 feet, to a depth of 3 feet, and they grew Kubanka, and Turkey Red wheat on this plot in comparison with the same wheats grown on the native Sacramento Valley gray silt loam soil. The Kansas soil was black in color, contained more nitrogen and potash and less sand than the California soil. The wheat grown on these plots during the seasons of 1907-1908, and 1908-1909 was analyzed, with the results shown in Table 44.

From these results Shaw and Walters conclude that a normal soil has little influence on the nitrogen content of the wheat kernel, but that climatic conditions are the controlling factors.

TABLE 44

Protein Content of Wheat Grown at Davis, California, on California and Kansas Soil (Shaw and Walters, 1911)

	Total Protein in Wheat	
	California Soil, Per Cent	Kansas Soil, Per Cent
Kubanka, 1907-1908	15.14	15.07
Velvet Don, 1908-1909	13.23	12.95
Turkey red, 1907-1908	18.29	18.02
" " 1908-1909	15.07	14.32

Another series of plots were prepared by Shaw and Walters, using the same Kansas and California soils as before, but in addition they added a plot of soil taken from Arlington, Virginia.[1] High gluten wheats were planted on all three plots, and the composition of the resulting wheat crop was determined. The Virginia soil was described as a light, yellow clay. As a check or control plot, an undisturbed California soil was included in the series. As the result of the analyses of these wheats they concluded that "In the light of the present data it seems quite certain that the soil nitrogen has little if any direct effect upon the nitrogen content of the grain grown upon such soil, and that some climatic factor is sufficient to entirely overshadow the soil factor." Also "It may be that certain physical factors, enabling the soil to hold moisture at certain periods of the plant's growth, are responsible for this difference." To the inability of the Virginia soil to hold moisture was attributed the low yield of grain, with pinched kernels and high nitrogen content, which was produced on that plot in 1909-1910.

[1] In Shaw and Walters' paper it was stated that the soils were from Arlington, Maryland, but the Arlington farm is in Virginia.

TABLE 45

CHARACTERISTICS OF WHEAT GROWN ON SOIL-EXCHANGE PLOTS AT DAVIS, CALIFORNIA (SHAW AND WALTERS, 1911)

Season 1908-1909

	Check California Soil	California Soil	Kansas Soil	Virginia Soil
Total protein, per cent	15.61	16.33	11.80	12.03
Typical kernels, per cent	99.10	99.82	79.75	97.47
Starchy kernels, per cent	.90	.18	20.25	2.53
Kernels in 10 grams	....	303.	269.	356.
Days to mature	272.	288.	208.	231.

Season 1909-1910

Total protein, per cent	13.45	14.27	11.23	17.44
Typical kernels, per cent	99.4	100.	64.7	100.
Starchy kernels, per cent	0.6	0.	35.3	0.
Kernels in 10 grams	347.	374.	300.	474.
Days to mature	238.	253.	204.	317.

A continuation of the work of Shaw and Walters was reported by Le Clerc and Yoder (1914). The completed series of studies included the results of analyses of wheats grown for four seasons on plots of California, Kansas and Maryland soils in each of the three states. There was a fair degree of uniformity in the protein content of wheat grown in any one locality, independent of the soil on which they grew. Four of the 42 cases, or less than 10 per cent, did not follow this general rule. These exceptions were of such an order as to result in a somewhat higher average protein content in the wheat grown on the Maryland soil in the three localities. The wheats grown on these Maryland soils were also somewhat more shriveled, having a lower average weight per 100 kernels than the wheats grown on the California and Kansas soils. Wheat grown on the several plots in Kansas averaged higher in protein than the wheat grown in the other two localities. It was also lower in average weight per bushel, and weight per 100 kernels. The averages of the tests and analyses of wheats grown in each locality on the several plots, and on plots of the three soils in all localities, are given in Table 46. Basing their conclusions on these data, Le Clerc and Yoder offer the opinion that "Climate is the principal factor influencing the protein content of wheat, and that soils, when used as in this experiment, have little or no influence."

Soil in Pullman, Washington, containing 0.204 per cent of total nitrogen was transferred to Ritzville, Washington, and soil from Ritzville containing 0.090 per cent of nitrogen was similarly transferred to Pullman, in an experiment reported by Thatcher (1913 b). Two crops were grown upon these plots, as well as upon the native soils at each location. There appeared to be no relationship between the composition

TABLE 46

Average Composition of Wheats Grown by Le Clerc and Yoder (1914)

	All Soils in Different States			Different Soils in Three States		
	California	Kansas	Maryland	California Soil	Kansas Soil	Maryland Soil
Physical properties						
Water, per cent	8.98	9.53	9.53	9.35	9.46	9.29
Weight per 1000 kernels, grams	30.2	19.	25.6	26.5	27.9	22.1
Weight per bushel, lbs	62.8	57.2	60.1	60.9	60.4	....
Flinty kernels, per cent	86.	99.	35.	71.	69.	85.
On water-free basis, per cent						
Nitrogen	2.42	3.30	2.18	2.48	2.52	2.75
Protein ($N \times 5.7$)	13.11	18.83	12.43	13.88	13.94	15.44
Alcohol-soluble nitrogen	.92	1.27	.90	1.00	.94	1.05
Gliadin in protein	41.	42.	40.	42.	41.	40.
Fat	1.97	2.00	1.94	1.93	1.98	1.97
Fiber	2.34	2.89	2.63	2.55	2.59	2.73
Pentosans	8.45	8.76	8.56	8.41	8.48	8.87
Sugars	3.61	3.32	3.03	3.33	3.48	3.30
Ash	1.90	2.30	2.22	2.13	2.08	2.16
Phosphoric acid	.90	1.02	1.18	1.04	1.03	1.05
Potash	.57	.68	.67	.64	.61	.66
Phosphoric acid in ash	47.	45.	53.	48.	48.	48.
Potash in ash	29.	30.	30.	30.	29.	29.

of the soil and the protein content of the wheat crop. Thatcher accordingly concluded (p. 19) that the nitrogen content of the soil has little, if anything, to do with the nitrogen content of the grain growing upon it in any given season.

The composition of wheat grown on "breaking" as compared with wheat grown on soil which had been cultivated for 10 years, at Valley River, Manitoba, in 1905, was determined by Shutt (1908). The breaking was cleared scrub land, characterized by a high percentage of organic matter and nitrogen. In potash and phosphoric acid the two soils did not differ materially; the breaking was higher in percentage of lime, however. The high percentage of humus apparently increased the water-holding capacity of the soil in the breaking over that which had been cropped. The composition of the two soils, as reported by Shutt, is given in Table 47.

The moisture content of these soils was determined at intervals during the growing season, and the resulting data, shown in Table 48, indicate that the water-holding capacity of the soil in the breaking was substantially higher than that of the old field.

The percentage of protein in the wheat grown on the old field proved to be much higher than that in the wheat grown on the break-

TABLE 47
COMPOSITION OF AIR-DRIED SOILS ON WHICH SHUTT (1908) GREW WHEATS AT VALLEY RIVER, MANITOBA

	Breaking	Old Field Cropped 10 Years
Moisture	2.98	2.06
Organic and volatile matter	20.90	12.84
Insoluble residue (sand, clay, etc.)	51.74	65.07
Oxide of iron and alumina	5.50	10.52
Lime	10.25	3.47
Magnesia	2.44	1.63
Potash	0.14	0.19
Phosphoric acid	0.15	0.13
Soluble silica	0.02	0.02
Carbonic acid, etc. (undetermined)	5.88	4.07
Nitrogen in organic matter	0.642	0.371
Available: Phosphoric acid	0.0067	0.0067
Potash	0.0166	0.0069
Lime	1.306	0.930

TABLE 48
MOISTURE IN SOIL OF THE BREAKING AND THE OLD FIELD

Date	Breaking	Old Field Cropped 10 Years
May 5	32.96	22.45
May 15	36.49	23.39
May 29	33.45	23.39
June 22	30.49	21.70
July 13	35.23	21.24
Aug. 2	30.37	13.24
Aug. 24	32.84	18.28

ing. The former contained 13.52 per cent of crude protein ($N \times 5.7$), while the latter contained only 10.01 per cent.

Jones and Colver (1916) pointed out that some of the conclusions relative to the correlation between climate and rainfall have been based on data gathered over so extensive an area that other factors might have come into play. They believed that the nitrate content of the soil might be of significance in influencing the protein content of the grain. Legumes in the rotation appeared to increase the percentage of protein in wheat subsequently grown in the rotation.

Tillage in its relation to the available nutrients of the soil was discussed by Swanson (1924), who indicated that the protein content of wheat is conditioned by the available nitrogen of the soil. When an ample supply of available nitrogen is maintained during the entire growing period, a large yield of high protein wheat may be produced, providing there is sufficient moisture in the soil and other favorable growth factors are maintained. A limited supply of moisture during the later stages of growth of the wheat plant results in a high protein wheat, but a low yield per acre. The opposite condition is usually encountered when wheat is grown under irrigation, when an ample supply of moisture in the soil means a low concentration of nitrates in the soil solution. Cool weather restricts the production of available nitrogen in the soil, which in turn is conducive to the production of soft, low gluten wheat, while the yield per acre may be high.

These extensive data resulting from the studies of the influence of environment on the composition of wheat appear to support the following general conclusions:

Cool summers with high rainfall, and a relatively long growth period result in the grain of wheat being high in percentage of starch and low in percentage of protein. Hot, dry summers, on the other hand, result in the production of wheat having a high percentage of protein. In general, the amount of rainfall during that period of growth which immediately precedes ripening is responsible for the rate of starch deposition in the kernel, and, consequently, for the relative protein content.

High protein seed, when planted in a region having a long growing season or a high rainfall, and consequently conducive to the production of low protein wheat, will yield a crop with a lower protein content than the original seed, and no better in this respect than the wheat raised from low gluten seed of the same variety and strain.

Climatic conditions are responsible for greater variations in the composition of wheat than are other factors, with the possible exception of irrigation.

Since varying rainfall results in variations in the soil moisture available to the plant, and this in turn induces variations in the protein content of the wheat grown on the soil, it follows that applications of irrigation water are similarly responsible for variations in protein content of irrigated wheat. The addition of irrigation water to the soil will, within certain limits, increase the yield of grain per acre, but the kernels, while heavier and plumper, will contain a lower percentage of protein than when no water is applied. The protein content of the grain varies inversely with the quantity of irrigation water applied.

Texture of soil is apparently responsible for greater variations in the composition of wheat than are variations in soil composition within the limits ordinarily encountered in agricultural practice in wheat-growing regions. Soils which are retentive of moisture, and at the same time have a reasonably high coefficient of availability of moisture, produce wheats lower in protein than when the opposite is true.

Applications of nitrogenous fertilizers, particularly inorganic forms, result, as a usual thing, in the production of wheat containing a higher percentage of protein than when no nitrogen is used. The time in the life of the plant when the nitrogenous fertilizers are applied determines, in a measure, the response to the treatment in terms of grain composition. In field trials, delaying the application of nitrates, up to the time of blossoming, effected the greatest increases in the percentage of protein in the grain. Phosphates applied to soils in liberal amounts usually resulted in the production of wheat lower in protein than when no phosphates were applied. Potash did not materially affect the composition of the grain, nor did mixed or "complete" fertilizers unless the proportion of phosphates in the mixture was high. It does not appear that the use of commercial fertilizers will effect sufficient change in the gluten content of wheat to justify their use from this standpoint alone, although increased yields of grain may result from their use. The use of legumes in the rotation affords a more practicable means of increasing the concentration of nitrates in the soil, and of thus indirectly increasing the protein content of subsequent wheat crops.

Chapter 5.

Defects of and Impurities in Commercial Wheat.

Commercial wheat as it reaches the miller frequently includes kernels which are not sound and normal, as well as admixtures of weed seeds, smut balls and spores, and other impurities. The unsound wheat kernels cannot ordinarily be removed from the mixture in which they are present, and are accordingly ground with sound kernels. Several kinds of damage may be represented in the unsound wheat fraction, including green or immature, frosted, scabby, bin-burned or heat damaged, weevily or weevil-cut, and sprouted kernels, and grains with "black-tip." Not all these forms of damage have been subjected to chemical study to an extent that puts us in possession of adequate information concerning the biochemical changes that have occurred in them in departing from the normal. All are viewed with suspicion by the practical miller. The great difficulty that arises in dealing with them in grading and merchandising wheat is the variation in the *degree*, or extent of damage manifested by individual kernels. Thus certain kernels may show evidence of damage, which, however, is so slight as not to impair their milling properties. In other instances the damage is so great that a small percentage of such kernels in a milling mixture may seriously affect the baking properties of the flour that is produced.

Composition and Quality of Frosted Wheat.

Frosted wheat, as the term is applied in the grain trade, includes those kernels which were produced by plants that were frozen before they ripened. The extent of damage is conditioned in large part by the relative time before ripeness that the plants were frozen. When nearly ripened, while in the stiff dough stage, the effect of freezing is slight. Kernels frozen at this stage exhibit a blistered appearance along the back. When the wheat kernels are in the milk stage at the time they are frozen, the effect on the milling quality of the grain is marked. The kernels in such cases appear blistered along the back, cheek and into the crease. The author's experience has indicated that when the blistered area is confined to the back of the kernel, no marked difference in the quality of the milled flour can be detected when compared with normal flour.

Harper (1889) states that "In my analyses the lowest percentage of protein found in the 'rusted and frosted' wheats is higher than the average for the graded wheats." By graded wheats in this connection was meant wheats free from such defects as rust and frost damage. Difficulty was experienced by Harper in separating crude gluten (by washing) from these "rusted and frosted" wheats, but unfortunately he does not distinguish between the two types of damage in his discussion or tabulation of data.

The percentage of protein in the shrunken, frosted wheat was found by Harcourt (1908) to be higher, the starch lower, and the woody fiber of the bran layers greater than in normal wheat.

The proportion of total nitrogen present in the form of "non-albuminoid" nitrogen was found by Shutt (1908) to be higher in frosted than in normal wheat. There was no substantial difference in this regard observed by Shutt on comparing the flours milled from the sound and frosted grain.

The forms of nitrogenous constituents of frosted and normal wheat were studied by Blish (1920), whose data, presented in part in Table 49, show that the severely frosted wheat contained two to three times as much total nonprotein nitrogen as the sound wheat. The increase in ammonia and amid nitrogen (in the acid hydrolysate) was proportional to the increase in nonprotein nitrogen. In the samples of frozen wheat a much larger percentage of the nonprotein nitrogen was in the α-amino form than in the samples of sound or normal wheat. Similar fractionation and study of flours milled from these wheats revealed corresponding differences. Thus in flours from samples 1, 7, and 8, the percentage of nonprotein nitrogen in the total nitrogen was 1.84, 4.40 and 10.56 respectively, and of α-amino nitrogen in the total nitrogen, 0.27, 1.29 and 4.85 per cent.

TABLE 49

EFFECT OF FREEZING ON NITROGEN COMPOUNDS OF IMMATURE MARQUIS WHEAT (1917 SERIES, REPORTED BY BLISH, 1920)

Date seeded	May 12	June 16	June 23	June 30
Date of first killing frost	Oct. 17	Oct. 17	Oct. 17	Oct. 17
Stage of development	Mature	Mature	Immature	Late milk stage
Percentage of total nitrogen in wheat	2.59	2.67	2.61	2.44
Percentage of nonprotein N in total N	4.17	4.34	7.05	13.98
Percentage of α-amino N in total N	0.55	0.65	1.84	5.05
Percentage of α-amino N in nonprotein N	13.52	15.18	26.08	36.06
Percentage of ammonia N in nonprotein N	....	3.45	3.04	3.28
Percentage of amid N in total N	0.54	0.52	0.88	1.51
Percentage of amid N in nonprotein N	12.96	12.07	12.48	10.82

Heavy frosting slightly lowered the percentage of flour produced in milling in the experiments of Whitcomb, Day and Blish (1921). The texture of bread baked from frosted wheat flour was not so good as that made from sound wheat flour, but the difference was only slight and barely discernible when the frost damage was light. The same general observations were made relative to the loaf volume of bread baked from frosted wheat flour and normal wheat flour.

Of nine samples of frosted wheat subjected to milling and baking tests by Wilhoit (1916) only one evidenced any impairment of its properties.

Wheat carrying varying percentages of frosted wheat kernels was subjected to milling and baking tests by Birchard (1920), who concluded that the most noticeable effect of frosted grain was to reduce the volume of the loaf. Where the damage was heavy, a gray or dark appearance was also imparted to the bread. When wheat was affected by light bran frost the quality of the flour was practically unimpaired, and excepting in those cases where the frost damage was severe, and the proportion of kernels so damaged fairly large, the texture and appearance of the resulting loaves was only slightly influenced.

Sharp (1925) froze immature wheat kernels containing 46.5 per cent of moisture, and on thawing them out observed an increase in the amino-nitrogen content of the grain. The amino-nitrogen of frozen kernels held at 25° C. for 36 hours increased 29 per cent, while that of kernels held at 35° C. increased 44 per cent. This is in contrast with the non-frozen kernels held under similar conditions, in which the amino-nitrogen decreased. The proportional increase in amino-nitrogen of frozen and thawed kernels was approximately the same during a period of 22 days of kernel development, but the actual increase was much smaller toward the close of the period. This is shown by the

TABLE 50

Change in Amino-Nitrogen of Wheat Kernels Harvested at Different Stages of Development, Frozen, Thawed and Held 48 Hours at Room Temperature in a Saturated Atmosphere (Sharp, 1925)

Time from Start, Days	Moisture, Per Cent	Amino-Nitrogen × 5.7 in Dry Matter	
		At Start, Per Cent	After Freezing, Thawing, and Holding 48 Hours in Saturated Atmosphere, Per Cent
0	69.4	1.71	2.38
12	50.6	0.77	0.48
14	46.5	0.42	0.59
18	45.0	0.48	0.59
20	43.5	0.36	0.51
22	38.7	0.27	0.43

data in Table 50. It thus appears that the amino-nitrogen content of the frozen kernels was related to the amino-nitrogen content of the kernels at the time of freezing. In another experiment it was observed by Sharp that the amino-nitrogen of flour milled from frozen and thawed wheat was about seven times that of flour milled from non-frozen heads.

Flour milled from wheat that was frozen when the plants were in an immature or unripe stage increased in hydrogen-ion concentration at a faster rate when stored than did normal flour, in the researches reported by Sharp (1924).

Fryer (1921) found that wheat kernels apparently increased slightly in weight on thawing out the plants after subjecting them to 13° (F.) of frost. Frosted kernels absorbed water more rapidly than normal kernels on immersion in water, and in sodium chloride solution.

Effect of Black Stem Rust Infections on the Composition and Quality of Wheat.

The exact manner in which the cereal rusts affect the metabolism of the host plants is not known. Tissue development of the rust organism is not ordinarily sufficiently extensive to involve any substantial drain upon the nutrients in the tissues of the host. It is possible that toxic substances are excreted by the pathogens which serve to poison the host in some manner not understood. The evidence offered by Weaver (1916) that transpiration is greatly increased when cereal plants are infected by rust, is the most tangible measure thus far afforded of the results of such infections. He states that a moderate infection may more than double the amount of transpiration. Such an acceleration of transpiration may not be due wholly to the rupturing of the epidermis, although presumably such a mechanical effect may be most important. However this may be, it is quite evident that infected wheat, with a doubled transpiration rate, will experience difficulty in retaining sufficient water within its tissues to maintain normal turgidity and afford a liquid medium for metabolism and transport. In consequence, the usual processes of carbohydrate synthesis, and translocation of nitrogenous and carbohydrate materials, will be interfered with in infected tissue. Infection of wheat plants with Puccinia graminis and P. triticina (leaf rust) was likewise observed by Weiss (1924) to result in lower water economy of the host. Sodium nitrate in the nutrient medium in which the host grew resulted in somewhat readier infection, but did not predispose to greater injury. That metabolism, as indicated by the rate of respiration, is interfered with, has been shown by Bailey and Gurjar (1920), in a study of the comparative respiration of normal and rust-infected wheat plants. The latter respired at a lower rate and were also observed to have a lower moisture content than the normal plants.

No correlation between the hydrogen ion concentration of the expressed juices of wheat plants, and susceptibility or resistance to rust, could be detected by Hurd (1923). Environmental factors produced greater differences in hydrogen ion concentration of the expressed juice than was found between varieties of plants of different ages grown under identical conditions. Liming the soil resulted in a lower hydrogen ion concentration in plants grown on such soil than in plants grown on unlimed soil. Hydrogen ion concentration of expressed wheat sap examined by Hurd, stated as pH, ranged around $pH = 6.0$. The pH averaged 0.1 higher when plants were cut at 1 P.M., than when cut at 9 A.M.

Analyses of wheat kernels from normal and rust-infected wheat plants, presented by Snyder (1905 a) are given in Table 51, and show the rusted wheat to have been high in its content of ash and crude fiber, and low in carbohydrates. The percentage of crude protein was also higher in the rusted wheat. In another section of the same publication Snyder also indicated that shriveled, light weight wheat kernels which he analyzed were higher in their content of protein than plump, heavy kernels.

TABLE 51

Composition of Normal Wheat, and Wheat from Rust-Infected Plants (Snyder, 1905)

Percentage of	Wheat from Rusty Plants	Wheat from Rust-Free Plants
Ash	2.87	1.92
Crude protein	16.37	13.34
Ether extract	2.41	2.24
Crude fiber	3.67	2.01
Carbohydrates	62.50	70.73
Moisture	12.18	9.67

Rust-infected and non-infected wheats harvested on the same day at East Selkirk, Manitoba, were subjected to chemical analysis by Shutt (1905), with the results shown in Table 52. Straw from the rusted wheat plants was substantially higher in protein content, showing the rust infection had interfered with the translocation of nitrogenous material from the straw to the head. The percentage of crude fiber, crude protein and ash was higher in the rusted grain than in the rust-free sample. Shutt does not present any data relative to the baking strength of flour milled from rusted wheat, but in his conclusions states: "The probabilities, from deduction, rather point to a deterioration or lack of 'strength' in the gluten."

Headden (1918 c) noted that "The very marked stoppage in the transference of the nitrogen from the leaves and stems to the head on

TABLE 52

Analysis of Rusted and Rust-Free Wheat, Grain and Straw (Shutt, 1905)

Straw	Weight of 100 Kernels, Grams	Moisture	Crude Protein	Crude Fat	Carbohydrates	Fiber	Ash
From rust-free wheat	.....	7.92	2.44	1.65	39.00	39.95	9.04
From rusted wheat...	.....	7.92	7.69	1.97	38.44	36.78	7.20
Grain							
From rust-free wheat	3.0504	12.26	10.50	2.56	70.55	2.29	1.84
From rusted wheat...	1.4944	10.66	13.69	2.35	68.03	3.03	2.24

the appearance of the rust indicated that a very large part of the total effect should be attributed to this cause." (Note p. 12.)

Ellis (1919) states that cutting rust-infected wheat on the green side will not result in larger yields than allowing it to ripen before cutting. Cutting prematurely resulted in shrunken berries and decreased weight, while cutting when the grain was "firm" gave the greatest yields and greatest weight per bushel. Grain cut in the late milk stage weighed 56 pounds per bushel, while that cut in the firm stage weighed 59 pounds per bushel.

While not all the light-weight, shriveled wheat that has reached the market owes these characteristics to infection of the wheat plants with rust, a considerable fraction of the shrunken hard spring wheat, grown and marketed in the northwestern states, is produced by infected plants. Bailey and Hendel (1923) found no correlation between plumpness of the wheat berry and protein, or gluten content of the flour, in working over the data resulting from a study of five wheat crops. Similar findings were reported by Mangels and Sanderson (1925). From this it would appear that the crude protein content of rusted wheat would depend, not upon the infection alone, but upon the environment of the wheat plant as well, and that grain from rust-infected plants might, and probably frequently does have, a low content of gluten.

Relation of Weight per Bushel to Yield of Flour.

While it has been shown by Bailey and Hendel (1923) and by Mangels and Sanderson (1925) that the plumpness and weight per bushel of wheat is not necessarily correlated with the protein content of flour milled from the wheat, it seems appropriate at this point to briefly refer to the correlation established between weight per bushel and yield of flour in milling.

Milling data resulting from testing spring wheat samples of the crops from 1911 to 1916 was summarized by Sanderson (1918). The yields of flour ranged from an average of about 72 per cent, in case of wheat weighing (average) 62.7 pounds per bushel, to 62.2 per cent of flour from wheat averaging 47.6 pounds per bushel. This represents a decrease of about 0.65 per cent of flour per pound decrease in weight per bushel.

A graph showing the correlation between yield of straight flour and weight per bushel of wheat was presented by Thomas (1917 b). The curve takes the form of a fairly straight line. Wheat weighing from 51.1 to 52 pounds per bushel yielded about 64.6 per cent of flour, while that weighing from 64.1 to 65 pounds per bushel yielded about 74.6 per cent of flour. The decrease in yield averaged about 0.75 per cent of flour for each pound decrease in weight per bushel.

Kansas State Board of Agriculture (1920) published a table prepared by the Howard Wheat and Flour Testing Laboratory of Minneapolis, which showed the average yield of straight flour from wheat of different weights per bushel. These data were compiled from the results of several thousand tests, and "are based upon the experience of well equipped mills grinding closely. Smaller mills not grinding closely could expect from 2 to 4 per cent lower yields." A portion of the data from this tabulation are given in Table 53.

TABLE 53

Relation Between Weight Per Bushel and Yield of Flour as Reported by the Howard Laboratory

Weight per Bushel, Pounds	Yield of Straight Flour, Per Cent
45	53.5
50	63.8
55	72.4
60	77.4
64	79.6

Data resulting from milling tests of wheat were compiled by Shollenberger (1923 a), from which it appears that wheat samples having a weight per bushel (after removal of dockage) between 63.0 and 63.9 pounds yielded 68.9 per cent of flour, while wheat weighing 47.0 to 47.9 pounds per bushel yielded 58.0 per cent of flour. Through this range of test weight (omitting wheats with a lower weight per bushel, since they are not usually regarded as milling wheat) the decrease in 16 pounds per bushel is accompanied by a decrease of 10.9 per cent of flour, or about 0.68 per cent of flour for each pound of test weight per bushel. From a graph presented in this report it would appear that hard red winter wheat yielded a higher percentage of flour than hard red spring wheat having the same weight per bushel through the range recorded (from 50 to 62 pounds per bushel). Shollenberger (1923 a) also presented in graphic form the correlation between color, texture, absorption, and volume of loaf, and the weight per bushel of wheat. There was a tendency toward a diminishing color score, and texture score of the bread, with decreasing weight per bushel of the wheat from which the flour was milled, but no significant correlation between absorption or loaf volume and weight per bushel.

Shollenberger (1925) reported the average flour yield from wheats grouped on the basis of weight per bushel, for the period of the ten crop years of 1915 to 1924 inclusive. In the hard spring class, a reduction in average test weight from 63 to 53 lbs. per bushel inclusive was accompanied by an average reduction of 0.85 per cent of flour per pound of test weight; with the hard winter wheats the average reduction was 0.60 per cent of flour per pound of test weight. The yield of flour from the hard spring and hard winter wheats was greater than from the soft red winter and white wheats, when samples weighing 58 pounds per bushel and over were compared.

TABLE 54

AVERAGE YIELD OF FLOUR FROM WHEAT OF VARIOUS TEST WEIGHTS PER BUSHEL, REPORTED BY SHOLLENBERGER (1925)

Test Weight per Bushel of Dockage-Free Wheat, Pounds	Class of Wheat				
	Hard Spring	Hard Winter	Durum	Soft Red Winter	White
63	73.8	73.3	72.9	72.4	70.3
62	72.8	73.5	72.4	71.6	70.3
61	71.8	72.5	71.0	70.7	70.8
60	71.0	71.8	70.3	69.6	70.4
59	70.8	71.3	69.2	69.6	70.3
58	69.7	70.8	68.7	68.3	69.7
57	69.0	70.7	66.8	67.9	69.2
56	68.0	70.5	65.1	67.3	68.3
55	66.4	69.1	64.3	67.0	66.9
54	65.8	68.3	62.5	66.3	66.3
53	64.5	66.6	...	...	...
52	63.6	67.1	...	...	...
51	62.8	65.5	...	...	...

Bailey (1924) summarized the results of two seasons' work at the Minnesota Experimental Flour Mill, showing the correlation between weight per bushel and yield of flour. Plotting weight per bushel as ordinates, and yield of flour as absissæ, the resulting curve is not a straight line, but a simple hyperbola. These observations are recorded in Table 55, and shown graphically in Figure 3. There was an average decrease of 0.78 per cent of flour for each pound decrease in weight per bushel.

It is of interest to compare these yields of flour from wheats of varying weight per bushel with the requirements of the U. S. Food Administration during the late war. (See U. S. Food Administration,

Special License Regulations No. 11, Wheat miller and manufacturers of mixed flours, M. S. 1000, July 22nd, 1918). The yields required in

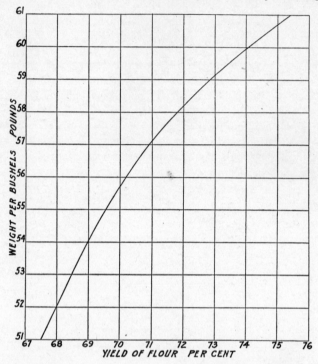

FIG. 3.—Graph showing the correlation between weight per bushel of wheat, and yield of flour. (Bailey, Minn. State Dept. Agr. Bul. 34.)

milling in the licensed flour mills in the United States during the period of Food Administration control are shown in Table 56.

TABLE 55

YIELD OF FLOUR FROM WHEAT SAMPLES GROUPED ON THE BASIS OF TEST WEIGHT PER BUSHEL

Group Range of Weight per Bushel, Pounds	Average Weight per Bushel of Each Group, Pounds	Average Yield of Flour Produced in Milling Each Group, Per Cent
51.0 to 51.9	51.00	67.37
52.0 to 52.9	52.00 [1]	63.46 [1]
53.0 to 53.9	53.50	69.09
54.0 to 54.9	54.12	68.98
55.0 to 55.9	55.17	69.74
56.0 to 56.9	56.22	70.28
57.0 to 57.9	57.22	71.53
58.0 to 58.9	58.34	72.01
59.0 to 59.9	59.23	73.55
60 and over	60.89	75.22

[1] Omitted in preparing the graph in Figure 3.

TABLE 56
Minimum Flour Extraction Permitted in the United States During the Food Administration Control (Regulations of July 22nd, 1918)

Test Weight of Wheat per Bushel	Pounds Wheat per 196 Pounds Flour	Percentage of Flour Extraction
58 lbs. or heavier	264 lbs.	74.3
57 lbs. or heavier	268 lbs.	73.2
56 lbs. or heavier	272 lbs.	72.1
55 lbs. or heavier	276 lbs.	71.0
54 lbs. or heavier	281 lbs.	69.8
53 lbs. or heavier	286 lbs.	68.6
52 lbs. or heavier	292 lbs.	67.2
51 lbs. or heavier	298 lbs.	65.8

Sprouted or Germinated Wheat.

A form of abnormal or "damaged" wheat frequently encountered in commercial lots of this cereal is sprouted or germinated grain. Sprouting of the kernel commonly results from the exposure of unprotected shocks or stooks to rains, which raise the moisture content of the grain to a sufficiently high level so that growth or germination can take place. Certain physiological aspects of germination have been discussed in the section dealing with the chemical life history of the wheat plant, and at this point the milling and baking properties of such grain will be detailed.

The loss in weight resulting from progressive germination of wheat was determined by Teller (1896). In three days this amounted to about 6 per cent of the original weight of the kernel, as shown in Table 57.

TABLE 57
Loss in Weight of Kernels on Sprouting, as Shown by Teller (1896)

Time Sprouted, Hours	Parts per 100 Lost
24	1.5
48	2.5
72	5.9
99	6.7
120	10.1

Three portions each of soft winter and hard winter wheat were sprouted by Harcourt (1911). In lot No. 1 the sprouts were allowed to grow to the length of the kernel before the grain was dried. In lot No. 2 the sprouts were twice the length of the kernel, and in lot No. 3 they were from 1 to 1½ inches long. The dried wheats, as well as a control or ungerminated sample, were milled into flour, the yields of flour not being reported. The resulting flours were subjected to baking tests, and an improvement in the quality of bread resulted from germinating the grain until the sprouts were the length of the kernel. (No. 1 sprouted.) When the sprouts were allowed to grow any longer than this, the baking quality of the flour was impaired.

TABLE 58

Tests of Flour Milled from Normal and Sprouted Wheat (Harcourt, 1911)

Sample of Wheat	Wet Gluten Per Cent	Water Absorbed Per Cent	Weight of Loaf Grams	Volume of Loaf cc.	Quality of Loaf		
					Color	Texture	Appearance
Ontario winter wheat							
Normal	28.00	47.65	469	1,670	100	100	100
No. 1 sprouted..	25.97	47.65	471	1,730	100	102	101
No. 2 " ..	22.03	47.65	473	1,450	97	poor	poor
No. 3 " ..	....	47.05	468	1,270	95	poorer	poorer
Manitoba spring wheat							
Normal	39.93	75.30	527	2,900	100	100	100
No. 1 sprouted..	39.47	67.65	496	2,900	104	102	104
No. 2 " ..	38.87	63.53	486	2,600	102.5	90	97
No. 3 " ..	....	63.53	489	2,430	100	80	93

The nitrogen and the soluble carbohydrate fractions of the sprouted wheat were also determined by Harcourt, with the results shown in Table 59. There was little change in the alcohol-soluble nitrogen, due, perhaps, to the appearance in this fraction of certain of the products of hydrolytic cleavage of the gliadin. The glutenin nitrogen decreased, and the salt-soluble nitrogen increased somewhat. The greatest proportional change was in the amide nitrogen fraction (after precipitation of proteins with Stutzer's reagent). This fraction increased progressively with the degree of sprouting; the percentage of sugars likewise increased considerably.

TABLE 59

Results of Analysis of Normal and Sprouted Wheats (Harcourt, 1911)

	Percentage of Nitrogen in Various Forms					Carbohydrates	
	Total Nitrogen	Alcohol Soluble Nitrogen	Glutenin Nitrogen	Salt Soluble Nitrogen	Amide Nitrogen	Total Sugars	Soluble Starch and Dextrin
Ontario winter wheat							
Normal	1.895	.833	.563	.499	.170	.90	3.46
No. 1 sprouted	1.891	.829	.547	.515	.172	1.44	3.22
No. 2 "	1.889	.824	.519	.546	.536	1.63	4.03
No. 3 "	1.890	...	...	...	.708	2.83	4.23
Manitoba spring wheat							
Normal	2.365	.928	.855	.582	.175	1.02	3.66
No. 1 sprouted	2.354	.930	.811	.613	.337	1.48	3.55
No. 2 "	2.358	.928	.763	.659	.539	2.39	4.62
No. 3 "	2.350	...	....	...	.700	3.02	5.22

Wheat was germinated in damp sand for varying lengths of time up to 5 days by Willard and Swanson (1911). There was a fairly regular increase in the loss in weight with increasing length of germination period, likewise an increased loss in scouring the wheat in the process of cleaning, and a decrease in the percentage of flour produced in milling the wheat. From the data presented, and the photographs of the loaves, it would appear that flour milled from the wheat sprouted for 1, 2 and 3 days was baked into satisfactory bread. A longer germinating period apparently resulted in impairment of baking strength. In Table 60 is shown the percentage of amino compounds ($N \times 5.7$ in the aqueous extract not precipitated by phosphotungstic acid), protein soluble in 1.5 per cent NaCl solution, and acidity of the patent and break flours. Increasing the length of the period of germination resulted in a tendency toward progressive increases in these components.

TABLE 60

Nitrogen Fractions and Acidity of Patent and Break Flours Milled from Sprouted Wheat by Willard and Swanson (1911)

Days Sprouted	Amino Compounds		Protein in Compounds Sol. in 1.5% NaCl Solution		Acidity of Flour	
	Patent, Per Cent	Break, Per Cent	Patent, Per Cent	Break, Per Cent	Patent, Per Cent	Break, Per Cent
0	.128	.276	2.19	3.10	0.90	2.00
1	.136	.225	2.14	2.71	1.10	2.10
2	.144	.208	2.14	2.50	1.15	1.55
3	.184	.240	2.26	2.48	1.30	1.95
4	.247	.327	2.26	2.63	1.62	2.25
5	.351	.384	2.42	2.83	1.55	2.45

In another experiment Willard and Swanson (1913) sprouted wheat until the sprouts were an inch long, dried the grain, and milled it into flour in the usual manner. Varying proportions of this flour were mixed with sound wheat flour, and the mixtures subjected to baking tests. In instances where the ratio of germinated wheat flour to sound wheat flour was as low as 1:12, there was an evident impairment in quality of the bread produced. Extracts of bran and shorts milled from germinated wheat did not effect any improvement in the quality of bread when used in the dough batch.

Another series of studies of germinated wheat was conducted at the Kansas Agricultural Experiment Station by Fitz and Swanson (1916), in which different lots of the same sample were allowed to sprout for varying lengths of time. On milling the germinated samples, the yield of flour was reduced in proportion to the length of the germination period. As in the previous studies, the percentage of amino compounds in both wheat and flour increased as the sprouting proceeded. The same was true of the acidity, and of the sugar content of the flour.

Difficulty was experienced in recovering the gluten from flours when the wheat was sprouted for more than 3 days. Baking tests of the flour showed some improvement resulting from germination up to 3 days. Prolonging the germination period beyond 2 days weakened the gluten and impaired the loaf texture. Certain of the analytical data presented by Fitz and Swanson are given in Table 61.

TABLE 61

TESTS AND ANALYSES OF GERMINATED WHEAT AND FLOUR MILLED THEREFROM
(FITZ AND SWANSON, 1916)

Hours of Germinated	Yield Total Flour	Wheat Amino Compounds	Wheat Acidity	Flour Protein	Wet Gluten	Dry Gluten	Pure Gluten	Flour Amino Compounds	Sugar	Acidity
0	....	....	....	9.99	28.29	9.79	8.27	.145	2.23	.180
28	70.20	.369	.488	9.66	26.62	9.08	7.97	.093	1.91	.178
43	71.08	.412	.483	9.65	24.49	9.02	7.90	.101	2.10	.192
67	65.60	.538	.633	9.51	22.90	8.87	7.69	.143	2.14	.235
76	61.64	.549	.846	9.72	20.34	7.87	6.89	.220	2.46	.350
91	62.36	.704	.843	8.97	16.56	6.24	5.51	.255	2.69	.281
99	53.36	1.127	.945	8.78	2.92	1.12	...	.265	3.69	.303
119	46.40	1.253	1.080	9.08	9.31	3.38	2.79	.376	3.93	.263

To determine the effect of small proportions of sprouted wheat flour in a blend, varying amounts of such flour were added to different lots of sound wheat flour. In general, additions of such sprouted wheat flour appeared to effect an improvement in loaf quality when not to exceed 25 per cent was used in the flour blend, and when the wheat from which it was milled was not sprouted more than 3 days. The authors note that this experience was somewhat different from that previously reported by Willard and Swanson (1913), and call attention to the fact that in this series a new wheat was used which yielded flour stronger than that milled from the old wheat used in the previous study.

Wheats which weighed 59.5 and 57 pounds per bushel, respectively, before sprouting weighed 53.5 and 51 pounds per bushel after being sprouted by Sanderson (1913a). Acidity of the flour milled from the wheat after sprouting was substantially higher than before sprouting. Likewise the protein content was somewhat higher, due to oxidation of carbohydrates. A comparison of a field-sprouted sample containing 4.2 per cent of sprouted kernels with a sample containing 14.4 per cent, showed the latter to give the best results on milling and baking except in the matter of color of the flour. Additions of sprouted grain to the sample containing only 4.2 per cent of sprouted kernels resulted in improvements in baking strength until the sprouted kernels exceeded 11 + per cent of the mixture.

The loss in weight of wheat with progressive elongation of the epicotyl in germination was traced by Olson (1917b). In the case of a sample of bluestem wheat, when the epicotyl was one-fourth (1), once (2), and twice (3) the length of the kernel, the losses in dry weight were (1) 0.60, (2) 2.70, and (3) 3.70 per cent, respectively. The yields of flour on milling the wheats were also reduced by (1) 7.05, (2) 14.60, and (3) 22.45 per cent, respectively, below the control or unsprouted sample of bluestem wheat, which yielded 74.15 per cent of total flour. Diastatic activity, expressed in terms of time, required to effect an autolysis of starch to a disappearance of the iodine reaction, was (1) 180, (2) 70, and (3) 55 minutes for the three lots of germinated wheat flour, as compared with 480 minutes in case of the control. Difficulty was experienced in recovering the gluten from the sprouted samples, the dry gluten found in the three flours being appreciably less than that found in the control or ungerminated. On fractionating the nitrogenous compounds Olson observed a tendency toward a decrease in the glutenin fraction, and an increase in the "amide nitrogen" fraction (nitrogen not precipitated by phosphotungstic acid from a 10 per cent NaCl solution extract).

TABLE 62

EFFECT OF GERMINATION ON YIELD OF DRY GLUTEN, AND ON NITROGEN FRACTIONS
(OLSON, 1917 b)

Length of Epicotyl to Length of Kernel	Dry Crude Gluten, Per Cent	Nitrogen Fractions in Percentage of Total Nitrogen			
		Gliadin	Glutenin	Albumin and Globulin	Amide Nitrogen
Control	14.19	46.66	26.44	26.82	0.08
¼ : 1	11.53	45.71	32.54	21.67	0.08
1 : 1	9.26	47.83	23.48	24.35	4.34
2 : 1	2.06	50.00	6.36	34.55	9.09

Loaves baked by Olson from the first two lots of germinated wheat flour (1 and 2) were larger than the control, but the crumb texture was inferior. The third lot (3) of sprouted wheat flour gave a loaf inferior in all particulars. Additions of 1, 5 and 10 per cent of germinated bluestem wheat flour to flour milled from normal bluestem wheat effected an improvement in loaf quality in proportion to the quantity added up to 10 per cent.

Flours milled from Montana hard winter wheats containing a small percentage of sprouted kernels were subjected to baking tests by Thomas (1917a). The general properties of the resulting bread, including loaf volume, color and texture, compared favorably with samples of the same class of wheat in which no sprouted kernels were present. Flour milled from sprouted wheat had a consistently lower water absorption than flour milled from normal or unsprouted wheat in the studies reported by Stockham (1917). The differences in this respect, in the freshly milled flours, are of the order of 2 per cent. On storing the sprouted wheat flour for 15 months its absorption increased only 1.2 per cent, while the control flour milled from the unsprouted grain increased in absorption 2.9 per cent.

The quantity of gas generated in dough made with flour from sprouted wheat was determined by Stockham (1920) and found to be greater than that produced from unsprouted wheat flour. The loaf baked from the sprouted wheat flour was only 1,270 cc. in volume, while that from the unsprouted wheat flour was 2,500 cc. Proteolytic activity of the sprouted wheat flour (p. 32) was also higher.

Milling and baking tests of Nigger wheat (1) unsprouted, (2) with sprouts to one-half the length of the kernel, and (3) with sprouts over one-half the length of the kernel, were conducted by Corbould (1921). The flour yield on milling was 70.8, 65.9, and 56.6 per cent, respectively, and the loaf volume on baking was 2,000, 2,125 and 1,975. It was concluded that if sprouting was checked before the plumule grew to not more than one-half the length of the kernel, a beneficial effect upon the flour for baking purposes would result.

The mixing of a small percentage of sprouted kernels with sound, normal grain that is low in diastatic activity was suggested by Sherwood (1925). The flour milled from such mixtures containing 2 and 3 per cent of sprouted kernels was found to have the same hydrogen ion concentration and titrable acidity as the flour milled from the normal grain. A further discussion of the comparative properties of these flours will be found in the section dealing with flour strength (Chapter 10).

Heat-Damaged or Bin-Burned Wheat.

Comparatively little data appears to be available which serves to indicate the modifications in the chemical characteristics of wheat effected in consequence of heat-damage or bin-burning. The author's observations lead to the conclusion that modifications of the flavor and odor of wheat and flour milled therefrom result before the chemical properties are altered to an extent that can be detected by methods now available. Frank (1922b) suggested that failure of the gluten of flour to agglutinate would serve to indicate the heat-damaged condition in individual wheat kernels. It is probable that substantial modifications of the baking properties of flour would be evident before the flour would react to Frank's test. A discussion of the phenomena giving rise to the heat-damaged condition will be found in the following chapter.

Effect of Admixtures of Weed Seeds and Rye on the Milling Quality of Wheat and Baking Quality of Flour.

Commercial wheat as it reaches the miller is frequently mixed with the seeds of a number of plants. This is true of spring wheat in particular, since it ripens at about the same time as the common weeds which grow in the wheat fields. Bailey (1922) indicated the relative prevalence of the more common weed seeds in northwestern spring wheat. Corn cockle was apparently more widely distributed in the spring wheat area than wild vetch and kingheads. The latter appeared more frequently in wheat grown in the northern portion of the Red River Valley, while wild vetch seed constituted a larger percentage of the mixture grown in the southern portion of the area. The seeds of certain weeds, such as the mustards, the foxtail grasses, wild buckwheat or black bindweed, and the like, are not particularly difficult to separate from the wheat before milling. A simple cleaning machine suffices to effect a complete separation of these from ordinary plump wheat kernels. Certain other weed seeds are much more difficult to separate from wheat, however, and special machinery is required for this purpose. Included in this group of difficultly separable weed seeds are corn cockle (Agrostemma githago), wild vetch (Vicia angustifolia, Vicia villosa), commonly known as "wild peas," giant ragweed or "kingheads" (Ambrosia trifida), wild rose (Rosa sp.), darnel (Lolium temulentum), and chess or cheat (Bromus secalinus). Bulblets of garlic and of wild onion are not uncommon in the soft red winter wheat raised in sections of Ohio, Pennsylvania and Maryland. Barley and rye kernels are often encountered, the latter being found particularly in winter wheat. In addition to seeds and bulblets, other impurities, such as the sclerotia of the fungus known as ergot, often referred to as ergotized grains, are occasionally mixed with wheat, especially of the durum varieties grown in sections of the Northwest. These cannot always be completely separated from such wheat in cleaning it for milling. Smut balls, resulting from the infection of the wheat plant with bunt or stinking smut, when mixed with wheat present a special problem in cleaning. Smut balls are no longer common in hard red spring wheat, but at present are found most commonly in wheat grown in the Palouse and Walla Walla districts of southeastern Washington.

The results of milling tests of wheat mixed with various impurities were reported by Miller (1915). Corn cockle seed, when present to the extent of 2 per cent or more in the wheat mixture, effected a reduction in loaf volume, texture and color of bread baked from the flour. An irritation of the skin of the hands was noticed on kneading and handling dough made from flour containing 3 per cent or more of corn cockle seed. Seeds of the hairy vetch (Vicia villosa), in proportions of 1 per cent or more in the wheat when milled, resulted in an impairment of the baking qualities of the resulting flour. The loaf volume, color, and texture of the bread were reduced in just the proportion that such seeds were present in the wheat mixture. Vetch seed also imparted a characteristic odor and flavor to the bread. Kingheads, the seed of giant ragweed, has less effect on the loaf volume than the other two seeds mentioned above, but its presence resulted in a greater reduction in color score, and a substantial impairment of the loaf crumb texture. It also served to reduce the yield of flour in milling more than the other seeds studied. Rye had much less effect than the weed seeds studied by Miller, 2 per cent producing a small effect upon any of the properties of the baked bread, and up to 5 per cent had no great effect, although the color score was reduced somewhat.

When the wheat mixture contained 3 per cent or more of rye it was found by Bailey and Sherwood (1924) that the color of the resulting flour was darker than the normal, while 4 per cent or more of rye appeared to reduce the strength of the flour, as shown by the impaired loaf volume and texture. Two per cent or less of rye did not apparently affect the flour properties, nor reduce the yield of flour in milling.

No appreciable effect on the volume of loaf, color score, odor or other properties of bread baked from flour which was milled from wheat to which 1 and 2 per cent of wild vetch ("wildpeas") seed had been added was observed by Wilhoit (1915) when compared with flour milled from clean wheat. When 3 per cent or more of wild vetch seed was added to the wheat there was an effect upon the properties of the loaf.

Varying percentages of wild vetch seed were added by Bailey (1916b) to clean wheat, and on milling the mixture a fairly regular decrease in the yield of flour was observed as the percentage of vetch was increased. Expansion of the dough and volume of the loaf of bread diminished with increasing amounts of wild vetch seed, and the color score was lowered. No abnormal odor of the flour or dough was noted when 0.5 per cent of vetch seed was mixed with the wheat, while with 1 per cent or more the characteristic odor was distinctly noticeable. The odor of hydrocyanic acid first became apparent when flour containing ground vetch seed was mixed into dough with warm water. This odor soon disappeared, or was masked by the succeeding odor of benzaldehyde, which persisted until the dough was baked. With the smaller percentages of vetch the odor was not observed in the baked and cooled loaf.

The presence in corn cockle seeds of a sapotoxin was indicated in investigations cited by Wehmer (1911). Stoecklin (1917) referred to the occurrence of githagine in the seeds of Agrostemma githago and described a hemolytic method for determining the proportion of this sapotoxin in flour. Wehmer also states that the studies of Bruyning and van do Harst, and of Bertrand showed the presence of about 0.9 per cent of a cyanogenetic glucoside, viciamin, in seeds of wild vetch. About 0.075 grams of HCN were evolved from 100 grams of the seed.

Hydrocyanic acid equivalent to 3.3 milligrams in each 100 grams of wild vetch seed was found by Gortner (1920). A lot of pigs weighing 143 pounds each were fed for 21 days on a ration including 1.5 pounds daily of wild vetch seed per head. These pigs made satisfactory gains in weight on this ration, and were healthy at the end of the period. Another lot was fed on a ration including 0.9 pounds of wild vetch seed per day, per head, and not only remained healthy, but gained steadily. A lamb was then fed on meal consisting of equal parts of barley and ground vetch, without harmful effect. From these studies it would appear that in reasonable quantities seeds of wild vetch may be safely fed to pigs, and possibly also to sheep, although there is some doubt about the desirability of feeding it to ruminants.

C. Louise Phillips (1922) reports the results of work conducted by the U. S. Bureau of Agricultural Economics to determine the milling yield of wheat and rye mixtures. It was found that each per centum of rye in the mixture effected a reduction of approximately one-third of 1 per cent of flour in milling. This was believed to be due to the difference in endosperm texture of the wheat and rye kernels. Rye has a softer kernel texture and, when mixed with hard wheat, is overtempered when water is added in tempering the mixture. Color score of the bread crumb was reduced in proportion to the quantity of rye in the mixture, 3 per cent of rye reducing the color score about one-half point.

In the same report by Miss Phillips it was indicated that when sweet clover seeds were mixed with wheat, and the mixture stored in tight containers for two months, their characteristic odor was imparted to the wheat. The mixtures containing 0.1 to 0.5 per cent of clover seed gave a faint odor of sweet clover at time of milling, and those containing 0.7 per cent or more gave a distinct odor. The odor was less apparent in the flour and feed than in the grain. Only when the grain had a pronounced odor of sweet clover could the odor be detected in the cold bread.

In the Agricultural Gazette of N. S. Wales (1923) appeared a report on the effect of Bokhara clover seed, which stated that bread baked from wheat contaminated with 3 per cent by count of such seed possessed the characteristic odor of the weed. With an admixture of 1 per cent it was doubtful whether or not the characteristic scent could be detected in the bread.

The U. S. Bureau of Markets and Crop Estimates (1921) examined 47 samples of Maryland wheat, and found an average of 40 garlic bulblets per bushel. One of the objections noted to the presence of garlic is the fact that it clogs up the corrugations of the rolls in the flour mill so these cannot function properly.

The average number of garlic bulblets per 1000 grams of Maryland wheat was found by Metzger (1922) to be 89.3 in the 1920 crop, and 79.4 in the 1921 crop. Metzger published a chart prepared by Shollenberger and Sommers of the U. S. Bureau of Agricultural Economics, showing the discount paid for garlicky wheat in Baltimore to range between 5 and 19 cents per bushel during the crop seasons of 1920 and 1921. During most of this period the discounts ranged between 8 and 16 cents. The greatest difficulty in marketing such wheat was, of course, the abnormal or garlicky odor which it possessed. Flour milled from garlicky wheat was purchased by blenders at substantial discounts and gradually worked off in flour blends or mixtures with normal flour.

Duvel (1907) and Cox (1916) indicate that garlic bulblets may be removed from wheat by drying the mixture with heated air, and then blowing out the light, desiccated garlic bulblets with an air current.

The presence of any considerable quantity of the spores of bunt or stinking smut in the wheat at the time of grinding usually results in the production of flour with a grayish hue, due to the brownish-black spore of this fungus. A peculiar "stale fish" odor often may be detected when the proportion of smut is large. The commercial value of such discolored flour is appreciably lowered, due to the preference on the part of the buying public for white or creamy tints. Millers accordingly resort to washing, or otherwise treating the wheat to remove such smut before grinding. The methods used in this connection will be described in the section in which the roller milling process is discussed.

Attention was called by Coleman and Regan (1918) to the occasional presence in bulk wheat of galls resulting from the infection of wheat heads by the nematode Tylenchus tritici (Steinbuck) Bastian. Wheat containing these galls originated chiefly in Virginia, although a trace was found in samples from California. The galls, on chemical analysis, were found to be high in the content of pentosans (20.4 per cent), and crude fiber (22.2 per cent). On subjecting a sample of wheat containing 2.5 per cent of nemotode galls to milling tests, it was observed that the flour possessed a characteristic disagreeable odor, and was darker in color than flour milled from the same wheat with galls removed. The volume and texture of the loaves baked from these two lots of flour were not appreciably different.

Yellow-Berry in Wheat, and Its Relation to Composition and Quality.

The condition known as yellow-berry has been referred to in the discussion of the influence of environment on the composition of wheat. Yellow-berry kernels are frequently encountered in samples of hard wheat varieties. They are distinguished by a softer, lighter-colored "starchy" endosperm, lacking the corneous or vitreous texture characteristic of the typically hard kernels. Yellow-berry kernels are variously described as "starchy" or "mealy," and as "piebald" when the entire kernel is not affected, as is frequently the case. As antonyms the terms "flinty," "hard," "corneous," "vitreous," "horny," and (in the case of durum wheat) "amber" are applied.

Bailey (1916a) discussed the observations of several workers which established that yellow-berry kernels possessed their characteristic appearance in consequence of the inclusion of a larger volume of vacuoles or air-spaces in the endosperm. This, in turn, modifies the refraction of light passing through such material, and occasions the lighter visual appearance of such structures. Hayes, Bailey, Arny and Olson (1917) emphasized the fact that the visual appearance of wheat is determined by the joint effect of two factors: 1st, the presence or absence of a pigment in the bran; and, 2nd, the physical condition or vitreousness of the endosperm cells. They proposed that the several gradations of endosperm density be designated as (a) corneous, (b) subcorneous, (c) substarchy, and (d) starchy, in order of decreasing hardness.

Samples of hard and soft wheat were separated by Snyder (1904a) into light or starchy, and dark or glutinous kernels. On analysis the average percentage of protein was found to be highest in the dark kernels, as shown by the data in Table 63.

TABLE 63

CRUDE PROTEIN IN LIGHT OR STARCHY, AND DARK KERNELS (SNYDER, 1904a)

	Protein, Per Cent
Miscellaneous samples	
Light or starchy kernels	12.68
Dark kernels	15.33
Selected seed wheat	
Light or starchy kernels	12.83
Dark kernels	14.93

Snyder (1905a) later observed consistently higher percentages of protein in "glutinous" wheat than in "starchy" (yellow-berry) wheat from the same seed. Lyon and Keyser (1905) also found the yellow-berry kernels to be lower in nitrogen content than the horny red kernels. They conclude that soil and climatic conditions previous to harvesting affect the proportion of yellow-berry kernels.

The yellow-berry problem was discussed at length by Roberts and Tillman (1908), and Roberts (1910). Their studies dealt with the influence of climatic conditions on the proportion of yellow-berry in hard wheat, the inheritance of yellow-berry, and the determination of hardness, rather than with the chemical composition of the grain. Roberts (1919) also measured the size of the starch granules in hard grains and soft grains and found them to average 0.0292 mm. in the hard and 0.0269 mm. in the soft grains. The specific gravity of the yellow-berry kernels averaged 1.369 and of the hard kernels 1.392. The average chemical composition reported for the two types of kernels is recorded in Table 64, and shows the hard, flinty wheat to contain a higher percentage of protein and lower percentage of starch.

TABLE 64

Composition of Hard, Flinty and of Yellow-Berry Wheat Kernels Reported by Roberts (1919)

		Yellow-Berry	Hard, Flinty Wheat
Moisture	per cent	7.85	7.63
Ash	" "	1.79	1.97
Crude protein	" "	10.51	11.99
Crude fiber	" "	2.25	2.37
N-free extract	" "	75.68	74.16
Pentosans	" "	7.56	7.88
Starch	" "	67.01	64.76
Ether extract	" "	1.97	1.88

The flour yield, specific gravity and total nitrogen content of hard or vitreous wheats was found by Bailey (1916a) to be higher than in the yellow-berry or mealy samples of the same type. This is shown by the average of several comparisons, given in Table 65.

TABLE 65

Results of Tests and Analyses of Vitreous and Yellow-Berry Wheats (Bailey, 1916a)

	Total Nitrogen, Per Cent	Flour Yield, Per Cent	Specific Gravity
Vitreous spring wheats	2.48	71.0	1.4207
Yellow-berry or mealy spring wheats	1.93	69.3	1.4063
Vitreous winter wheats	2.27	71.0	1.4227
Yellow-berry or mealy winter wheats	1.60	67.6	1.4034

Soft red winter wheats grown in Ohio, Indiana and Maryland had a lower specific gravity, as well as a lower nitrogen content, than the hard wheats of Minnesota.

Density, as indicated by specific gravity, contributed to the ease of milling, the yellow-berry samples yielding lower percentages of flour than the hard samples which were otherwise similar in physical characteristics.

The yellow-berry condition was discussed by Headden (1915a, 1916a, 1916b, 1918a, 1918b) in connection with his studies of the influence of various factors on the composition of Colorado wheat. He found (1918b, p. 42) the yellow-berry kernels to be invariably lower in percentage of crude protein than the flinty kernels grown under the same conditions. The average protein content of the former was 9.03 per cent, and of the latter 10.83 per cent.

Hard and yellow samples of Wis. No. 45 wheat of the 1917 crop were subjected to milling and baking tests by Leith (1919), who found the hard samples to be superior in flour yield and volume of loaf. Chemical analysis of yellow-berry and hard kernels of durum and hard winter varieties showed the former to contain from 1.3 to 3.55 per cent more crude protein than the latter.

In the case of 44 samples separated into yellow-berry and dark hard portions by Frank (1922a, 1923) the protein content of the two fractions was 13.40 per cent and 14.04 per cent, respectively. Only a fair degree of success attended the effort to estimate protein content from the percentage of dark, hard and vitreous kernels in a sample, and it was concluded that determinations of the percentage of hard kernels are of little value in classifying hard wheats on the basis of protein content. The correlation of the percentage of dark hard vitreous kernels and protein content of wheat was determined by Mangels and Sanderson (1925). Samples of the 1922 and 1924 crops showed a positive coefficient correlation of $+ 0.660 \pm 0.041$, and $+ 0.453 \pm 0.030$, while in the 1923 crop the coefficient correlation dropped to $+ 0.067 \pm 0.047$.

The author has observed that, in a general way, the flour milled from wheats high in yellow-berry possessed a more creamy or yellow color than flour milled from dark and vitreous kernels.

Chapter 6.

Storage and Handling of Wheat.

After the wheat is harvested and threshed a considerable time may elapse before it reaches the miller. It may remain in the hands of the producer or be shipped to a terminal and stored there in an elevator, it may travel long distances in freight cars and the holds of vessels; and then be stored by the miller for an extended period before it is actually ground into flour. During this period the grain frequently undergoes certain changes which substantially modify its milling and baking properties. In wheat which has been threshed from the shocks almost immediately after the harvest a process known as "sweating" occurs, which is evidently due to deep-seated changes in the grain that are possibly accelerated by enzymes. The same changes can occur in shocked wheat, and it has been observed in stacked grain. In the latter case it is evidenced by the appearance of a moist condition, which is responsible for the term sweating. A slight rise of temperature may result from sweating, either in stacks or bins, which persists for a few days and then falls to normal if the grain is not damp. Several views have been expressed concerning the occasion for this phenomenon, certain of which have been summarized by the Kansas State Board of Agriculture (1920). It seems to be generally agreed that this is a periodic change, occurring as one phase of the normal life cycle of wheat. The writer advanced the hypothesis that sweating is one manifestation of the process of after-ripening, which cereals are known to undergo. It will probably occur in normally ripened and reasonably dry grain, regardless of its location, but will be observed by gross physical modifications only when the bulk is large, as in a stack or bin.

During the sweating, and for a time thereafter, wheat undergoes a modification which is reflected in improved baking properties. The rate and extent of these changes on aging of wheat is indicated by the data published by Fitz (1910). A lot of wheat was threshed and 4 days later (August 31) a sample was milled (No. 398). A portion of the same lot of wheat was stored for 18 days in a bin and then milled (398A), at which time it appeared to be going through the sweat. A third sample (398B) was milled after 57 days' storage in the bin. Part of the wheat was stacked (460), and this was threshed 57 days later and at once milled, at the time 398B was milled. The results of baking tests of these four lots of wheat are given in Table 66. From these it appears that there was a fairly steady improvement in baking quality in so far as water absorption and loaf volume were concerned, although the color score of the freshly harvested wheat flour was the highest.

TABLE 66

RESULTS OF BAKING TESTS OF FLOURS MILLED FROM SHOCK-THRESHED AND STACK-THRESHED WHEAT (FITZ, 1910)

Sample	Treatment	Water Absorption, Per Cent	Volume of Loaf, cc.	Color
398	Shock-threshed and milled at once	51.5	2,440	102
398A	Shock-threshed and stored in bin 18 days	51.5	2,500	100
398B	Shock-threshed and stored in bin 57 days	54.1	2,650	100
460	Stacked and later threshed and milled when 398B was milled	54.7	2,710	99

After the period of sweating is concluded the wheat may continue to improve for an extended period so far as baking qualities of the flour milled from it are concerned. Saunders (1908) stored Pringles Champion wheat from 1907 to 1908. In midwinter 1907 its baking strength was 80, and a year later it was 91. Baking strength of Red Fife wheat increased during the same period from 95 in 1907 to 101 in 1908. This tendency toward improvement in baking strength on aging was confirmed by further studies (1910, 1911).

A sample of hard, red spring wheat was stored in the experimental mill elevator at North Dakota Agricultural Experiment Station, and portions of it were milled and baked at intervals over a period of nine years. The resulting data as reported by Stockham (1920) show a significant increase in baking strength during the first year. For five years thereafter there was little change in the loaf volume and texture score of bread baked from the samples as successively milled. After the sixth year there appeared to be a decrease in strength. A portion of this lot of wheat was milled at the end of eight years, and the flour baked at once, the loaf volume being 2,550 cc. and the texture score 94. The flour was then stored for 4 months and again baked, with unsatisfactory results, the loaf volume being 1,800 cc. and the texture score 85.

Saunders, Nichols and Cowan (1921) report an extensive study of the baking strength of flour milled from wheat stored from September, 1907, until December, 1912. Huron and Yellow Cross samples tended to increase progressively in baking strength throughout the entire period, while lots of other varieties studied reached the maximum strength a year earlier, and then receded somewhat. The Red Fife sample increased in strength from 98 in September, 1907, to 108 in December, 1909, and then decreased somewhat during the three years following.

In the wheat storage experiments which Mangels (1924) conducted with common spring and durum wheats the changes in baking qualities were not so consistent. Common wheats of the 1921 crop exhibited little change in baking strength as indicated by loaf volume, such change as was observed being in the direction of improvement. Of the durum wheat samples certain of them exhibited improvement, while others were impaired in baking quality.

A tendency toward an increase in the percentage of total sugars in wheat stored unground for 2 years was noted by Leavitt and Le Clerc (1909). The average percentage of sugar, dry basis, of 7 lots of wheat was 3.06 per cent at the beginning, and 3.49 per cent at the end of the 2-year storage period. Two samples increased .82 per cent, while two others did not change in sugar content. The alcohol-soluble protein of wheat tended to decrease slightly during two years of storage.

Wheat was analyzed at the beginning and end of a storage period of 16 months by Shutt (1909b), and a slight increase in the percentage of protein was observed, which he suggests might be accounted for by the slow oxidation of carbohydrates during storage. A survey of Shutt's data indicates that in most instances the increase in protein was only .3 per cent or less. Shutt (1911c) extended this study for two years more, and noted a continuance of the tendency in the direction of increasing protein content.

Much of the wheat grown in North America, particularly in the districts east of the Rocky Mountains, is handled in bulk. The treatment accorded grain in bulk handling was described by Bates and Rush (1922). With the relatively large volumes of grain thus handled, ranging from a carload of 1,200 bushels to cargoes of over 400,000 bushels, great care must be taken to prevent spoilage and heavy losses. Large quantities of wheat and other cereals go out of condition every year, the greatest losses being sustained in grain handled in bulk.

The heating of stored wheat doubtless results from the respiration of the grain tissues and of organisms on the surface of the grain. The germ or embryo of the kernel respires more vigorously than the endosperm or storage tissue, as shown by the researches of Karchevski (1901), and Burlakow (1898), and is probably the seat of most of the biological oxidation occurring in the grain. Respiration of this kernel structure apparently starts the process of heating in a bulk of wheat. As the temperature rises organisms on the surface of the grain, notably thermophyllic bacteria, become active, and carry the heating process along until the grain reaches a comparatively high temperature unless the process is arrested by cooling or drying.

The relative rate of respiration is influenced by a number of factors, most prominent of which is the moisture content of the grain. Bailey (1917b) advanced an hypothesis to account for the relation of moisture content to rate of heating. It was suggested that in dry grain there is not sufficient water to produce a gel; that is, the colloidal material does not have a continuous structure. The exact percentage of moisture at which a gel passes into a discontinuous structure is not known. It probably varies with the percentage of gluten in the grain, since gluten possesses a greater water-imbibing capacity than raw starch. Increasing the moisture content above the maximum at which discontinuity exists results in the formation of an elastic gel through which diffusion can occur. Uninterrupted diffusion is essential to respiration, since the

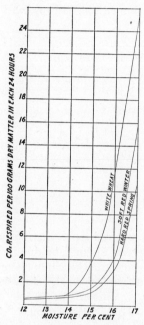

FIG. 4.—Graphs showing the comparative rate of respiration of hard spring, soft red winter, and soft white winter wheat. (Bailey and Gurjar, 1918.)

material to be oxidized (largely sugars) must diffuse to, and the gases and other products of respiration must diffuse away from the actual seat of respiration in the cells. Conditions such as a diminishing rigidity of the gels constituting the cell protoplasm facilitate diffusion, which, in turn, may accelerate the rate of respiration. The same conditions may likewise result in an acceleration of other enzyme phenomena, such as diastatic activity, thus, possibly, increasing the concentration of the substratum or sugar which is oxidized in respiration. Since this would tend to increase the rate of respiration, the effect of increasing moisture content thus becomes cumulative.

Bailey and Gurjar's (1918) observations on the relative rate of respiration of hard and soft wheats are recorded in Table 67, and shown graphically in Figure 4. It is evident from these data that the critical moisture content at which a sharp rise in respiratory rate of sound wheat is encountered is in the neighborhood of 14.5 per cent in the case of the hard wheat; this observation was confirmed by Dendy and Elkington (1920). The critical moisture content appears to be somewhat lower in the case of the soft wheats. Differences in the respiratory rates of the two types of wheat are attributed to differences in gluten content. Hard wheat contains a higher percentage of gluten, which imbibes more water, and thus yields a more rigid, viscous gel than soft wheats of the same moisture content.

TABLE 67

INTERPOLATED QUANTITY OF CARBON DIOXID RESPIRED PER UNIT OF TIME AND MATERIAL, AT EVEN PERCENTAGES OF MOISTURE

Class of Wheat	Carbon Dioxid Respired per 24 Hours for Each 100 Gm. of Dry Matter					
	12 Per Cent Moisture, Mgm.	13 Per Cent Moisture, Mgm.	14 Per Cent Moisture, Mgm.	15 Per Cent Moisture, Mgm.	16 Per Cent Moisture, Mgm.	17 Per Cent Moisture, Mgm.
Hard spring	0.50	0.58	0.68	1.13	2.72	10.73
Soft red winter..	...	.63	.81	1.37	3.84	15.51
White winter ...	.49	.60	.83	4.15	9.85	25.18

Shriveled wheat also respired more vigorously than plump wheat of the same moisture content. This may be due in part to the larger proportion of germ or embryo tissue in a unit volume of shriveled wheat. The greater proportion of surface in the shriveled kernel may likewise contribute to the increased rate of respiration, because this facilitates diffusion of the gases involved in respiration.

Damaged grain, such as frosted kernels, and, as shown in a later paper by Bailey and Gurjar (1920), sprouted kernels also have a higher rate of respiration than sound wheat. The relative difference depends upon the extent of damage; thus wheat sprouted for 48 hours, 24 hours, and the unsprouted wheat respired in the ratios of 100:53:14. Studies of the respiration of shelled corn conducted by Bailey (1921) showed badly heat-damaged kernels to have a higher rate of respiration than sound or normal corn of the same moisture content. It is thus apparent that damaged or unsound grain constitutes a greater hazard in bulk storage than does sound material.

Increasing the temperature of wheat accelerated respiration until the temperature exceeded 55° C. The relative increase per unit rise of temperature was not uniform, as shown by the data in Table 68, which are presented graphically in Figure 5. Similar observations were recorded by Dendy and Elkington (1920). Respiration is such a complex phenomenon that a cumulative effect resulting from several contributing factors may be occasioned by increasing the temperature. From the practical standpoint it is evident that as the temperature of heating grain

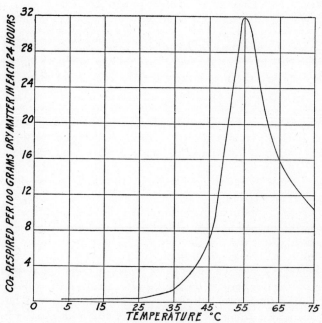

FIG. 5.—Graph showing the correlation between temperature and the rate of respiration of stored wheat. (Bailey and Gurjar, 1918.)

rises the hazard increases, to the extent that the rate of change becomes greater at the higher temperatures. Thus damp grain at 20° C. may increase in temperature no more in a week than it will in a day after the temperature reaches 30° C.

TABLE 68

RESPIRATION OF HARD SPRING WHEAT AT DIFFERENT TEMPERATURES

Temperature, °C.	Carbon Dioxid Respired per 24 Hours for Each 100 Gm. of Dry Matter, Mgm.	Temperature, °C.	Carbon Dioxid Respired per 24 Hours for Each 100 Gm. of Dry Matter, Mgm.
4	0.24	55	31.73
25	0.45	65	15.71
35	1.30	75	[1] 10.28
45	6.61		

[1] A part of this carbon dioxid may have resulted from roasting the grain.

Accumulated carbon dioxid in the space between kernels which results from respiration was shown to depress the rate of respiration. Wheat stored for milling purposes should, accordingly, not be turned or ventilated unless by so doing the temperature can be reduced. Storage in an oxygen-free atmosphere, that is, in nitrogen, reduced the respiratory rate as well, it being about 40 per cent of the rate in ordinary atmospheric air.

The exact temperature which wheat may reach in consequence of respiration without being damaged for milling purposes has not been determined. Observations of the writer indicated that discoloration resulted when the grain reached temperatures in excess of 100° F. (37.8° C.). The first effect of heat damage upon baking properties of the flour is to be observed in the rather intangible and difficultly measurable qualities of flavor and odor. No satisfactory correlation between the chemical composition of heat-damaged wheat and baking qualities have been published, so far as the author is aware. Frank (1922b) proposed a test for heat-damaged kernels, based upon the difficulty of agglutinating the crude gluten from a crude flour expressed from such kernels. Wheat mixtures might contain a limited quantity of kernels so badly heat damaged as to react to this test, but if all the kernels were heat damaged, the extent of damage would have to be much less if the wheat were to yield acceptable flour.

The conclusions drawn from these respiration studies respecting the influence of various factors on the keeping qualities of wheat are supported by experiments on a large scale with bulk wheat. Thus Bailey (1917a) examined a number of carlots of wheat which arrived at Minneapolis, Minn., in a heating condition during the unusually hot summer weather of August, 1916. No plump spring wheat containing less than 14.4 per cent of moisture was found to be in a heating condition, excepting a few lots which contained damaged kernels. Numerous lots of heating spring wheat were encountered, the kernels of which were shrunken and shriveled, and which contained less than 14.5 per cent of moisture. A few lots of wheat, with a fairly high test weight per bushel, but containing damaged kernels, such as frosted or sprouted grains, and which contained less than 14.5 per cent of moisture were found to be in a heating condition. This indicates the increased liability of shriveled or of damaged wheat to go out of condition. A considerable number of carlots were examined that were in a heating condition and which contained 14.4 to 15.0 per cent of moisture, and were comprised of fairly plump, heavy kernels, as indicated by the weight per 1000 kernels.

Bailey (1917b) conducted storage experiments in elevators at Duluth, Minn., over a period of nearly two seasons. Carlots of wheat of various types, and containing different percentages of moisture, were under observation during this period. The grain was stored in bins equipped with temperature indicating devices attached to a vertical cable at 10-foot intervals. Of two lots of wheat stored on September 11-12, one containing 16.5 per cent of moisture kept only 49 days before it was heating sufficiently to necessitate turning it, while one containing 15.5 per cent of moisture kept 333 days before it reached the same temperature (80° F.), and required turning to cool it. It was shown by these experiments that (1) atmospheric temperatures, (2) initial temperature of the grain, (3) insulation afforded by the container and its surroundings, bear an important relation to the actual time involved in effecting a unit change of temperature in consequence of respiration. It was further noted that as the temperature rises in consequence of heating, the rate of temperature increase becomes greater the higher the temperature, within the limits studied.

Birchard and Alcock (1918) equipped a steamship with temperature indicating devices, and wheat containing 13.2 to 15 per cent of moisture was loaded into the vessel at Vancouver, B. C., on November 9-13, 1917. Observations of temperature in different locations in the grain cargo were made en route, via the Panama Canal, to Norfolk, Va. On discharging the cargo at London on February 18, 1918, no appreciable damage to the grain was observed except in the neighborhood of the stokehold bulkhead, and in locations where condensation from cold hatches dripped back onto the grain. Temperatures as high as 100° F. were recorded on December 29th, near the bulkhead. It was concluded from these observations that, while they did not definitely determine how much moisture wheat may safely carry, it appeared that a moisture content of more than 14.5 per cent should be regarded as dangerous.

Sharp (1924) so treated different portions of the same sound wheat sample as to modify their moisture content, and three such lots containing 4.8, 9.3, and 14.1 per cent of moisture, respectively, were further subdivided and stored under different conditions, namely (1) out of doors, (2) room temperature about 22° C., and (3) in a thermostat at 35° C. The hydrogen ion concentration of the wheats was determined at intervals over a period of 52 weeks, and but a slight change was observed in pH during this time, even in the wheat of the highest moisture content maintained at 35° C. This lot changed in pH from 6.40 to 6.26 during the year. A lot of frosted immature wheat was also studied. The moisture content of the several lots was 5.9, 8.8 and 14.8, respectively, and these were stored under the same conditions as the sound wheat sample. Only the lots with the highest moisture content changed in pH appreciably. In the case of the samples containing 14.8 per cent of moisture and held at room temperature, and at 35° C., the initial pH of 6.55 had changed to 6.07 and 6.06, respectively, at the end of 52 weeks.

Graphs and tabulated data showing the percentages of wheat found to be out of condition, when the samples were grouped on the basis of their moisture content, were presented by Phillips (1921). In the case of hard spring wheat the percentage of samples of musty, sour and heating grain encountered when the moisture content was less than 14 per cent was comparatively small. With wheat containing 15 per cent of moisture the proportion of samples in a heating condition was more than twice as great, and with 16 per cent of moisture about 8 times as great as the wheat containing 14 per cent. A much larger proportion of the soft winter wheat samples were in a musty, sour or heating condition than of the hard spring wheats containing the same percentage of moisture. Thus, about 10 per cent of the soft red winter wheat samples containing 14 per cent of moisture were out of condition, while with hard spring wheat of the same moisture content only about 2 per cent were out of condition. Hard, red winter wheat was approximately intermediate between the hard spring and the soft winter wheat in this respect. These data support the conclusion previously reached by Bailey and Gurjar that with the softer wheat, of lower gluten content, the storage hazard was greater than with the hard and more glutenous wheats.

The average moisture content of wheat grown in Germany was 16-17 per cent, according to Buchwald (1916). In dry seasons it fell to 14 per cent, while in a damp season it increased to 18-20 per cent. He regarded 16 per cent as the maximum limit for warehousable wheat, and maintained that artificial drying should be effected at temperatures not above 45° C. Several fungi were identified as actively developing in damp wheat, including Aspergillus glaucus and Penicillium crustaceum, when the moisture content exceeded 16 per cent.

The moisture content of heating wheat which arrived in Port Arthur, Canada, during the summer months was determined by Birchard (1920). The highest moisture content of any of the wheat which was cool and in good condition on arrival was 16.5 per cent, while all the cars examined which had a higher percentage of moisture showed signs of heating. That bulk wheat does not dry rapidly, even when heating, was indicated by the fact that wheat sampled from a car in a heating condition contained as much moisture 3 inches below the surface, as in the center of the bulk. These data, given in Table 69, show that no substantial drying of the grain had occurred except at depths of one or two inches.

TABLE 69

Moisture Content of Wheat at Various Depths from the Surface (Birchard, 1920)

Depth in Car	Moisture, Per Cent
Surface of grain	14.0
1 inch below surface of grain	16.6
3 inches " " " "	20.5
1.5 feet " " " " "	19.4
4 " " " " "	19.2

Observations of the temperature changes in carlots of damp wheat stored in an elevator at Port Arthur were made by Birchard over a period from early August to the latter part of September. Sound wheat containing 16.5 per cent of moisture or less remained sound and cool during this period, while a lot of wheat grading No. 5 tough, frosted and green which contained 15.5 per cent of moisture showed signs of incipient heating after 25 days of storage, but the temperature of the grain gradually dropped, and after 25 days more was again as cool as when placed in storage.

The average percentage of moisture in the several grades of Canadian wheat as recorded by Birchard are given in Table 70. The "dry" wheat carried no side notations, while the wheat with excessive moisture was graded with a side notation of "tough," or of "damp," depending upon the percentage of moisture in the wheat.

TABLE 70

Average Percentages of Moisture in Manitoba Inspection of Northern Spring Wheat from September, 1916, to March, 1917, as Recorded by Birchard (1920)

	No. 1 Northern	No. 2 Northern	No. 3 Northern	No. 4	No. 5	No. 6
Straight grade	13.2	13.4	13.2	13.1	13.1	12.7
"Tough"	15.1	15.2	15.3	15.3	15.3	15.4
"Damp"		17.4	17.6	17.5	17.5	17.4

The advantages of cooling stored grain to reduce the rate of respiration, and thus decrease the liability of heating, were stressed by Dienst (1919). Hoffman (1918) presented an equation for use in computing the rate of change in temperature with each increment of temperature increase.

Wheat containing an average of 11.3 to 12.6 per cent of moisture which was infested with weevil was observed by Baston (1921) to exhibit a tendency to heat in bins at Buffalo, while noninfested wheat remained sound and cool. He concluded that the rising temperature in the weevily wheat was due to the activities of the weevils.

C. Louise Phillips (1921) abstracted the results of an experiment recorded in the U. S. Department of Agriculture Weekly News Letter, vol. 3, No. 6 (Sept. 15, 1915), which demonstrated that moisture is rapidly transferred from damp wheat kernels to dry wheat kernels when they are mixed. A lot of white wheat containing 9.7 per cent of moisture was mixed with a lot of red wheat containing 15.1 per cent. These were mixed August 3rd, and transferred on August 6th to another bin. At the time of transferring the wheat the red wheat contained 12.9 per cent and the white wheat 12.2 per cent of moisture. On August 10 the grain was again transferred, and at that time the red wheat contained 12.5 per cent and the white wheat 12 per cent of moisture. Thus, with an initial difference of 5.4 per cent, the difference 3 days after mixing was 0.7 per cent, and 7 days after mixing was 0.5 per cent of moisture. It is apparent that the mixing of damp and dry wheat will facilitate the handling of the former by promptly reducing the moisture content of the damp material.

The drying of wheat for milling purposes was studied by Cox (1916), who found that when wheat containing 16.3 per cent of moisture was dried for one hour at a temperature of approximately 140° F., the baking qualities of flour milled from it were equal to natural or unheated wheat. This heating reduced the moisture content of the wheat to 12.9 per cent. In another series of tests conducted by Cox, three lots of wheat were artificially dried at an average air temperature of 140° F., with similar results. Samples of wheat were drawn from the drier during the drying operation, placed at once in thermos bottles, and their temperature found to range from 110° to 132° F., with an average of 125°.

The effect of artificially drying wheat at elevated temperatures on its baking quality was studied by Birchard (1920). Drying small samples in an air oven at 149° F. for 2 hours did not result in any damage to the wheat, as evidenced by baking tests of flour milled from the damp and the dried grain. Subsequent studies with wheat samples containing varying percentages of moisture and dried at different temperatures for 3 hours indicate that a safe temperature limit for drying depends upon the moisture content of the grain before drying. Thus the limit with wheat containing 14 per cent of moisture was about 176° F., but with wheat containing 17 per cent of moisture it was 158° F.

The Australian Advisory Council of Science and Industry (1917) studied the protection of wheat against the attack of weevils and mice by mixing it with freshly calcined lime (quicklime). Wheat which was thus mixed with 1 per cent of ground quicklime and allowed to stand for two weeks was not damaged, as shown by milling and baking tests. With mouse-tainted wheat the taint was removed by this treatment.

Beyer (1913) was granted a German patent covering the freshening of moist and musty grain by spraying with a freshly powdered mixture of CaO or MgO, and $NaHCO_3$, allowing to stand for 5 days and removing the powder mechanically after 5 days. Moisture extracted from the grain forms hydroxides of Ca and Mg, which in turn are converted into harmless carbonates by the CO_2 liberated from the $NaHCO_3$.

The observations of Duvel (1909), Duvel and Duval (1913), Shanahan, Leighty, and Boerner (1910), and Boerner (1919) respecting the keeping qualities of shelled corn have been reviewed by Bailey (1921), to which review the reader interested in the storage and transportation of corn is referred.

The hygroscopic character of wheat was established by Brewer (1883), who reported his investigations, and those of Hilgard. Hilgard exposed wheat to an atmosphere over a free water surface, and it gained 18.8 per cent in weight during an 18 day period. In a dry atmosphere another portion lost 6.2 per cent in weight during the same period. Thus dried wheat may gain 25 per cent in weight when exposed to an atmosphere saturated with water vapor.

Wheat was weighed into paper boxes by Brewer and exposed to the air of his office. A loss in weight was registered from September, 1880, to February, 1881, amounting to 5 + to 7 + per cent, which was followed by a gain in weight to September, 1881. At the end of the year these wheats weighed more than at the outset. Hickman (1892) buried boxes filled with wheat in a bin of the same wheat. These remained undisturbed for 3 years, at the end of which time the average shrinkage of 14 lots so studied was 2.32 per cent. The losses were quite variable, however, ranging from 0.00 to 4.94 per cent. A later series of studies from January to July resulted in gains of weight during the period.

About 3,000 pounds of dry, threshed Clausen wheat placed in bins by Smith (1901) lost less than one-half of one per cent in weight after 332 days. A lot of Buda Pesth wheat lost less than one-tenth of one per cent in weight in the same length of time.

Changes in weight of sacked wheat and oats during storage in Utah were determined by Harris and Thomas (1914). Wheat containing 6.5 to 8.7 per cent of moisture in August, 1911, gained about 3 per cent in weight during the period ending the following April. During the period from May to September, 1912, a loss of approximately 1 per cent occurred, followed by a gain of about the same amount during the following winter. A second loss was experienced during the summer of 1913 until the experiment was terminated in August, 1913.

Changes in weight of 34 lots of wheat stored in the elevator of the experimental mill at Fargo, N. D., were reported by Sanderson (1913b). There was an average loss of 0.7 per cent in storing hard spring wheat for an average period of 421 days. Stockham (1917) exposed spring wheat to a saturated atmosphere and found that after 2 weeks the moisture absorbed exceeded 30 per cent and reached about 34 per cent after 24 days. At this time the grain was moldy.

A 195-pound sack of wheat was weighed every day for 2 years by Guthrie, Norris and Ward (1921). Between weighings the bag was suspended in the air of their laboratory at Sydney, N. S. Wales. It was observed that the weight of the grain did not adjust itself immediately to a change in the humidity of the atmosphere. From early February, 1918, to mid-May, 1918, there was a gradual increase in weight, amounting to 1.4 per cent of the original weight. By November, 1918, the weight had decreased and then remained fairly constant in weight until February, 1919, when changes with variations in atmospheric humidity were recorded. There was again an increase in weight until May, 1919.

In general, periods of low humidity effected decreases in moisture content, and vice versa.

The variations in moisture content of wheat in sacks which resulted from fluctuations in atmospheric humidity were followed by Sutton (1921) in Western Australia. The average moisture content of the three lots of wheat used was 9.33 per cent, in January, 1919, when harvested. This increased to an average of 15.21 per cent in September, 1919, receded to 11.15 per cent in January, 1920, and again increased to 16.04 per cent in June, 1920. Curves are presented which indicate that the moisture content of wheat varied with the humidity of the interval just preceding the testing of the wheat. When the relative humidity exceeded 70 per cent the moisture in the wheat was about 15 per cent, and the moisture content fell to 11 per cent during the period when relative humidity was somewhat less than 40 per cent.

Samples of all classes of American wheats were exposed to atmospheres of different relative humidities (at $25°$-$28°$ C.) by Coleman and Fellows (1925) until the hygroscopic moisture of the grain reached equilibrium with the aqueous vapor pressure of the atmosphere. The moisture content of the wheat increased fairly regularly from 7.0 per cent to 17.3 per cent as the relative humidity of the atmosphere increased from 15 to 75 per cent. A further increase in humidity to 90 per cent increased the hygroscopic moisture of the wheat to 24.8 per cent, while in a saturated atmosphere the wheat contained 34.4 per cent of moisture. In comparing wheats of varying protein content, no difference in the hygroscopicity was observed. The same was true when the percentage of vitreous kernels was correlated with hygroscopicity.

The several studies indicate that the degree of exposure of commercial grain determines the rate of response to fluctuating humidity. When exposed in sacks the response is fairly prompt, while bulk grain, especially when the quantity is large, may change in weight comparatively little in consequence of losses or gains of hygroscopic moisture. Substantial variation in moisture content of wheat may result from protracted exposure to atmospheres varying in humidity through the range which may be encountered in the regions in which wheat is handled in commerce.

Chapter 7.

Chemistry of Roller Milling.

In modern roller milling an effort is made to separate the branny covering, or pericarp and other fibrous structures, and the germ from

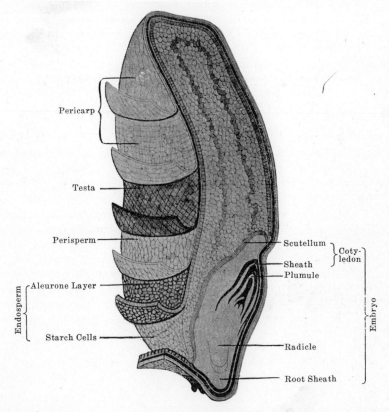

Fig. 6.—Diagram of the wheat kernel showing the several kernel structures. (Guthrie, 1902.)

the endosperm of the wheat caryopsis or grain, and to then pulverize the endosperm to particles of uniform size. It is these particles which constitute the refined white flour. From the biological standpoint the bran is the protective covering of the grain, the germ or embryo is the structure from which growth starts in germination, while the endosperm is a relatively large reservoir of reserve food which is utilized in the germination of the kernel and growth of the seedling. Each of these structures or groups of structures is characterized by certain physical peculiarities, of which advantage is taken in milling. Thus the bran is tough, due to its high content of fiber, particularly when the grain is properly conditioned or tempered for milling. The embryo or germ contains a large amount of fat, which facilitates flaking between rollers. The starchy endosperm, on the other hand, is more or less friable, and tends to fracture and pulverize when ground between rollers.

The position occupied by these structures in the wheat berry is indicated in Figure 6, reproduced from Guthrie's (1902) bulletin. The pericarp, testa and perisperm constitute the fibrous protective covering of the endosperm and embryo. The testa contains the reddish-brown pigment which confers the characteristic color of "red" wheats. The aleurone layer (sometimes erroneously termed the "gluten" layer) commonly adheres to the perisperm in milling and is thus removed with the bran. Characteristics of the individual starchy parenchyma cells are shown in Figure 12.

In the gradual reduction process of roller milling much of the endosperm is separated from the bran and germ in a fairly pure condition, to the extent that the highly refined grades of patent flour contain little other than endosperm fragments. The separation is not quantitative, however, in spite of the best efforts of millers and milling machinery designers. Increasing proportions of bran and germ fragments appear in the flours in progressing from the most highly refined to the lower grades, while substantial losses of endosperm are sustained in the offals or by-products, such as bran and shorts. The average roller mill probably does not recover as merchantable flour more than 72 to 75 per cent of the potential 82 to 85 per cent of the flour in the endosperm. Thus about 10 per cent of the kernel which, in a perfect separation would be returned as flour, appears in the lower priced offals. The ideal of milling would involve a peeling of the bran, degermination, and finally a pulverizing of the decorticated endosperm; but such an ideal has not been reached as yet.

The proportion of the three groups of kernel structures was determined by Fleurent (1896a) in wheats from three widely separated districts, with the results shown in Table 71.

Details of the Milling Process.

In this discussion of the roller milling process only brief reference can be made to the mechanical aspects. For detailed consideration of the technology of flour production, the texts by Kick (1888), Amos (1912),

TABLE 71

PERCENTAGE OF ENDOSPERM, EMBRYO AND BRAN IN WHEAT KERNELS AS DETERMINED BY FLEURENT (1896 a)

	Average Weight of Kernels, Grams	Endosperm, Per Cent	Embryo, Per Cent	Bran, Per Cent
Russian	0.030	84.95	2.00	13.05
Algerian	0.048	84.99	1.50	13.51
Canadian	0.037	84.94	2.05	13.01

Oliver (1913), Kozmin (1917), and Dedrick (1924) should be consulted. The description of the process which follows is designed to tie into the discussion of the chemistry of the mill streams with which this section is chiefly concerned.

1. *Cleaning the Wheat.* Reference has already been made to the characteristics of certain of the impurities found in commercial grain, and their effect upon the flour when ground with the wheat. Each wheat district presents its peculiar problems in so far as the cleaning department of the mill is concerned. In the Pacific Northwest particular attention must be given to the separation of the "smut balls" and spores of bunt or stinking smut. Ordinary dry scouring is not wholly effective in separating these spores, and the grain is either mixed with fine, bolted air-slaked lime before scouring, or is washed with water in special washing machines. In the hard spring wheat mills special machines must be available for the separation of the seeds of corn cockle, wild vetch and kingheads, or giant ragweed. In the eastern United States garlic bulblets are frequently encountered in the wheat. These can be separated only after vigorously heating the bulk grain which dries up the bulblets so they can be subsequently lifted out by an air current in an aspirator.

2. *Conditioning the Wheat.* The conditioning or tempering process immediately follows the cleaning process, or becomes a part of the cleaning when the wheat is washed. In tempering, a suitable addition of water is made, the wheat and water thoroughly mixed, and the wetted grain allowed to stand for several hours. The quantity of water added per unit weight of wheat is varied, depending upon the condition of the wheat. The three principal variables which determine the tempering treatment, so far as the wheat is concerned, are: (a) moisture content, (b) relative plumpness of the grain, and (c) relative hardness or vitreousness of the wheat kernels. Increasing proportions of water are used in tempering with decreasing moisture content, increasing plumpness and increasing hardness of the grain. When substantial alteration of the properties of the bran is desired, a moderate heating of the grain (to about 30°C.) may advantageously precede the addition of the tempering water. Kent-Jones (1924) states that

temperatures as high as 45°-50° C. may be reached by the wheat, but such practices are not usual in America. This conditioning or tempering imparts an added toughness to the bran, thus facilitating its subsequent separation when the wheat is ground and bolted. The lapse of time between the addition of the tempering water and the grinding is varied so as to produce the maximum toughness of the bran, with the minimum of penetration of water into the endosperm. If the latter absorbs much water, it likewise tends to become tough, and will flake rather than pulverize between the rollers, with consequent losses into the feeds or offals. When the moisture content of the wheat is somewhat above normal a short tempering period of from 1 to 2 hours is usually sufficient. With wheat of a normal moisture content, 2 to 4 hours is commonly allowed, while with dry wheat, particularly hard wheat containing less than 13 per cent of moisture, the water is ordinarily added in two portions. Thus with abnormally dry hard wheat containing 10 per cent of moisture or less, the best results are usually secured by adding sufficient water to bring the moisture content up to the normal and allowing the wetted wheat to stand for about two days, or until the water penetrates to the center of the endosperm. A second and smaller addition of water is then made at the end of this first and long tempering period, and a second tempering period of a few hours' duration is allowed before grinding. The duration of the first tempering period is reduced as the wheat approaches the normal in moisture content.

3. *Breaking the Wheat Berry.* The first stages of grinding, known as the breaks, are the work of corrugated rolls, mounted in pairs, and rotating toward each other with a differential of about $2\frac{1}{2}$ to 1. The grooves or corrugations vary in "sharpness" in different mills, depending upon the rate of pulverizing that is desired. In the first break the rolls are set some distance apart, and the grain is crushed into a few coarse fragments, most of which are returned to the second break rolls, where they are crushed finer. The coarse fraction bolted from each break chop returns to the next set of break rolls, until after the last break, of which there are commonly 5 or 6 in a large modern mill, the coarse fibrous flakes that remain are spouted to the bran.

4. *Bolting or Sifting Process.* After each grinding on the break rolls the ground material or chop is bolted to effect a separation of the particles of varying degrees of fineness. Three general classes of material are thus separated: (a) coarse fragments, which are reground until they become bran; (b) fine fragments or flour, which are spouted directly to the flour bins; and (c) granular particles, intermediate in size between (a) and (b), which are further classified in the purifiers, and are then known as "middlings" or "semolinas." Several types of sifters are employed in separating the break chop into these fractions. The hard wheat mills commonly use boxes fitted with sloping sieves,

which are rapidly shaken with a gyratory motion. A cross-sectional diagram of such a sifter is shown in Figure 7. Bolting of the reduced or ground middlings is effected in much the same manner, except that finer sieves are used. Wire screens are used as the coarse sieves in the break system, while silk, or occasionally phosphor-bronze sieves are used

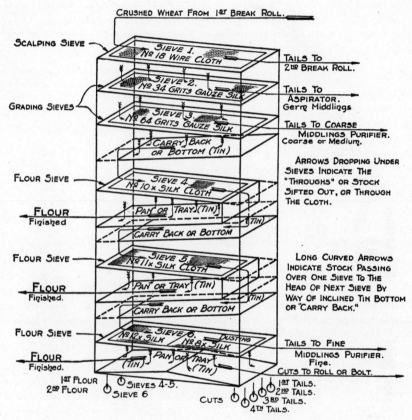

FIG. 7.—Section of sifter with inclined sieves. (Dedrick, "Practical Milling.")
Reproduced by permission of the National Miller, Chicago.

as the finer sieves in both break and middlings reduction systems. Cylindrical or hexagonal reels of bolting cloth are used in bolting many of the streams in soft wheat mills, and in hard wheat mills toward the tail of the process, where the reground material has lost its granular character and bolts less freely.

5. *Purification.* In the first chapter reference was made to the introduction of the purifier into modern milling practice. While various

types of purifiers are in use, those most commonly employed consist of a box which houses a long, sloping sieve, which sieve is caused to vibrate lengthwise with a jigging motion. The sieve has graduated openings, with fine meshes at the head, and meshes of increasing size or coarseness as the tail or discharge end is approached. As the particles travel down the vibrating sieve, a classification is effected on the basis of size. In addition to this sifting process, the granular particles are aspirated by means of air currents set up either by a fan mounted in the upper part of the sifter box, or by an external fan

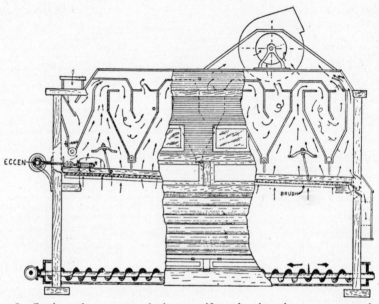

Fig. 8.—Section of one type of sieve purifier, showing the movements of air currents. (Dedrick, "Practical Milling.")

Reproduced by permission of the National Miller, Chicago.

communicating thereto by an air duct. It is this aspiration which accomplishes the purification suggested by the name of the machine, since properly controlled currents of air will lift out the flakier and more fibrous particles from the mixture in the middlings stream. The purified middlings are thus freer from fibrous material, and on subsequent reduction will yield a somewhat more highly refined flour than when they are not subjected to this treatment. A vertical section of one type of purifier is shown in Figure 8.

6. *Reduction.* The purified middlings are reduced to the desired fineness by grinding them vigorously between smooth iron rollers. A system of springs on each pair of rolls exerts a pressure which facili-

tates grinding. Flour produced by the first grinding of any stream of purified middlings is more highly refined and freer from fiber than that produced by subsequent grinding of the coarse residues. The coarse particles bolted from the chop resulting from the initial reduction of the middlings still have endosperm or floury material present, and on regrinding these some flour can be recovered. As the proportion of branny material in these fragments becomes greater with each reduction, the quantity of bran fragments in the flour which is produced likewise becomes greater.

Early in the milling process a stream of coarse, granular fragments, known as sizings, is separated from the break chop. These sizings contain most of the germ, present in an almost intact condition, and easily recognized by the shape of the particles and their lemon-yellow color. They are torn from the grain by the shearing action of the corrugated break rolls. When these sizings are subsequently reduced by grinding them between smooth rolls, the wheat germs become flat flakes and are separated from the flour on bolting the chop. In certain mills a portion of the germ flakes is bolted off and sold as wheat germ for use in making vitamin preparations. Most of the germ appears in the shorts or standard middlings, and this is used as cattle feed.

7. *Flour Bleaching.* Since the chemistry of flour bleaching is discussed in a later section (chapter 9), only mechanical facilities used in treating flour will be mentioned here. When the bleaching reagent is in the form of a gas, this is mixed with a large volume of air before bringing it in contact with the flour, and blowers are provided for effecting this dilution. The mixture of gas and air is then blown into an agitator where the flour is treated. Most of the agitators now in use in American mills are of the vertical type. In certain of these the flour is dropped into a funnel, thence over a cone, which spreads the flour in a thin sheat and tends to expose each flour particle to the action of the bleaching agent. In other agitators a thin sheet of flour is produced by spouting the flour onto the center of a flat metal plate, which is revolving rapidly in a horizontal plane. Centrifugal force causes the flour to move toward the periphery of this plate and the gas is blown into it at the point where it drops over the edge of the plate.

When solid reagents are used in bleaching, special precautions must be taken to insure thorough mixing of the relatively small amount of the reagent with the flour. Usually the reagent is first mixed with several times the weight of the flour to be treated; this mixture is then fed in the proper proportions into the flour stream, and this in turn is spouted into a high-speed mixer where the flour and reagent are stirred for a time. The treated flour is then discharged into a flour dresser or rebolt reel, which completes the mixing process.

In this discussion of flour milling, only the fundamental and usual processes have been mentioned. Certain auxiliary devices, such as granulators or middlings mills, scrolls, bran and short dusters, and the like, have been omitted from consideration in the interest of simplicity. Their action and application to the milling process are treated in the works on milling technology to which reference has been made.

Rôle of Moisture in Roller Milling.

While water is, from the chemical standpoint, the simplest constituent of the wheat kernel, it plays an exceedingly important rôle in flour milling because of its profound effect upon the physical properties of other constituents of the grain. Most of these other substances are in a colloidal condition and imbibe water freely with consequent swelling. Incidental to swelling the dry materials on wetting become appreciably less friable, and acquire certain characteristics of an emulsoid gel. While the moisture content of milling wheat is never increased sufficiently to produce as moist a gel as those with which the colloid chemist usually works, a small increase of, say, 2 or 3 per cent above the normal moisture content of wheat will reduce the friability of the grain to a surprising extent. Thus the practical miller characterizes wheat containing 16 per cent of moisture as "tough," because of the reduction in vitreousness which he observes on comparing such wheat with the same grain dried to a moisture content of 13 per cent or less. When dry wheat is moistened, as in the tempering or conditioning process, the added moisture is rapidly absorbed by the outer or branny layers of the kernel, and slowly penetrates into the endosperm. The writer's observations lead to the conclusion that it requires about 72 hours for equilibrium to be reached in this process of distributing the added moisture throughout the berry. Consequently the major portion of the water added in tempering is in the bran for several hours after wetting the grain.

When the grain is abnormally dry before tempering, the best results in milling generally follow if the wheat is allowed to stand in a bin for at least 2 days to permit the water to penetrate into the endosperm. With normally dry grain about 4 to 6 hours will usually give the desired penetration, with the result that the bran is still distinctly moist and tough while the endosperm is but slightly affected. In this condition there is a diminished tendency for the bran to pulverize, and it emerges from the crushing rolls in broad flakes which are readily separated from the middlings and flour in the sifters. An excessive addition of water in tempering may not only result in an increased penetration of water into the endosperm, but will also tend to increase the aqueous vapor pressure of the atmosphere in the roller mill housing and attached spouting. In this more humid atmosphere the middlings particles, because of their hygroscopic nature, will absorb increased quantities of moisture, and their moisture content will be thus raised.

The characteristic most desired in the bran—toughness—is objectionable in the endosperm, which should pulverize when ground between the same rolls that are flaking the bran. If the endosperm be unduly tough it also will tend to flake, with consequent losses of this valuable material into the by-products or feeds. A nice adjustment of these physical characteristics is demanded by the miller, however, since an abnormally dry endosperm will possess too high a crushing strength and require an undue amount of power to grind. Thus Dedrick (1924) (Note p. 218) states that tempered wheat required 2 to 10 per cent less power to grind than dry wheat.

While statements are frequently made to the effect that substantial changes are effected in wheat during the tempering process in consequence of the accelerated activity of the enzymes of the kernel, little evidence in support of such a contention has been advanced. On the other hand, the findings of Tague (1920) indicate that such biochemical changes do not occur. Tague tempered wheat, brought to a moisture content (determined in an air oven at 110° C.) of 15.5 per cent, at temperatures of 5°, 20° and 40° C. for 24, 48 and 72 hours. Raising the temperature and prolonging the time of tempering increased the hydrogen ion concentration of the wheat slightly. The other chemical changes observed were of small magnitude, and it would appear from this study that the improved milling quality of tempered wheat is due chiefly to changes effected in physical properties. The optimum conditions were (1) a temperature of 20-25° C., (2) a moisture content of 15½ per cent, and (3) a tempering period of 48 hours. Hard wheats and dry wheats were improved more by the treatment than either soft wheats or wet wheats.

It is generally agreed by millers that in tempering hard wheats the moisture content should be raised higher than in similarly conditioning soft wheats. Miller (1921) reported that in the mill which he operated the best results were obtained with Kansas hard winter wheat when its moisture content was raised to 14.6-14.8 per cent in tempering, while with Missouri soft winter wheat the moisture content of the tempered wheat was .2-.5 per cent lower.

More water is added to plump than to shriveled wheat in tempering. Thus in the milling operations of the Minnesota State Experimental Flour Mill, reported by Bailey (1924), the spring wheat of the 1921 crop, when the average weight per bushel was 54.3 pounds, was tempered to a moisture content of 15.15 per cent (determined by the Brown-Duvel method), while that of the next crop, having an average weight per bushel of 58.9 pounds, was tempered to a moisture content of 15.5 per cent.

Losses of Moisture in Milling.

In ordinary roller milling practice there is a progressive loss of moisture in the process of grinding, purifying and bolting, occasioned by the evaporation of water from the exposed material, when the comparisons are based on the tempered wheat. In the case of three mills in which studies were conducted by Shollenberger (1919), the moisture content of the mill products (determined by drying in a water oven at the temperature of boiling water), was invariably less than that of the tempered wheat. As the process of milling progressed through the several stages from the breaks to the tailings and low grade reductions, there was a fairly regular decrease in the moisture content of the stock on the various rolls. The exceptions to this rule were the higher percentages of moisture in the 2nd, 3rd, and 4th break stocks when the tempered wheat was rather dry (less than 14.7 per cent), due to the removal of the drier endosperm and the concentration of the moist bran in these stocks. Wheat tempered to a moisture content of 16 per cent lost sufficient moisture in the break system so that the 2nd and each succeeding break stock contained less moisture than the tempered wheat. The moisture content of these break streams average about 0.5 per cent higher before grinding than after grinding, in each case. The moisture content of the flours produced in successive stages of the process also tended to diminish. Thus the percentage of moisture in the break flours averaged higher than in the middlings flours, and the latter in turn contained more moisture than the tailings and low grade flours.

In any particular mill two major factors operate to influence the percentage loss in milling due to evaporation when losses are calculated on the basis of the weight of cleaned wheat before tempering. These are (1st) the moisture content of the wheat before tempering, and (2nd) humidity of the atmosphere to which the streams are exposed in the machines and spouts.

The significance of the moisture content of the wheat before tempering in determining losses due to evaporation was indicated by Ladd and Bailey (1911), and further emphasized by Bailey (1914c). Recent investigations conducted in the Minnesota State Experimental Flour Mill, reported by Bailey (1923), are more acceptable from the quantitative standpoint, because the operations were on a larger scale. No loss or gain in milling the 1921 crop spring wheat was experienced when the moisture content of the untempered wheat averaged 13.1 per cent (Brown-Duvel method). For each one per centum of moisture under 13.1 per cent which the wheat contained there was a net gain in weight of total products of about .75 per cent. A corresponding ratio of loss was experienced when the moisture content of the wheat exceeded 13.1 per cent. This correlation between moisture content and yield of total products is shown graphically in Figure 9. The following year, as shown by Bailey (1924), the heavier wheat permitted of the addition of more water in tempering, with the result that a yield of total products approximately 0.8 per cent greater could be expected in milling wheat of any particular moisture content. With the dry crop of the season of 1922, containing 11.9 per cent of moisture, the yield of total products was 101.85 per cent of the cleaned wheat milled, the gain of 1.85 per cent being the water added in tempering which was not subsequently evaporated in milling. Shollenberger's data, previously referred to,

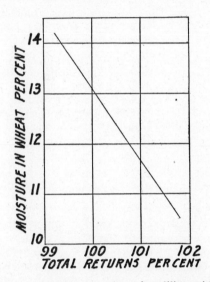

Fig. 9.—Correlation of total yield of products in milling with the original moisture content of the untempered wheat, crop of 1921. (Bailey, Minn. State Dept. Agr. Bul. 21.)

showed a gain of 0.3 per cent in milling wheat containing 12.4 per cent of moisture, and losses of about 1 per cent in milling wheat containing 14 per cent of moisture.

The moisture content of freshly milled patent and straight grade flours produced from hard winter and soft winter wheats as reported by the Association of Operative Millers (1923-24) ranged around 13.0 per cent. This is shown by the data in Table 109. The analytical methods employed by the numerous chemists, whose data are averaged in this table are not indicated in their reports.

Mill data showing the invisible loss in milling (largely evaporation of moisture) per barrel of flour were secured by Bowen (1914). Divided into six months' periods, the greatest loss was reported between July and December in each of the two calendar years covered by the report.

The average loss was 1.23 pounds per barrel of flour of 196 pounds, equivalent to .63 per cent. These data appear in Table 72.

Humidity of the atmosphere to which the mill streams are exposed in the machines and spouts is likewise correlated with the evaporational losses in milling. It has not proven feasible to control the atmospheric humidity within the large mill building, but when this was done by Shollenberger (1921a) in a small experimental milling unit, the data resulting from his studies showed that each 10 per cent increase in relative humidity above 35 per cent resulted in an average increase of about 0.5 per cent in the total weight of the products obtained. In Figure 10, which shows Shollenberger's data graphically, this correla-

TABLE 72

Amount of Wheat Ground in a Certain Mill to Make One Barrel of Flour and the Invisible Loss per Barrel by Six Months' Periods (Bowen, 1914)

Date	Amount of Wheat Ground to Make 1 Barrel of Flour, Bushels		Invisible Loss per Barrel of Flour Produced, Pounds
		Pounds	
January to June, 1906	4	24.3	0.96
July to December, 1906	4	27.7	1.67
January to June, 1910	4	30.3	1.01
July to December, 1910	4	31.5	1.27

tion between relative humidity and losses or gains in milling is clearly demonstrated. Since the various samples studied are grouped on the basis of their moisture content (determined by drying in an air oven at 108° C. for 5 hours), these data accordingly further emphasize the relation of the moisture content to milling returns which was discussed in the preceding paragraph.

At relative humidities ranging from 35 to 39 per cent the average moisture content of the flour produced by Shollenberger was 12 per cent (water-oven method), and at 65 to 69 per cent relative humidity the average moisture content was 13.3 per cent, or an increase of 1.3 per cent of moisture for a difference of 30 per cent in relative humidity.

That the quality of flour produced in humid atmospheres is superior to that milled in an atmosphere of low humidity is indicated by data compiled by Shollenberger (1923b). The ash content of the flour diminished regularly with increasing humidity, while bread scoring highest in color and texture was baked from the flours milled in atmospheres the relative humidity of which ranged between 60. and 69.9 per cent.

While it would be difficult during winter months to substantially increase the humidity of all the air in a large mill situated in the northern part of the temperate zone, efforts have been made to introduce water vapor into the air in certain milling machines, notably purifiers and aspirators. The Humphries Process, described in the bulletins of

Henry Simon, Ltd., (Manchester, England), and referred to in the *Millers Gazette* (1911), has been employed in certain British and Continental mills to introduce moisture in the form of a fine mist or spray. This is done to compensate for the normal losses by evaporation. The spraying is done at suitable points in the process, the amount introduced at any one point being comparatively small. In this way the moisture content of the stocks is fairly constant throughout the process.

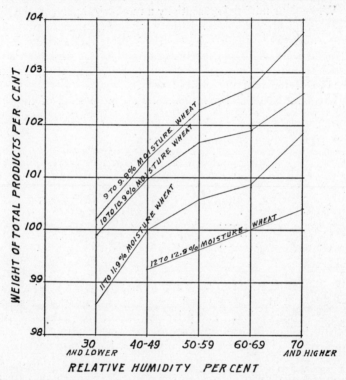

FIG. 10.—Correlation between the relative humidity of the mill atmosphere and the weight of total products, using wheat of varying moisture content. (Shollenberger, U. S. Dept. Agr. Bul. 1013.)

Several extensive papers have appeared in the German journals, notably those by Buchwald and Neumann (1913), M. Miller (1912), Mohs (1921a), and others, which describe the process in some detail. In addition to tap water, solutions of various salts, malt extract and other materials have been used, which serve to improve the baking qualities of the flour.

A process was originated by Woolcott (1921) in which an air conditioning machine was provided capable of supplying humidified air to the roller mills and purifiers. This not only reduced evaporation, but, it was claimed, produced whiter flour as well. In using this process the moisture content of the tempered wheat as spouted to the first break rolls was reduced from 1.5 to 2 per cent. It was further claimed that the dust explosion hazard was reduced, since propagation of flame does not proceed so readily in a humid as in a dry, dusty atmosphere.

The Composition of Roller Mill Streams.

While flour milling is essentially a mechanical process, the differences in the chemical composition of the wheat kernel structures result in substantial differences in the composition of the several streams intermediate in the process of milling as these structures are separated and appear in the various streams. The chemical as well as the physical characteristics of the partially ground and the finished products are accordingly conditioned in large measure by the proportion of the bran, germ

TABLE 73

COMPOSITION OF THE FIRST FOUR MIDDLINGS FLOURS FROM THE C. A. PILLSBURY & CO. MILL, OF MINNEAPOLIS, MINN. (RICHARDSON, 1884)

Flour Stream	Ash, Per Cent	Oil, Per Cent	Nitrogen, Per Cent	Fiber, Per Cent
1st middlings flour	0.39	1.58	1.93	0.25
2nd " "	0.44	1.66	2.02	0.33
3rd " "	0.38	1.36	1.79	0.28
4th " "	0.40	1.42	2.13	0.38

and endosperm which make up these materials. In viewing the gradual transition from wheat to flour and offals, the intermediates can be characterized not alone by physical properties and visual appearance, but by chemical composition. The principal rôle of the chemist in the control of flour milling is not to prepare reagents, or determine the course of a reaction, as in many industrial processes, but rather to supply a quantitative measure of characteristics which otherwise would perforce be described in vague terms. The measure thus supplied may then serve to suggest modifications in mechanical procedure which will effect a more exact separation of the several parts of the wheat kernel.

One of the earliest studies of the chemistry of American flour milling was reported by Richardson (1884). Practically all of the important streams from the roller mill of C. A. Pillsbury and Company (Minneapolis, Minn.), were analyzed and the data published. It is interesting to note the low ash content of the flour resulting from the reduction of the purified middlings at this early period in the use of the roller milling process in America. The composition of the first four middlings flours, shown in Table 73, indicates that these flours were as highly refined, and did not contain a materially higher percentage of ash than many of the corresponding streams in mills operating 40 years later.

TABLE 74

Composition of Mill Streams and Products, Analysed by B. R. Jacobs of the U. S. Department of Agriculture, Bureau of Chemistry

P. C.	Description	H₂O, Per Cent	Ash, Per Cent	P₂O₅, Per Cent	Fat, Per Cent	Fiber, Per Cent	Pentosans, Per Cent	Total Sugar, Per Cent	Total N, Per Cent	Nitrogen Soluble in Dist. Water, Per Cent	Nitrogen Soluble in 5% K₂SO₄ Solution, Per Cent	Nitrogen Soluble in Alcohol 70%, Per Cent	Crude Gluten Wet, Per Cent	Crude Gluten Dry, Per Cent	P₂O₅ in Ash, Per Cent	Ratio P₂O₅ to N,	P. C.
4098	1st break flour	11.80	.660	.342	1.14	0.16	3.34	1.42	1.91	.288	.267	.996	30.24	14.00	51.8	5.58	4098
99	2nd " "	11.32	.558	.282	1.36	.10	3.34	1.33	1.99	.323	.274	1.08	34.12	14.24	50.5	7.06	99
4100	3rd " "	11.52	.488	.258	1.45	.10	2.59	1.42	2.08	.379	.281	1.17	37.24	15.64	52.9	8.06	4100
1	4th " "	11.21	.637	.348	2.20	.13	2.94	1.52	2.29	.324	.309	1.28	36.96	16.40	54.6	6.58	1
2	5th " "	10.87	1.029	.549	2.61	.11	2.82	1.65	2.35	.351	.365	1.13	38.40	16.88	53.3	4.28	2
3	1st middlings flour	11.49	.363	.194	1.02	.06	2.64	1.23	1.80	.351	.264	1.01	30.36	11.28	53.4	9.28	3
4	2nd " "	10.98	.372	.184	.97	.06	3.02	1.33	1.77	.554	.253	.996	29.40	11.32	49.5	9.62	4
5	3rd " "	11.07	.379	.204	1.09	.05	2.99	1.37	1.80	.505	.270	1.01	31.20	11.88	53.8	8.82	5
6	4th " "	10.76	.415	.258	1.10	.10	2.82	1.49	1.85	.456	.291	1.09	39.92	12.28	62.2	7.17	6
7	5th " "	10.67	.435	.236	.90	.10	3.04	1.52	1.89	.421	.302	1.07	33.04	12.40	54.2	8.01	7
8	6th " "	11.08	.519	.280	1.43	.13	3.59	2.10	1.96	.372	.351	1.05	32.98	12.92	53.9	7.00	8
9	7th " "	11.08	.647	.356	1.41	.14	3.41	2.50	1.96	.337	.400	1.05	33.68	13.00	55.0	5.51	9
4110	8th " "	10.83	.614	.326	1.44	.16	3.06	2.00	1.91	.330	.379	.996	30.96	12.00	53.1	5.86	4110
11	9th " "	10.75	.606	.320	1.48	.18	2.95	2.04	1.84	.320	.365	.968	30.96	12.20	52.8	5.75	11
12	1st tailings (coarse) from purifier.	9.77	3.670	1.85	5.44	4.40	11.52	4.28	2.57	.657	1.211	.540	....	....	50.4	1.39	12
13	2nd tailings (fine) from purifier.	9.14	1.780	.872	1.65	1.93	6.64	2.62	2.30	.400	.561	.873	....	....	49.0	2.64	13
14	1st germ middlings (coarse)	10.91	.482	.254	1.48	.13	2.70	1.42	1.70	.337	.285	.925	29.28	11.20	52.7	6.69	14
15	1st germ flour (fine)	10.30	1.087	.606	2.56	.25	2.92	2.01	1.91	.351	.488	.870	28.24	11.82	55.7	3.15	15
16	Pure germ	8.46	4.80	2.729	11.87	1.83	6.15	15.09	4.84	1.830	3.030	.660	....	....	56.8	1.76	16
17	Dust from purifier.	9.86	1.1118	.529	1.79	1.63	6.06	2.05	2.09	.344	.414	.909	29.44	12.20	47.3	3.95	17
18	Dust from rolls	11.03	.509	.248	1.08	.29	3.21	1.45	1.79	.349	.257	.937	31.12	12.20	48.7	7.22	18
19	Dust from scoured wheat	7.65	4.16	.860	2.26	18.91	25.16	3.07	1.70	.526	.530	.327	....	....	20.7	1.98	19
4120	1st patent flour	11.52	.403	.216	.99	.16	2.94	1.32	1.82	.467	2.88	1.039	30.64	11.64	53.6	8.43	4120
21	1st clear flour	10.98	.807	.436	1.71	.24	3.57	1.75	2.13	.337	.393	1.091	34.76	14.08	54.0	4.89	21
22	2nd clear flour	10.44	1.341	.704	2.04	.31	3.36	2.10	2.33	.358	.470	1.099	36.44	15.52	52.5	3.31	22
23	Red dog	9.22	3.15	1.632	5.40	2.37	8.45	6.45	2.87	.751	1.347	.702	....	....	51.8	1.76	23
24	Bran	8.79	6.38	3.15	4.06	10.84	25.07	5.35	2.33	.558	.863	.400	....	....	49.4	.74	24
25	Shorts	8.87	4.10	2.23	5.21	8.43	16.31	6.03	2.47	.691	1.210	.393	....	....	54.4	1.11	25
26	Wheat (cleaned) before tempering	10.29	1.73	.848	2.07	....	5.13	2.57	2.05	.390	.547	.811	....	....	....	....	26

In a series of 6 patent flours milled in Minnesota and North Dakota, reported by Richardson, the ash content ranged between .39 and .45 per cent.

An excellent digest of the data accumulated during the last half of the nineteenth century which related to flour was compiled by Wiley et al. (1898). Richardson's data were included in this compilation, together with the results of the examination of a large number of commercial flours by the U. S. Dept. of Agriculture Bureau of Chemistry.

Jacobs analyzed all of the flour streams from a large Minneapolis flour mill about ten years ago. His data, hitherto unpublished, are presented in Table 74 through the courtesy of the Bureau of Chemistry of the U. S. Department of Agriculture. The first break flour was somewhat higher in its content of ash than the second and third break flours. It may be presumed that it was also substantially darker in color, although Jacobs did not include the color score of these flours in the data available to the writer. The fourth and fifth break flours were less highly refined than the second and third break flours, and in this particular may be rated in the order of their production in the mill. The first, second and third middlings flours were the most highly refined of the flour streams and were quite similar with respect to ash content, as is usually the case in most five break mills. Next in order were the fourth and fifth middlings flours, which resembled each other closely; the sixth middlings flour contained about 0.10 per cent more ash than the fourth and fifth middlings flour; while the seventh, eighth and ninth middlings averaged about 0.10 per cent more than the sixth middlings flour. The tailings and dust collector flours contained the highest percentage of ash. The percentage of P_2O_5 in the flours paralleled the percentage of ash except in the case of the dust from the scoured wheat, as shown by the fairly constant ratio of P_2O_5 to ash recorded in the second column from the right. There was a fairly high positive correlation between the content of ash, and of fat and fiber in these flours, particularly when account was taken of the experimental error in the determination of fiber in refined flours. The fat content of the first break flour was lower than might be anticipated from its ash content, due, no doubt, to the fact that at this stage of milling the germ was still uncrushed and its oil had not been expressed to the extent that was the case in later grinding operations. The "pure" germ was notably rich in fat, while the first five middlings were lowest in this constituent. The percentage of pentosans tended to parallel that of the fiber, but because of difficulties in the quantitative determination of the fiber in refined flours, the pentosan content was doubtless a more exact measure of the proportion of pericarp or bran in the flour than the fiber content.

While the wheat contained in excess of 2 per cent of nitrogen (which multiplied by the factor 5.7 would be recorded as percentage of crude protein), the first, second and third middlings flour contained only 1.8 per cent of nitrogen, since they were derived from the central portion of the endosperm, which is lowest in protein content. The percentage of nitrogen tended to increase in progressing through the middlings system. The break flours were still higher in nitrogen content, and their inclusion in the first clear flour served to increase its nitrogen content above that of the patent, which was made up largely of the refined middlings flours. The percentage of dry crude gluten paralleled the nitrogen content rather closely, as might be expected. Probably little significance can be attached to the percentage of nitrogen soluble in distilled water, but the nitrogen soluble in 5 per cent K_2SO_4 solution afforded a fair measure of the refinement of the flours. This nitrogen fraction was low in the first, second and third middlings flours, and high in the fourth and fifth break flours. It is doubtful if this fraction can be satisfactorily employed as a criterion of flour grade, however; not alone because of the comparative lack of precision in the method, but particularly because, as shown in the discussion of sprouted and frosted wheat, the soluble nitrogen fraction may be considerably higher in grain so damaged.

An interesting series of data relating to flours milled from hard red winter wheat in a five break mill were reported by Swanson, Willard and Fitz (1915). These include the results of baking tests and of chemical analyses, which are presented in Tables 75 and 76. The same general relations between flour streams originating at different points in the milling process that were observed in Jacobs' work are found in the results of these analyses of winter wheat flours. In general, the percentages of ash are somewhat higher in the latter. The percentage of "amino compounds" approximately paralleled the percentage of ash, though this correlation might disappear if flours milled from different wheats were compared, particularly if sprouted or frosted kernels were included in the wheat mixture.

The results of the baking tests included in these studies are of special interest, because of the relation of composition to baking qualities which can thus be observed. Loaves of surprisingly good quality were baked from the second and third break flours, with regard to volume, texture and color of crumb. The gluten quality rating assigned these flours was considerably lower than that of the second, third, fourth and fifth middlings flours. This gluten quality factor is the product of the maximum expansion of dough, rise in oven, and volume of loaf multiplied by 10^{-4}. The color score of the first break flour was lower than would be anticipated from its ash content. Thus it contained only .02 per cent more ash than the third break flour, but scored 8 points lower in color. This is usually the case, since the break flours pick up considerable of the dirt on the surface of the cracked wheat kernel, which darkens the color without increasing its ash content to a like

TABLE 75

Composition of Flour Mill Streams from a Five Break Mill (Swanson, Willard and Fitz, 1915)

MILL STREAM	Moisture, Per Cent	Ash, Per Cent	Protein, Per Cent	Fat, Per Cent	CRUDE GLUTEN		ACIDITY		Amino Compounds, Per Cent	PHOSPHORUS			
					Wet, Per Cent	Dry, Per Cent	25° C., Per Cent	40° C., Per Cent		Water Soluble at 25° C., Per Cent	Water Soluble at 40° C., Per Cent	Total, Per Cent	Per Cent of Total Phosphorus Soluble in Water at 40° C.
Wheat	12.08	1.70	13.04	1.86			.137	.420	.466	.058	.192	.424	45.5
Patent, 70%	12.16	.46	11.17	1.01			.065	.114	.142	.018	.025	.097	25.7
Clear, 27-27½%	12.10	.73	13.65	1.40			.149	.215	.237	.062	.086	.164	52.4
Low grade, 2½-3%	11.40	.96	14.06	1.67	49.9	14.43	.190	.300	.348	.086	.135	.238	56.7
First break	13.00	.72	11.50	1.12	40.4	13.23	.118	.208	.240	.046	.071	.141	50.3
Second break	12.97	.70	12.66	1.96	42.3	12.73	.120	.184	.208	.052	.072	.144	50.0
Third break	12.72	.70	13.77	1.17	48.6	15.03	.110	.192	.207	.054	.073	.150	48.6
Fourth break	12.86	.77	14.88	1.29	49.4	15.06	.139	.220	.249	.061	.090	.178	50.5
Fifth break	12.48	1.12	18.10	1.73	66.8	20.20	.206	.324	.318	.100	.164	.259	63.3
First middlings	12.10	.46	10.97	.92	33.2	10.73	.062	.089	.171	.015	.020	.083	24.1
Second middlings	12.26	.44	11.12	.83	34.8	10.56	.059	.096	.175	.015	.022	.082	26.8
Third middlings	11.82	.48	11.53	.92	32.9	10.80	.061	.110	.182	.023	.029	.094	30.8
Fourth middlings	11.81	.51	12.14	1.07	38.2	9.95	.071	.125	.177	.023	.037	.104	35.5
Fifth middlings	11.53	.55	11.97	1.00	39.1	12.25	.077	.131	.192	.027	.039	.106	36.8
Sixth middlings	11.33	.55	12.21	1.21	40.1	17.92	.083	.145	.202	.035	.047	.118	39.8
Seventh middlings	11.32	.84	13.67	.97	50.4	14.18	.146	.244	.297	.071	.091	.191	47.6
First sizings	12.35	.56	10.80	1.10	37.2	12.11	.080	.141	.170	.037	.045	.111	40.5
Chunks	12.09	.90	11.67	1.47	41.7	12.10	.140	.254	.274	.074	.104	.198	52.5
Second sizings	11.78	.78	11.41	1.35	36.5	11.11	.128	.208	.245	.048	.080	.164	49.4
B middlings	11.69	.64	11.41	1.05	41.6	12.35	.114	.184	.215	.050	.068	.151	22.5
First tailings	11.98	1.01	13.33	1.57	48.9	14.43	.150	.277	.279	.067	.128	.230	55.6
Second tailings	11.68	1.20	14.59	2.19	47.1	14.96	.164	.349	.392	.083	.155	.289	53.6
Second low grade	11.08	.90	12.65	1.50	44.2	12.48	.161	.271	.302	.058	.108	.204	52.9
Bran-duster flour	11.06	1.13	15.62	1.72	58.6	16.25	.196	.329	.350	.072	.143	.268	53.3
Ship-duster flour	11.32	1.40	13.44	2.14	40.9	12.50	.212	.413	.448	.062	.169	.331	51.0
Roll-suction stock	11.57	.70	10.83	1.40	35.1	10.66	.115	.413	.211	.060	.069	.145	47.5
Stock going to 4th middlings from 2nd sizings	12.17	.40	10.99	1.13	41.4	11.85	.057	.117	.067	.020	.069	.099	69.6

TABLE 76
Baking Tests of Flour Streams from a Five Break Mill (Swanson, Willard and Fitz, 1915)

Serial No.	Mill-Stream Flours	Percentage Moisture	Percentage Absorption	Flour for Loaf, Grams	Water for Loaf, Grams	Maximum Expansion of Dough, cc.	Rise in Oven, cm.	Weight of Loaf, Grams	Volume of Loaf, cc.	Gluten Quality Factor	Texture of Crumb, 100=Perfection	Color of Loaf, 100=Perfection
390	Patent, 70%	12.16	60.0	342.0	205.5	2,650	4.75	499.0	1,845	2,322	100	100
391	Clear, 27–27½%	12.10	60.0	341.3	204.8	2,350	5.20	501.5	1,875	2,291	88	85
392	Low grade, 2½–3%	11.40	61.6	342.4	211.2	2,175	3.65	515.0	1,665	1,321	78	70
393	First break	13.00	55.0	348.8	191.8	2,700	5.90	494.0	1,980	3,154	86	80
394	Second break	12.97	55.0	348.8	191.8	2,500	4.50	495.0	1,845	2,076	93	86
395	Third break	12.72	55.0	343.6	189.0	2,475	4.70	491.5	1,845	2,146	94	85
396	Fourth break	12.86	55.0	342.2	189.3	2,400	4.35	495.0	1,840	1,920	93	84
397	Fifth break	12.48	56.6	343.0	194.4	2,350	3.15	499.0	1,700	1,258	75	67
398	First middlings	12.10	60.0	341.3	204.8	2,575	4.50	502.0	1,845	2,137	98	100
399	Second middlings	12.26	60.0	342.0	205.2	2,800	4.90	507.0	1,910	2,620	99	99
400	Third Middlings	11.82	61.6	340.1	209.7	2,675	4.85	505.5	1,888	2,471	100	100
401	Fourth middlings	11.81	63.3	340.1	215.7	2,575	5.10	509.0	1,885	2,475	98	99
402	Fifth middlings	11.53	61.6	339.0	209.0	2,550	5.05	506.0	1,885	2,427	99	99
403	Sixth middlings	11.33	61.6	338.3	203.0	2,475	4.65	505.0	1,805	2,077	100	82
404	Seventh middlings	11.32	60.0	338.3	208.6	2,200	2.75	505.0	1,660	1,004	90	94
405	First sizings	12.35	58.3	342.0	199.5	2,400	4.20	501.0	1,775	1,789	96	76
406	Chunks	12.09	58.3	341.3	199.1	2,375	4.65	501.5	1,885	2,081	80	90
407	Second sizings	11.78	60.0	340.0	204.8	2,375	4.15	499.5	1,825	1,798	95	92
408	B middlings	11.69	60.0	339.7	203.8	2,375	4.05	499.5	1,770	1,702	96	74
409	First tailings	11.98	60.0	340.9	204.6	2,350	4.05	503.5	1,770	1,684	79	71
410	Second tailings	11.68	60.0	339.8	203.9	2,200	3.15	501.5	1,700	1,178	82	80
411	Second low grade	11.08	58.3	337.4	196.8	2,150	3.15	498.5	1,665	1,127	77	66
412	Bran duster	11.06	60.0	337.4	202.5	1,875	2.40	502.0	1,585	713	75	60
413	Ship duster	11.32	60.0	338.3	203.0	1,850	1.00	503.0	1,483	274	70	69
414	Roll-suction stock	11.57	58.3	339.0	198.0	1,950	3.70	494.0	1,710	1,233	77	
415	Stock going to 4th middlings from 2nd sizings	12.17	55.0	341.7	188.0	2,625	4.30	494.0	1,795	2,026	99	97

extent. The first to the sixth middlings flours scored highest in color and texture, and the loaf volume was large. The sizings flours were inferior to the middlings flours in loaf volume, color and texture, while the tailings flours rated still lower in these particulars.

It should be noted that the 70 per cent patent referred to in tables 75 and 76 included the following flour streams: First, second, third, fourth and fifth middlings, and first reduction of chunks and first sizings. The clear flour, 27 to 27½ per cent, included all five break flours, the sixth and seventh middlings, and the flour from second sizings, B middlings, and first and second tailings.

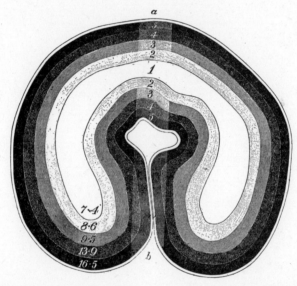

FIG. 11.—Diagram of a cross section of a wheat grain showing the gluten content of five arbitrary "zones." (Cobb, "Universal Nomenclature of Wheat.")

The relation of the percentage of ash and gluten in roller mill streams was discussed by Weaver (1921). In the crush or middlings flours there was a tendency toward an increasing gluten content with an increasing percentage of ash. This relation did not maintain exactly in the break flours, the first break flour being higher in ash than the third break, but lower in gluten. The best bread was baked from the streams lowest in ash, as shown in Table 77. Poor milling was held by Weaver to have more effect on the ash content than the percentage of ash in the wheat.

Weaver also compared the ash and gluten content of a number of flours of the same grade milled from Kansas hard winter wheat, and failed to find any correlation between the percentage of these two constituents.

TABLE 77

PERCENTAGE OF ASH AND GLUTEN AND THE LOAF VOLUME AND COLOR SCORE OF BREAD BAKED FROM ROLLER MILL STREAMS, REPORTED BY WEAVER (1921)

Flour Stream	Ash, Per Cent	Gluten, Per Cent	Volume of Loaf, cu. in.	Color Score
BREAK FLOURS				
First	.54	10.32	156	96
Second	.41	10.42	156	98
Third	.48	13.42	178	99
Fourth	.64	15.21	140	90
Fifth, No. 1	.73	16.21	96	50
Fifth, No. 2	.96	16.88	85	50
CRUSH OR MIDDLINGS FLOURS, COARSE				
First	.33	11.55	154	99
Second	.34	11.73	152	99
Third	.38	12.07	160	100
Fourth	.39	12.07	160	100
Fifth	.45	12.68	170	98
Sixth	.54	13.07	162	97
Seventh	.72	13.21	146	96
CRUSH OR MIDDLINGS FLOURS, FINE				
First	.35	11.81	158	100
Second	.37	11.81	153	100
Third	.40	11.81	152	99
Fourth	.45	12.55	172	98
Fifth	.44	11.98	160	98
Sixth	.61	12.42	150	97
Seventh	.60	14.21	156	96
SIZINGS				
No. 1	.44	11.91	152	99
No. 2	.42	11.37	152	100
No. 3	.58	11.55	148	98
FINISH FLOURS				
First, No. 1	.92	15.60	140	85
First, No. 2	.89	14.50	144	85
Second, No. 1	.63	15.00	150	88
Second, No. 2	.68	12.60	148	87
Third	.75	13.30	146	86
Germ tailings	.67	12.02	158	94
Coarse tailings	.92	13.30	148	87
Fine tailings	.95	16.08	144	83
Bran and shorts duster, No. 1	.82	14.00	144	86
Bran and shorts duster, No. 2	.68	12.60	148	87

It might be deduced from the consistently lower protein content of the refined middlings and flours produced therefrom, as contrasted with the break and tailings flours, that the central portion of the endosperm, from which the refined middlings are derived, is lower in its content of protein than the outer region of that structure. That such is the case has been shown by Cobb (1905), who painstakingly scraped the endosperm of plump Purple Straw wheat kernels in such a way as to effect a separation into 5 concentric zones. This was done with a sufficient number of kernels as to accumulate several centigrams of the flour. The flour from each zone was then analyzed, and it was found that the percentage of gluten decreased progressively from the outer to the central part of the endosperm. This is shown by the data in Table 78, in which the zones are listed in order from the central (No. 1) to the outer zone (No. 5). The order of zoning and gluten content of each is also shown in Figure 11, reproduced from Cobb's paper.

TABLE 78

GLUTEN CONTENT OF THE DIFFERENT PORTIONS OF THE ENDOSPERM, AS SHOWN BY COBB (1905)

Zone Number	Weight of Flour	Gluten, Per Cent
1	.1308	7.4
2	.1254	8.6
3	.1315	9.5
4	.1267	13.9
5	.1114	16.5

Gluten was found by Cobb in all of the endosperm cells. The illustration of a typical endosperm cell, as drawn by him, appears in Figure 12, and the gluten is shown to be present in the form of a reticular network, in the interstices of which large and small starch granules are embedded. The nucleus was found to be of stellate form, due to the pressure of the starch granules. It was formerly believed, and the older literature frequently contained statements to the effect that the gluten is contained in a single layer of cells, commonly referred to as "gluten cells," which constituted the outer zone of the endosperm. These cells, known more properly as aleurone cells, are now known to be devoid of gluten, and they could not contribute to the protein or gluten content of patent flour, since they are almost entirely removed with the bran in the process of roller milling.

The proteins of wheat bran were studied by Jones and Gersdorff (1923), who identified three proteins. These were an albumin, a globulin and an alcohol-soluble protein, which constituted 16.64, 13.62 and 31.01 per cent respectively of the total protein of the bran. The alcohol-soluble protein precipitated on cooling its hot solution, and redissolved on warming. It could be precipitated from its alcoholic solution by adding aqueous sodium chloride solution.

The quality of gluten, as indicated by the viscometric procedure of Sharp and Gortner (1923), was determined in the flour streams of the Minnesota State Experimental Flour Mill by Hendel and Bailey (1924). A low gluten quality factor was encountered in the break flours, the average for the five being 1.85, the fourth and fifth break flours being particularly low in this respect, although the high per-

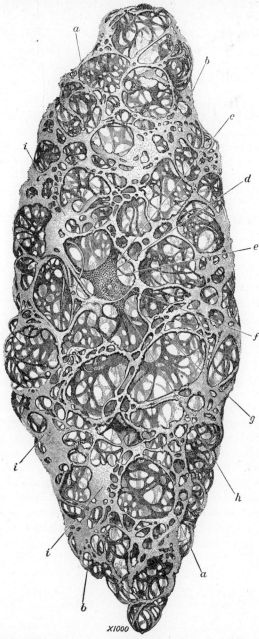

FIG. 12.—A cell dissected from the starchy endosperm of a wheat kernel, showing the protoplasmic network (a, b) and the stellate nucleus (c, d, e, f, g, h). The starch granules have been removed. (Cobb, "Universal Nomenclature of Wheat.")

centage of gluten in these flours compensated in part for the low quality. The sizings flour was intermediate, while the quality factor of the middlings flours was uniformly high, averaging 2.65.

Criteria of Flour Refinement or Grade.

Since the miller endeavors to separate the bran and germ from the endosperm in the process of milling, it would appear that the most direct method to determine the success of his efforts would be to ascertain the quantity of bran and germ fragments in the flour that is produced. Several efforts have been made to classify or grade flours on the basis of the results of such determinations. Girard (1895) counted under the microscope the number of bran fragments in a cubic millimeter of flours representing different percentages of extraction, and found them to increase in number fairly regularly with the extraction. This is shown by certain of his data in Table 79.

TABLE 79

INCREASE IN THE NUMBER OF BRAN FRAGMENTS WITH INCREASING PERCENTAGE OF FLOUR EXTRACTION (GIRARD, 1895)

Percentage of Extraction	Number of Bran Fragments per gram
45	3,400
60	10,700
70	32,300
80	44,100

Vinassa (1895) separated the fibrous materials from flour by digesting with HCl, neutralizing with alkali, and precipitating the fibrous residue by centrifuging. The residue was then dyed, and examined with a microscope to determine the relative amount, and the presence of adulterants. Buchwald (1913) indicated a correlation between the number of cellular elements in wheat flour and grade or refinement of the flour. Certain of his data are shown in Table 80. The seed coat fragments and wheat hairs were obtained from a unit quantity of flour in each case, and the number increased with increasing ash content and decreasing refinement.

TABLE 80

RELATION OF NUMBER OF SEED-COAT FRAGMENTS, AND WHEAT HAIRS TO ASH CONTENT OF WHEAT FLOUR (BUCHWALD, 1913)

Flour	Ash, Dry Basis, Per Cent	Seed-Coat Fragments	Wheat Hairs
0	.42	23	14
I	.92	149	168
II	2.36	549	512

CHEMISTRY OF ROLLER MILLING

The number of hairs and bran fragments in a series of flours of varying degree of refinement, as produced in an English flour mill, were determined by J. Hume Patterson, and reported by Moore and Wilson (1914). During the progressive milling of wheat by the roller process the number of microscopic offal particles in the flour streams

TABLE 81

Hair and Bran Fragments in 3 Milligrams of Each of the Flour Streams, as Determined by Patterson (Moore and Wilson, 1914)

Milling Machine	Wheat Hairs	Bran Fragments	Total
A	0	0	0
B	0	0	0
C	0	0	0
D	5	0	5
E	7	0	7
F	13	5	18
G	18	26	44
H	25	26	51
Break (1, 2, 3)	28	24	55
Break (4)	38	30	81

was found to increase. Wheat hairs and hair fragments were not found in the most refined flour streams, while considerable numbers of such particles were counted in the flour streams produced near the tail of the mill. The count was also high in the break flours, sifted off after the initial breaks or grinding of the wheat. This is shown by the data in Table 81.

A microscopical method for examining flour to determine its refinement and freedom from non-endosperm structures was developed

TABLE 82

Bran Particles and Wheat Hairs in a Unit Weight of Flour of Different Classes and Grades (Keenan and Lyons, 1920)

Commercial Grade of Flour	Bran Particles Variation	Bran Particles Average	Hairs Variation	Hairs Average
Hard wheat patent, 40-90%	15-72	30	2-45	18
Soft " " 34-90%	19-133	49	1-34	20
Hard wheat straight, 92-100%	33-121	64	17-87	43
Soft " " 90-100%	34-153	82	22-81	45
Hard wheat clear, 6-52%	65-331	174	43-223	109
Soft " " 5½-50%	126-308	218	30-167	86
Hard wheat low grade, 2-10%	169-353	273	88-335	182
Soft " " " 2-10%	143-402	302	27-261	140

Experimental Flours

Patent, 70%		22		13
" 90%		40		28
Straight, 97½%		44		30
Clear, 27½%		58		49
Low grade, 2½%		320		124

by Keenan and Lyons (1920). A portion of the flour, weighing 5 milligrams, was placed on a ruled microscopic slide, and 4 drops of chloral hydrate solution (1 : 1) was mixed with the flour, the preparation being "cleared" by gentle heating. After cooling, the bran particles and wheat hairs in the mount were counted. A series of commercial flours were examined, and the following counts reported for the several classes and grades.

An examination was made by Keenan and Lyons of the mill streams which were combined in the manufacture of a straight flour milled from soft wheats, with the results shown in Table 83.

TABLE 83

Bran Particles and Wheat Hairs in the Flour Streams Combined to Produce a Soft Wheat Flour (Keenan and Lyons, 1920)

Flour Stream	Bran Particles	Hairs	Total
First break	113	38	151
Second break	75	38	113
Third break	131	53	184
Fourth break	228	106	334
Fifth break	368	173	541
First middlings	21	8	29
Second middlings	48	27	75
Third middlings	26	7	33
Fourth middlings	29	2	31
Fifth middlings	55	12	67
Sixth middlings	60	18	78
Seventh middlings	143	23	166
Eighth middlings	264	38	302
First germ flour	50	5	55

Such a procedure is especially useful in judging of the grade of flour used in compounding a self-rising flour. In such flours the ash content has been modified by the addition of chemical leavening agents, and accordingly cannot be used as an index of the flour grade.

Keenan (1923) later suggested that a count of wheat hairs presents less difficulty to the untrained microscopist than the count of bran particles. Data were presented graphically which showed the break flour to be consistently high in hair count, ranging between 40 and 123 (per 5 milligrams of flour), while the hair count of the middlings flours ranged from 9 (average of third middlings) to 37 (average of eighth middlings).

The color of a suspension of flour in alcohol, when compared with a standard prepared with starch-free bran particles, was used by Liebermann and Andriska (1911) in determining the grade of flour. The lower grades, which contained more of the branny particles, gave a darker coloration.

The fibrous residue collected on a filter after digesting a unit weight of flour with alkali was employed by Fornet (1922a) as a simplified method for determining the "degree of milling," or relative refinement of flour. This residue can be compared with a standard, and the quantity and appearance of the fiber thus conveniently and accurately estimated. Fornet (1922b) stressed the usefulness of this procedure in determining the grade of flour from which a specimen of bread was baked. In the case of bread, the usual criteria of flour grade, such as content of ash, fats, sugars and starch, are either modified by other ingredients of the bread or by the treatment to which the dough is subjected in fermentation and baking, while the fiber is not sufficiently altered to interfere with its separation and estimation by Fornet's procedure.

A somewhat more elaborate procedure for the separation of the branny particles was suggested by Wiedmann (1921). The flour was digested with a soda lye solution, and then with saturated bromine water. The residue was washed, collected in a graduated Gerber tube, and measured volumetrically. By testing a series of flours of known extraction the normal residue for flours of various grades can be determined and used as a basis of comparison.

Salvini and Silvestri (1912) evolved a test of the refinement of flour, which consisted of shaking 5 grams of flour with 25 cc. of a mixture containing 5 per cent of glycerol and 2.5 per cent of formaldehyde (40 per cent) in water. The suspension was allowed to settle for 30 minutes in a graduated cylinder 20 mm. in diameter. The volume of precipitate increased with decreasing refinement of the flour, and the data thus secured correlated with the ash content of the flour.

Calendoli (1918) described a simple test for the grade of flour. A pinch of flour was added to a few ccs. of concentrated HCl, and the color noted. Refined flour free from branny particles gave a violet coloration, while the color was reddish-brown when bran was present. The preparations were compared with a set of standards, and the relative grade of flour could thus be approximated.

The principal difficulty with these methods for estimating the content of branny fragments lies in the lack of precision. In the hands of different analysts varying results are often secured when an effort is made to state the findings quantitatively. Considerable experimental error is likewise evident in many instances, necessitating numerous replicate determinations. For these reasons such methods, while often convenient in confirming other observations, have not come into common use as the sole index of flour grade. It might appear at first thought that the percentage of crude fiber, as determined in an ordinary proximate analysis, would afford a quantitative measure of the proportion of branny particles, but, as a matter of fact, the content of fiber is so small in proportion to the experimental error in its determination as to render the results of doubtful value. The fine state of division of refined flour in itself complicates the situation, since the fibrous particles which are to be weighed in the final operations of the determination are so short as to make their collection upon the filter a difficult procedure.

It might be anticipated that the relative quantity of branny particles in flour would be correlated with the number of bacteria and other fungi. Jago and Jago (1911, note p. 550) indicated that such was actually the case, and that while the patent or highly refined flours were practically sterile, the lower grades contained considerable numbers of such organisms.

Ash content has been more extensively used as a criterion of flour grade than any other characteristic of flour, except possibly the color or visual appearance. About the middle of the last century Mayer (1857) found that the ash content diminished with increasing refinement. In 100 parts of dry substance he found:

	Ash, Per Cent	Nitrogen, Per Cent
No. 0 Superfine wheat flour	0.58	2.01
No. 4 Wheat flour	1.29	2.19
Wheat bran	5.72	4.32

Dempwolf (1869) called attention to the variable character of the ash of different wheats, and presented analyses of the ash of different grades of flour. The total ash content of the different grades was

TABLE 84

Percentage and Composition of the Ash of the Various Mill Products as Reported by Dempwolf (1869)

Grade and Mark	Percentage of Output	Total Ash, Per Cent	Fe_2O_3, Per Cent	CaO, Per Cent	MgO, Per Cent	KO, Per Cent	NO, Per Cent	PO_5, Per Cent
Kochgriese								
A	.489	.398	.525	7.296	6.899	34.663	.988	49.721
B		.386	.583	7.718	6.857	34.669	.891	49.218
Auszugmehle								
0	3.144	.380	.630	8.057	7.008	35.482	.744	48.896
1	2.635	.416	.643	7.946	7.105	35.285	.675	48.976
2	5.291	.452	.627	7.454	7.795	34.254	.678	48.519
3	7.165	.481	.635	7.092	8.343	33.876	.690	49.306
Semmelmehle								
4	14.757	.586	.596	6.798	9.924	32.715	.650	50.056
5	17.925	.611	.570	6.791	10.574	32.239	.726	50.187
Brodmehle								
6, 7	15.419	.764	.334	6.626	10.870	30.386	.946	50.146
Schwarzmehle	6.805	1.176	.425	5.536	12.234	30.314	1.260	50.204
8	2.576	1.549	.484	4.741	12.947	30.299	.974	50.173
Kleien, 9	9.516	5.240	.208	2.747	16.861	30.672	.701	50.152
10	9.000	5.680	.426	.502	17.349	30.142	1.080	49.112
Kopfstaub								
11	1.290	2.648	1.671	.203	13.023	31.489	2.144	44.054

shown to increase with diminishing refinement, but the proportion of the different ash constituents was fairly constant in the flours as produced (No. 0 to 7). The percentage of MgO tended to increase in proceeding from the higher to the lower grades, while the CaO decreased.

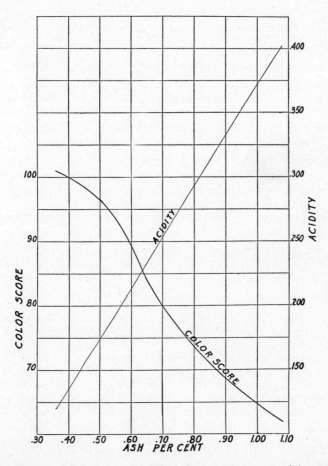

FIG. 13.—Graphs showing the correlations between ash and acidity content of flour, and ash content and color score of flour. (Bailey, 1913c.)

Vedrödi (1893) established a correlation between ash content and the grade of Hungarian flours. A fairly consistent increase in the percentage of ash with decreasing refinement was observed on analyzing flour samples from any individual mill. There was considerable range, however, in the ash content of flour samples bearing the same numerical designation, but produced in different mills. A system of classification was proposed by Vedrödi in which the percentage limits of ash content for flours of each numerical designation is given.

TABLE 85

Vedrödi's Proposed Classification of Flours on the Basis of Ash Content

Numerical Designation	Average	Ash Content in Per Cent, Limits Proposed
0	.31	.20– .34
1	.36	.35– .39
2	.41	.40– .43
3	.47	.44– .52
4	.58	.53– .60
5	.66	.61– .70
6	.98	.71–1.16
7	1.53	1.17–1.80
8	2.24	1.81–3.15

A positive correlation between percentage of ash and of acidity, and a negative correlation between the percentage of these constituents and the color score was observed by Bailey (1913c). In arranging the data presented in Table 86 the samples were grouped on the basis of their ash content, with a narrow range in each group. The mean ash content is given in the first column, while in the second and third columns are recorded the average percentage of acidity, and the average color score of the flours in each group. The same data are presented graphically in Figure 13.

TABLE 86

Relation Between the Percentage of Ash and Acidity, and the Color Score of Flour (Bailey, 1913 c)

Mean Percentage of Ash	Average Acidity, Per Cent	Color Score
.36	.120	101.0
.40	.139	99.7
.44	.156	98.3
.48	.166	97.6
.52	.172	95.9
.56	.193	92.1
.60	.209	88.5
.65	.232	83.5
.70	.264	80.9
.77	.279	73.7
.85	.314	70.2
.95	.354	66.9
1.08	.402	61.8

It is generally recognized by flour mill chemists that the ash content of flours cannot be invariably accepted as an index of the percentage of total flour represented in the flours under examination, or even of the absolute degree of refinement of these flours. Different flour mills will produce flours of the same percentage extraction which vary in ash content. The same mill may produce flour of the same extraction which varies somewhat in ash content from day to day, such variations being due to differences in tempering treatment of the wheat, modifications of the grinding, and fluctuations in temperature and humidity of the atmosphere. The type of wheat from which the flour is produced is reflected in the ash content of the flour. Thus, flours milled from soft winter wheat are usually lower in ash content than flours of the same percentage extraction milled from hard wheat. This is shown by the average composition of flours milled from the two classes of wheat, as recorded in Table 109, which is included in the consideration of flour grades. Even in the same class of wheat, variations in the ash content of the grain are reflected in the ash content of the flour. This is shown by data reported by Bailey (1924), resulting from the examination of four typical lots of spring wheat. As the ash content of the wheat increased, a corresponding increase in the percentage of ash in the straight grade flour was observed. These data are recorded in Table 87.

TABLE 87

PERCENTAGE OF ASH IN WHEAT, AND IN FLOUR MILLED THEREFROM (BAILEY, 1924)

Class and Grade of Wheat	Ash Content	
	Wheat, Per Cent	Straight Flour, Per Cent
No. 1 dark northern spring	1.44	0.45
No. 3 northern spring	1.58	0.46
No. 1 dark northern spring	1.98	0.55
No. 3 dark northern spring	2.14	0.60

Certain varieties of wheat yield flour characterized by an ash content greater than the average for the same percentage extraction. This has been observed in the examination of Kota wheat, a hard red spring variety, by Waldron, Stoa, and Mangels (1922), and by Clark and Waldron (1923).

The inorganic ingredients of the ash of flour are not, of themselves, directly responsible for the properties of the flour with which the percentage of ash is correlated. An increasing ash content is ordinarily assumed to indicate a diminishing degree of refinement of the flour, and the latter in turn usually implies a progressive impairment of baking qualities, color and keeping qualities. The substances found in the ash remaining after incineration of the flour might be substantially increased, however, without necessarily impairing any of the flour properties mentioned. In fact, it is likely that an improvement of baking quality might result from superimposing an equivalent quantity of the constituents of the ash upon a short patent flour. The ash content of flour is of significance, therefore, because it is correlated with the flour properties in question, rather than because it is directly responsible for those properties. The fibrous envelope or branny portion of the wheat kernel and the germ contain a much higher percentage of ash than the starchy parenchyma of the endosperm or floury portion of the same kernel, and ash content accordingly serves as an index of the relative proportions of bran or germ present in the commercial flour. The impairment of certain flour properties resulting from increasing the proportions of pulverized bran and germ in the flour is due to other constituents of those structures, rather than to the inorganic substances which appear in the ash.

In the discussion of Dempwolf's data it was pointed out that the percentage of MgO in the ash increased in proceeding from the higher to the lower grades, while the CaO decreased. The ratio between these two ash constituents in several of the flour grades was calculated by the author, with the following results:

Flour Grade	$\dfrac{CaO}{MgO}$ in the Ash
0	1.15
2	0.96
4	0.66
6	0.61
8	0.37

Similar ratios of CaO to MgO in the ash of wheat flours was shown by the results of Teller's (1896) analyses, and by those of Grossfeld (1920). In the latter instance the ash of fine flour contained calcium and magnesium in about equal proportions while the ash of the lower grade of flour contained nearly twice as much magnesium as of calcium.

Gerum and Metzer (1923) observed that the ratio of P_2O_5 to total nitrogen in dry crude gluten increased with increasing percentage of flour extraction.

The percentage of manganese, zinc, copper and iron was found by McHargue (1924) to be very low in refined flour, while much larger

TABLE 88

Concentration of Certain Metals in Wheat Products, as Reported by McHargue (1924)

Wheat Product	Parts per Million			
	Copper	Iron	Manganese	Zinc
Bran	16	210	125	75
Germ	46	270	150	160
Flour	Tr	24	10	22

proportions were found in bran and germ. The quantity of these metals in wheat products is shown in Table 88.

Iodin content of flour was about half that of wheat, as shown by McClendon and Hathaway (1925) and much less than that of bran and shorts. It has not been established that the iodin content of flour would serve as an index of grade, since it might vary with the environment in which the wheat is grown. In the run of wheat products which McClendon and Hathaway examined, the iodin content in milligrams per metric ton of wheat, straight flour, bran, shorts and red dog was, 6.6; 3.5; 15.5; 9.6; and 3.7 respectively.

Pentosans may be considered the most characteristic constituent of the branny coat of the wheat kernel. This was emphasized by Koning and Mooj (1914), who describe a method for determining the relative proportions of bolted and unbolted flour used in bread, based upon the content of pentosans. The pentosan content of a sample of bread prepared from unbolted flour was 6.23 per cent. When varying proportions of this flour were mixed with bolted flour, the pentosan content of bread baked from the mixture conformed closely to calculated values based upon the pentosan content of flours in the mixture.

The quantity of phloroglucide precipitated per unit weight of dry flour from the distillate prepared as in the determination of pentosans was determined by Prandi and Perracini (1917). The comparative quantity of phloroglucide from three grades of flour is given in Table 89, which shows that with longer extraction and lower grades of flour the quantity of phloroglucide increased appreciably.

TABLE 89

QUANTITY OF PHLOROGLUCIDE PRECIPITATED FROM THE ACID DISTILLATE OF VARIOUS GRADES OF FLOURS (PRANDI AND PERRACINI, 1917)

Percentage of Flour Extracted	Phloroglucide		
	Minimum	Maximum	Mean
80	3.10	3.31	3.25
85	3.81	4.61	4.09
90	6.70	7.18	6.95

Gerum (1920), working with Bavarian flours, found an increase in the pentosan content of flour with increases in percentage of extraction, but concluded that starch and ash determinations afford a more satisfactory index of the degree of extraction. This is shown by Gerum's data, certain of which are included in Table 90.

The fact that pentosans are the characteristic component of the bran, fat of the germ and starch of the endosperm was stressed by Jacobs and Rask (1920). This is shown by their data in Table 91 resulting from their analyses of flour, bran and germ.

TABLE 90

RELATION BETWEEN DEGREE OF EXTRACTION AND CHEMICAL COMPOSITION OF FLOUR AS SHOWN BY GERUM (1920)

Wheat Flour Extraction Per Cent	On Dry Basis			
	Ash, Per Cent	Starch, Per Cent	Fiber, Per Cent	Pentosans, Per Cent
50	.48	76.1	.38	1.86
60	.70	77.4	1.43	3.70
70	.85	75.6	1.60	3.55
70	.95	75.3	1.64	3.32
70	.94	75.1	1.61	3.36
70	.80	75.1	1.66	3.66
70	1.09	75.9	...	2.07
90	1.65	68.5	3.87	5.40
94	1.89	61.2	2.50	7.51
94	2.00	62.6	2.50	8.61

When Jacobs and Rask cut wheat kernels transversely, and analyzed the germ or lower ends, and the brush or upper ends separately, it was found that the percentage of fat in the germ portion was decidedly

TABLE 91

CHARACTERISTIC CONSTITUENTS OF ENDOSPERM, BRAN AND GERM OF THE WHEAT KERNEL. (JACOBS AND RASK, 1920)

Wheat Product	Starch, Per Cent	Fat, Per Cent	Pentosans, Per Cent
1st middlings flour (endosperm)..	68.78	.75	2.25
Commercial bran	9.07	3.58	22.72
Commercial germ	7.30	10.02	5.38

higher than in the other portion. This is shown by the following data, resulting from such an examination of Marquis hard red spring wheat.

ETHER EXTRACT (on dry basis), PER CENT IN

Entire kernel	2.32
Brush end of kernel	.91
Germ end of kernel	5.34
Percentage of kernel represented by germ end	31.8
Percentage of total fat contained in germ end	73.4

From data available it was then computed that the actual fat content of the germ was 12.5 per cent, and of the starchy endosperm, 0.75 per cent.

As one means of calculating the yield of flour, the following formula, based on an actual examination of the wheat and mill products, was suggested by Jacobs and Rask.

$$\text{Flour yield} = \frac{(\text{Starch in wheat} - \text{starch in offal}) \times 100}{\text{Starch in flour} - \text{starch in offal}}$$

Potential flour yield of wheat was likewise computed from the pentosan content of wheat. This was possible, due to the relatively constant percentage of pentosans in the bran, which contains most, if not all, of the pentosans in the kernel. The average percentage of pentosans found in the bran was 28.5, and the conversion factor accordingly becomes $\frac{100}{28.5} = 3.51$. The offal content of wheat will be the difference between the pentosan content of wheat and that of the endosperm, multiplied by this factor. Since the pentosan content of endosperm is likewise fairly constant, averaging 3.13 per cent, this value may be inserted in the equation, which thus becomes:

Percentage of offal = (Percentage of pentosans in wheat — 3.13) x 3.51, and the percentage of offal thus computed, subtracted from 100, gives the potential flour yield.

Since the starch content of the mill offals examined ranged between 28 and 37 per cent, averaging 30.3 per cent, it is obvious that much of the endosperm is finding its way into these by-products of milling rather than into the flour. Jacobs' and Rask's data indicate that in this particular the mills studied are only about 90 per cent efficient, 10 per cent of potential flour appearing in the feed or offals.

A formula for calculating the percentage extraction of flour from the starch content of the mill products was developed by Scholler (1922). This took the form: $A = 100 \left(2 - \frac{P}{G}\right)$, when A = the degree of extraction of flour, P = weight of starch in units per 100 units of mill products, and G = starch content of the flour. With wheat of normal characteristics, and milled in the usual manner, a simpler formula: $A = 2(100 - P)$ may be employed. When variations in properties of the wheat or processes in milling are encountered, the more accurate formula $A = \frac{100(G-R)}{P-R}$ should be employed, in which R = the starch content of the mill offals.

The ratio of starch to fiber in flour was regarded by Grossfeld (1920) as a convenient index of grade or degree of refinement. Using Kalning and Schleimer's (1913) data, the ratio became progressively narrower in three grades of diminishing refinement. This is shown by the data in Table 92.

TABLE 92

RATIO OF STARCH TO FIBER IN FLOUR GRADES AS CALCULATED BY GROSSFELD (1920)

Extraction	Ratio: $\frac{\text{Starch}}{\text{Crude Fiber}}$
0–31.5	661
31.5–73.5	456
73.5–78.7	89

This ratio is of particular significance in determining the grade of flour from which bread has been baked, since it is not greatly altered by baking.

An objection to the ash determination as a criterion of fineness in the case of flours ground between mill stones was registered by Cerkez (1895) because of the possibility of particles of the stone appearing in the ash and increasing the percentage. The fat content was deemed preferable as an index of grade, and the basis of classification shown in Table 93 was proposed:

TABLE 93
LIMITS OF FAT CONTENT OF FLOUR GRADES PROPOSED BY CERKEZ (1895)

Numerical Designation	Limits of Fat, Per Cent
0	0.60–0.75
1	0.96–1.05
2	1.06–1.15
3	1.16–1.25
4	1.26–1.45
5	1.46–1.62
6	1.63–1.84
7	1.85–2.50
8	2.51–3.45

Determinations of the percentage of ether extract in each of the flour streams of the Minnesota State Experimental Flour Mill, reported by Bailey (1923), indicate a close correlation with ash content. A few exceptions were noted, but in general the fat content was as satisfactory an index of refinement as the percentage of ash. The fat and ash content of a few representative flours from this series are shown in Table 94.

TABLE 94
PERCENTAGE OF ASH AND ETHER EXTRACT OR CRUDE FAT IN CERTAIN FLOUR STREAMS FROM THE MINNESOTA STATE EXPERIMENTAL FLOUR MILL (BAILEY, 1923)

Flour Stream	Ash, Per Cent	Ether Extract, Per Cent
1st break	.69	1.69
3rd break	.51	1.64
5th break	1.00	2.06
1st middlings	.40	1.47
5th middlings	.51	1.70
Low grade	1.81	3.32

Acidity is usually acceptable as a supplementary check on the grade or degree of refinement of flour. Reference has already been made to the correlation between ash content and acidity established by Bailey

(1913c). Such a correlation was likewise observed by Swanson (1912), who further indicated that the acidity of a water extract of flour could be accounted for by assuming the phosphorus compounds of the extract to be monosodium di-hydrogen phosphate. Instances have been noted by various workers where the acidity of particular samples of flour was higher than might be anticipated from their ash content, however, particularly when the flours were milled from sprouted or otherwise damaged wheat. This has already been indicated in the section dealing with sprouted wheat, in which is presented the data of Willard and Swanson (1911), who found that patent flour milled from normal or unsprouted wheat had an acidity of 0.90, while the same grade of flour milled from wheat sprouted for 4 days had an acidity of 1.62.

The percentage of albumin and globulin was found by Teller (1896) to increase in progressing from the higher to the lower grades of flour in roller milling. The difficulty in using the "soluble protein" fraction as an index of grade arises from the fact that an increase is encountered in any flour grade when the flour is milled from sprouted or frosted wheat. This is evident from the work of Willard and Swanson (1911), and might be deduced from the data published by Harcourt (1911).

Certain of these general relations of composition to flour grade and milling practice discussed in the preceding paragraphs are illustrated in the results of analyses published by Kalning and Schleimer (1913) which appear in Table 95. In Germany, where these studies were made, it has evidently been the practice to manufacture a more highly refined or shorter patent than has been the custom of late years in milling hard wheat in America. In the case of the three grades of flour with which Kalning and Schleimer worked, the finest, known as "Auszugmehl," represented 31.5 per cent. of the wheat. The next 42 per cent separated as flour in milling bears the designation "Semmelmehl" in German trade phraseology, and is less highly refined than the first separation, or Auszugmehl. After these two flours are taken off (which if combined would constitute about what is known in American practice as straight grade), an additional 5.2 per cent of low grade or "Nachmehl" was separated. It is interesting to note that the 42 per cent which was extracted after the removal of the superfine flour or Auszugmehl contained only .88 per cent of ash, and .20 per cent of fiber, the following 5 per cent of flour contained 2.36 per cent of ash and 1.09 per cent of fiber. This indicates that the streams of flour-like material produced after a certain limit has been reached (as in the manufacture of straight grade) will lack the characteristics of refined white flour. This break in quality is evidently quite sharp after the milling limit has been reached in the case of any particular lot of wheat.

TABLE 95

Composition of Flours Milled in Germany from 40 Per Cent Domestic and 60 Per Cent Foreign Wheat, as Reported by Kalning and Schleimer (1913)

	Wheat	Flour 0–31.5%	Flour 31.5–73.5%	Flour 73.5–78.7%
Total protein, per cent	15.49	13.24	15.08	19.36
Water soluble protein (12 hrs. extraction), per cent		2.37	3.00	5.27
Alcohol (60%) soluble protein		6.53	6.83	5.91
Fat, per cent	2.29	1.14	1.86	4.04
Starch, per cent	66.25	79.29	74.69	61.13
Glucose, direct reduction, per cent		1.46	2.18	4.31
Pentosans, per cent	7.94	2.59	3.37	5.52
Fiber, per cent	2.51	0.12	0.20	1.09
Ash, per cent	1.92	0.49	0.88	2.36
Phosphoric acid, per cent	0.97	0.23	0.50	1.27
Diastatic activity (Lintner)		19.	39.	58.

Enzymic Activity of Wheat Flour Grades.

Since the starchy endosperm of the wheat kernel is essentially a storage organ, in which metabolic activity is at a low level, while the germ and, to a lesser extent, the aleurone layer are the tissues in which metabolism reaches the highest level at various periods in the life cycle, the enzymic activity of the disintegrated germ and aleurone should be greater than in the pulverized endosperm. This has been indicated by the difference in respiratory activity of germ and of endosperm found by Burlakow (1898) and Karchevski (1901), the energy of carbon dioxid respiration being about twenty times as great in the germ as in the endosperm. It should accordingly follow that those mill streams which contain little germ and aleurone tissue should be comparatively low in enzymic activity. That this is true in so far as diastatic activity of the flours is concerned is indicated by the data of Kalning and Schleimer presented in the preceding section. Thus they found the diastatic activity of the highly refined flour or Auszugmehl to be equivalent to 19 units (Lintner), while that of the next lower grade was 39 units, and of the low grade or Nachmehl it was three times as high as in the Auszugmehle. The diastatic activity in these instances parallels the fat content of the flours, and it has already been indicated that the percentage of fat is the chemical criterion of the proportion of germ in flour.

Enzymic activity and gas-retaining power of flours produced on the breaks and from the successive reductions of semolinas or middlings was determined by Martin (1920a). The enzymic activity was expressed in terms of the carbon dioxid produced in 24 hours by a mixture of 10 grams of flour, 5 grams of water, 1.2 per cent of salt, and 1 per cent of yeast. Gas-retaining capacity was represented by the volume of the fermented dough, so prepared that by the addition of the requisite amount of starch all the doughs contained the same quantity of gluten, and stated in terms of flour sample "a1" as unity. These data, given in Table 96, indicated, in the opinion of Martin, that the enzymic activity as shown by the evolution of carbon dioxid on fermentation increased progressively from the interior to the exterior of the endosperm, since the first reductions were presumably from the central region of that structure.[1] Percentage of gluten increased in the same manner, while gluten quality diminished progressively in flour from regions approaching the cortex. These compensating gradations accordingly tend in the direction of producing flours from various parts of the endosperm having a common or uniform strength.

TABLE 96

Enzymic Activity and Gas-Retaining Capacity of Flours Produced by Successive Reductions of the Wheat Berry (Martin, 1920)

	Enzymic Activity- CO_2 Evolved, cc.	Relative Gas-Retaining Capacity of Gluten	Dry Gluten, Per Cent
REDUCTION FLOURS			
a1. From A B C reductions corresponding to the central portion of the grain	85	1.00	13.8
a2. From D E G reductions corresponding to a region exterior to the central portion	93	.93	15.9
a3. From K reduction corresponding to a layer nearer the cortex than D E G	115	.86	17.4
a4. From M N reductions corresponding to the region nearest the cortex	122	.77	17.9
BREAK FLOURS			
b1. From B'M₂ a flour from the central portions of the grain	58	.86	15.9
b2. From Br₄, a flour having its origin near the cortex	83	.64	21.6
STRAIGHT GRADE FLOUR representing the entire endosperm	88	.90	15.6

The rate of gas production in fermentation was at a higher level in clear flour dough examined by Bailey and Johnson (1924b) than in the dough made from patent flour milled from the same wheat. The clear flour dough was less effective in retaining the carbon dioxide of fermentation than the patent flour dough.

[1] The author doubts whether Martin's data indicate a difference in the enzymic activity of the various parts of the endosperm. It seems more probable that the differences noted may be accounted for by variations in the proportion of germ and bran fragments in the flours produced by successive reductions.

The proteolytic activity of different grades of flour, in terms of the time required to liquefy a 1.5 per cent gelatin jelly, was determined by Stockham (1920). There was little difference noted in the proteolytic activity of the patent and clear flours, but that of the second clear was appreciably higher, as shown by the data in Table 97.

TABLE 97

PROTEOLYTIC ACTIVITY OF DIFFERENT GRADES OF FLOUR MILLED FROM HARD SPRING WHEAT (STOCKHAM, 1920)

Grade of Flour	Baking Strength of Flour		Hours to Liquefy 1.5% Gelatin
	Expansion Loaf Volume Minus Dough Volume	Texture Score	
Patent	100	100	354
1st clear	87.5	95	330
2nd clear	75.	92	144

The brownish coloration of bread was attributed by Bertrand and Mutermilch (1907b) to an oxidizing enzyme of the nature of tyrosinase. The action proceeds in two stages, a glutenase first producing the material which is oxidized through the action of the tyrosinase. The glutenase was found to be inactive in an alkaline medium, somewhat active in a neutral, but more active in an acid medium. The oxidizing enzyme was found by Bertrand and Mutermilch (1907a) to be chiefly present in the bran of the wheat kernel. It was not laccase, as postulated by Boutroux, as it did not react with guaiacol to produce tetra-guaiacoquinone. In contact with trypsin it produces a coloration ranging from rose to brown.

Catalase, another enzyme concerned with oxygen transfers, which, according to Löew (1901) has the function of destroying the harmful peroxides found in living cells, is encountered in varying degrees of activity in the several structures of the wheat kernel. The varying activity of catalase in the different grades of flour was first studied by Wender and Lewin (1904), who measured the quantity of oxygen liberated in one hour from 35 cc. of hydrogen peroxide solution when mixed with 100 grams of flour and 200 cc. of water. A high grade flour, No. 0 was found to liberate 64 cc., while a low grade, No. 7½, liberated 246 cc.

The inadequacy of the Pekar and similar color tests in distinguishing grades of flour was stressed by Wender (1905), in a later discussion of the usefulness of the catalase activity as an index of flour grade. On modifying the method to involve a measurement of the volume of oxygen liberated in one hour from 10 cc. of H_2O_2 solution (12%) in 100 cc. of water by 25 grams of flour, the data given in Table 98 were secured. A fairly regular increase in catalase was observed in progressing from the more to the less highly refined flours. The range within each type, in the case of flours from three sources, was comparatively narrow. (Characteristics of these types of flours in so far as chemical composition is concerned, are indicated by the data of Dempwolf, Vedrödi, and Cerkez, presented in the preceding section.)

TABLE 98

CATALASE ACTIVITY OF VARIOUS FLOUR TYPES FROM THREE SOURCES (WENDER, 1905)

Flour Type	Oxygen Liberated per 25 Grams Flour in 1 Hour		
	Czernowitz Mill, cc.	Budapest Mill, cc.	Vienna Exchange, cc.
0	37	39	21
1	43	44	37
2	46	51	51
3	72	73	62
4	80	84	71
5	82	85	77
6	98	92	82
7	106	108	80
7½	117	112	118

In another experiment Wender found no diminution of the catalase activity of corn (maize) flour after 3 years' storage.

That the quality of flour can be judged by the quantity of oxygen which it liberates from hydrogen peroxide was found by Liechti (1909).

Miller (1909) compared various methods for determining the degree of refinement of flour, including the ash content, the percentage of fat, catalase activity, percentage of starch, pentosans and phosphoric acid. The phosphorus was evidently largely in organic combination, since little inorganic P_2O_5 was found in the water extract when enzymatic action was inhibited. A method was suggested for the determination of catalase activity. Instead of placing the flour in contact with the H_2O_2 solution, an extract of flour was prepared by digesting flour with water 1:10 for 4 hours; 5 cc. of the extract was then mixed with 15 cc. of H_2O_2 solution. The latter contained an equivalent of .075 grams of H_2O_2, as determined by titration with standard potassium permanganate solution. After 2 hours the residual H_2O_2 in contact with the flour extract was determined by permanganate titration. A comparison of the H_2O_2 decomposed by an extract equivalent to 1 gram of flour, with the quantity of gaseous oxygen evolved by a unit weight of flour in 30 minutes is shown in Table 99. It is interesting to note that while the oxygen evolved by No. 5 flour was only 150 per cent of that evolved by No. 1 flour, the quantity of H_2O_2 decomposed by the extract of the former was 880 per cent of that decomposed by the latter.[1]

TABLE 99

COMPARISON OF THE GASEOUS OXYGEN EVOLVED PER UNIT OF FLOUR, AND THE QUANTITY OF HYDROGEN PEROXIDE DECOMPOSED BY A FLOUR EXTRACT (MILLER, 1909)

Flour Number	Oxygen Evolved in 30 Minutes, cc.	Hydrogen Peroxide Decomposed by Flour Extract in 2 Hours, mgm.
00	32	6.5
0	38	6.8
1	61	8.8
2	63	9.5
3	66	16.6
4	70	32.6
5	90	76.5
6	102	109.0

The catalase activity of American wheat flours was studied by Bailey (1917), and it was found that in general the volume of oxygen liberated from hydrogen peroxide in contact with a flour suspension paralleled the ash content of the flour which was tested. A typical series of such tests, in which the patent, straight and clear flours produced in a certain mill were compared, is shown in Table 100. It will be noted that while the ash content of the clear grade is only double that of the 80 per cent patent, the quantity of oxygen evolved in 60 minutes was practically four times as great.

TABLE 100

CATALASE ACTIVITY OF PATENT, STRAIGHT AND CLEAR GRADE FLOURS PRODUCED IN A CERTAIN AMERICAN MILL (BAILEY, 1917 d)

Flour Grade	Ash, Per Cent	Catalase Activity cc. of Oxygen Evolved in 30 mins.	in 60 mins.
80 per cent patent	.43	5.5	11.5
Straight	.48	12.6	19.9
20 per cent clear	.86	30.1	43.4

Marion (1920) compared three flours of different grades, and found that as the degree of extraction was increased, with the consequent increase in ash content and fibrous material, the quantity of oxygen liberated from hydrogen peroxide per unit of time and material was likewise increased. These comparisons are shown in Table 101. When

[1] Author's note. Possibly some of this difference is to be attributed, not to any peculiar property of the extract in decomposing hydrogen peroxide, but rather to the relatively large quantity of O_2 dissolved in the reaction medium and not accounted for when gaseous oxygen is determined.

the first and second flours were mixed, the increase in catalase activity was in proportion to the ratio of the second flour in the mixture.

The effect of several variables upon the catalase activity of flour was studied by Merl and Daimer (1921). In contact with phosphate, acetate and lactate buffer mixtures the optimum pH = ± 7. Toluol was found to be without significant effect on the activity of flour catalase, and is thus suitable as an antiseptic material in preventing decomposition during digestion of the flour. Chloroform effected little change in activity, hydrocyanic acid diminished it somewhat, while benzol and alcohol exerted a considerable inhibiting effect increasing in the order named. Heating the dry material at 40° C. for 2 hours did not reduce the subsequently observed catalase activity. Similar heating at 60° and

TABLE 101

Composition and Catalase Activity of Different Grades of Flour (Marion, 1920)

Grade of Flour	Acidity as Sulfuric Acid	Ash, Per Cent	Fat, Per Cent	Fiber, Per Cent	O_2 Liberated at 15° C. in 5 mins. per Gram of Flour, cc.
First flour	.0196	.52	.80	.22	1.10
Second flour	.0319	.96	.96	.76	6.10
"Pour betail"	.0539	1.86	1.86	6.40	13.70

above reduced the catalase activity markedly. When one gram of the dry material was rubbed up with 50 cc. of water and then heated at various temperatures for 2 hour periods, there was a reduction in catalase activity at temperatures above 30° C. in the case of wheat flour and bran. The temperature coefficient of acceleration in activity from 0-10°, and 10-20° was 1.5; from 20-30°, 1.43; and in the next 10° interval there was a decrease in the observed activity. The data, plotted as a curve, indicated an optimum temperature of about 36-37° during the 20 to 30 minutes' duration of the test.

That catalases are localized near the periphery of the grain was the conclusion of Bornand (1921). Rammstedt (1910), in a critique of the methods employed in determining the quality of the flour, stressed the lack of knowledge concerning catalases, and called attention to the fact that ignited sea sand liberated oxygen from hydrogen peroxide. Fernandez and Pizarroso (1921) failed to find a correlation between catalase activity and the percentage of total P_2O_5 soluble in .4 per cent acetic acid, or phytin P_2O_5. Marotka and Kaminka (1922) likewise noted that the oxygen evolved from hydrogen peroxide in contact with flour did not correlate with the degree of bolting of the flour unless controls with the actual grain from which the flour was milled was included in the determinations. Thus the determination of catalase activity as an index of flour grade has been subjected to criticism, and the possibility of arriving at incorrect conclusions has been pointed out. This method, like others that have been employed for the same purpose, does not always yield identical results in the hands of different analysts, and as yet no standard procedure has been agreed upon. As a confirming test, in connection with certain other determinations, the measurement of catalase activity may be of considerable value, and in the control laboratory is especially useful because of the short time required to complete the determination.

Physico-Chemical Methods for the Determination of Flour Grade.

Specific conductivity of the water extract of wheat flour was suggested by Bailey (1918) as a criterion of flour grade, it having been observed by him that this was positively correlated with the ash content. In a more extensive study conducted by Bailey and Collatz (1921) the effect of several variables upon the conductivity of the extract of a certain flour sample was determined. It was found that increasing the temperature at which the extraction was conducted increased the conductivity of the extract until a temperature of about 55° C. was exceeded. Above this temperature the conductivity was diminished. With a patent flour there was an increase in conductivity with the lapse of time during the first 240 minutes, after which the increase was slight except when the extraction was conducted at 0° C. At this temperature, while the greatest change in conductivity was recorded during the first 240 minutes, there was an appreciable increase during the period from the 240th to the 960th minute. In the instance of a clear flour studied in a similar manner, it required about 240 minutes for the same proportional change in conductivity as was observed in half this time with the patent grade. After 240 minutes the relative change was slight when extractions were conducted at 25°, 40° and 60°, but at 0° a large increase occurred between the 240th and the 960th minute. In the interest of conserving time, a procedure was adopted which involved extracting for exactly 30 minutes at 25°. This temperature was selected primarily because it was easiest to maintain in an accurate manner under ordinary laboratory conditions. When a series of flours was extracted in this way, and the conductivity of the extract at a temperature of 30° plotted as absissas, with the ash content as ordinates, the resulting curve, although nearly a straight line, took the form of a simple parabola, as shown in Figure 14.

From the similarity of the response of flour extracts to temperature changes, and that of phytin solutions in contact with the enzyme phytase studied by Collatz and Bailey (1921), it appears that the conductivity of water extracts of flour is due chiefly to the presence in solution of dissociated inorganic salts of phosphoric acid resulting from the hydrolysis of phytin in contact with phytase present in the flour suspension.

The specific conductivity of the water extract of flour was unaffected by treatment with nitrogen peroxide in the proportions that this reagent is used in ordinary bleaching of flour. As shown by the work of Bailey and Johnson (1922), flour treated with chlorine yielded an extract with an increased content of electrolytes, however, due not alone to the

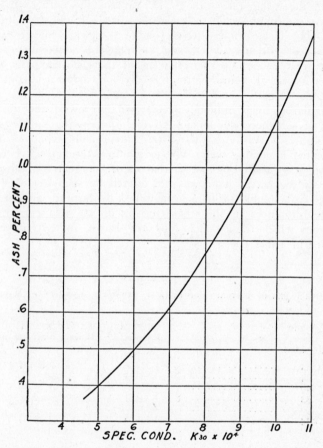

FIG. 14.—Graph showing the correlation between ash content of flour and the specific conductivity of the water extract. (Bailey and Collatz, 1921.)

hydrochloric and hypochlorous acids formed in the flour, but also, perhaps, to the acceleration of the enzyme phytase in consequence of increasing the hydrogen ion concentration. The extent of change in conductivity on treating flour with the dosage of chlorine ordinarily used in bleaching bread flours averaged about 0.48×10^{-4} (K_{30}), which was equivalent to a difference of about $.04 +$ per cent of ash when untreated or natural flours were compared.

No significant change in the conductivity of the extract of flours was observed when fresh flours were stored for a period of five months, according to Bailey and Johnson (1924a). In this particular the properties of both chlorine treated and natural flours were stable over the period embraced in the study in question.

With the collaboration of S. D. Wilkins the author studied the refractive indices of water extracts of patent, straight and clear grade flours. The clear grade extracts had a somewhat higher refractive index than the patent grade extracts, although it is doubtful if the difference is great enough to make a classification on this basis possible. When a series of flours with gradually increasing ash content was studied, the refractive index of their extracts did not likewise increase in a regular manner, although as between the lowest and the highest content of ash there was an appreciable difference in this property of their extracts. When extracts from patent and second clear flours were heated in boiling water for 30 minutes, the refractive index of the latter was reduced to that of the patent flour extract, although the extract of the patent flour was not altered by such treatment. This suggests that soluble proteins, coagulated by heating, are responsible for the differences in this characteristic of the two extracts. The data from which these conclusions are drawn, hitherto unpublished, are given in Table 102.

TABLE 102

REFRACTIVE INDEX OF EXTRACTS OF PATENT, STRAIGHT, AND CLEAR GRADE FLOURS

Ash in Flour, Per Cent	Refractive Index	
	Original Extract	Extract After Heating in Boiling Water 30 minutes
.43	1.33477	1.33474
.44	1.33493	
.49	1.33467	
.53	1.33474	
.90	1.33550	
2.14	1.33628	1.33474

The refractive index of a series of flour samples was studied by Stockham (1920), who likewise found little difference on comparing the several grades.

Buffer values of the various grades of flour were determined by Bailey and Peterson (1921) who found the lower or clear grades to be more heavily buffered than the patent grades. It accordingly required more acid (or alkali) to effect a unit change in the hydrogen ion concentration of a clear flour extract than was necessary with a patent flour extract. The buffer value was positively correlated with ash content, when the "value" is computed in terms of quantity of acid or alkali

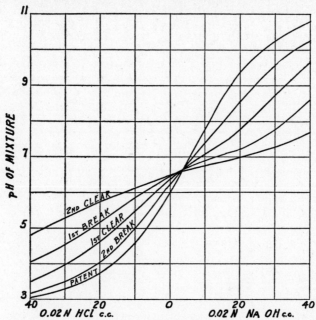

Fig. 15.—Electrometric titrations of five typical grades of flour of varying ash content. (Bailey and Peterson, 1921.)

required to effect a unit change in pH of the extract. This is shown by the data resulting from an electrometric titration of extracts of flours with varying ash content recorded in Figure 15.

Subsequent to the publication of these data Van Slyke (1922) suggested that buffer value be defined by the ratio $\frac{dB}{d\,pH}$, and that a solution has a buffer value of 1 when a liter will take up 1 gram equivalent of strong acid or alkali per unit change in pH. Substituting the data of Bailey and Peterson in Van Slyke's differential ratio, the buffer values of the 5 flours referred to in Figure 15 have been calculated by the author and are given in Table 103. It will be observed that the buffer value is correlated with ash content of the flours.

TABLE 103

Buffer Value of 5 Typical Flours Studied by Bailey and Peterson (1921)

Grade or Mill Stream	Ash, Per Cent	Buffer Value $\frac{dB}{d\,pH} \times 10^4$
Patent	0.40	1.30
2nd break	0.58	1.80
1st clear	0.83	2.04
1st break	1.34	3.54
2nd clear	2.38	6.00

Bailey and Peterson (1921) detailed a study of the influence of several variables on the buffer value of the flour extract. The buffer value increased somewhat with time of extraction, and quite appreciably increased as the temperature during extraction was increased from 0° to 40° C. Since the buffer value paralleled the specific conductivity of water extracts of the flours examined, and varied with modified conditions in the same direction and to about the same degree as the activity of phytase varied in hydrolyzing phytin, it appeared that phosphates produced by the hydrolysis of phytin during extraction with water, may be the principal buffering substances in the extract.

Bailey and Sherwood (1923) also showed that the hydrogen ion concentration of doughs made from low grade flours changed at a lower rate than doughs made from high grade flours during fermentation because of the higher buffer value of the former.

Bailey and Peterson (1921) found that the hydrogen ion concentration of the highly refined flour mill streams was higher than that of the lower grade streams even though the titrable acidity of the former is almost invariably lower. This is likewise attributed to the lower buffer value of the highly refined flours. These observations have been confirmed by Ostwald (1919), and by Kent-Jones (1924).

Lüers and Ostwald (1919) distinguished grades on the basis of the increased viscosity resulting from adding 2 per cent of the flour to water and heating the mixture. The relation of the percentage of extraction to viscosity as derived from the formula $\eta\Delta = 1000\ (\eta - \eta_0)$, where $\eta\Delta$ = increase in viscosity, η = viscosity of the paste, and η_0 = viscosity of water.

TABLE 104

Relation of Grade or Extraction to $\eta\Delta$ or the Increase in Viscosity Effected by Addition of 2 Per Cent of Flour in Making a Paste (Lüers and Ostwald, 1919)

Extraction Percentage	$\eta\Delta$
40	5839
60	5631
75	5328
90	3403
94	3272
Nachmehl or clear grade, 60-90%	715

Lactic acid was observed to increase the viscosity of high grade flour solutions, but had little effect on the lower grades. This is shown in Table 105.

Hendel and Bailey (1924) examined the flour streams resulting from the gradual reduction process employed in roller milling. The gluten "quality constant" determined by the method described by Gortner (1924) was found to be the highest in the middlings flours. Thus it averaged 2.65 in the instance of 1st, 2nd, 3rd, and 4th middling flour,

TABLE 105

Effect of Lactic Acid on Relative Viscosity (Water Taken as Unity) of Wheat Flour Suspensions (Lüers and Ostwald)

Grade or Extraction, Per Cent	20 Per Cent Flour Solutions	
	Viscosity in Water	Viscosity in 0.05 Normal Lactic Acid
0–40	3.54	16.16
0–60	4.24	22.29
0–75	4.54	11.04
0–90	4.71	7.67
0–94	5.13	7.70
60–90 (clear)	9.67	9.79

and only 1.65 in the five break flours. The lowest "quality constant" 1.57 was encountered in the duster flour, which represented fine material removed from the bran and shorts in the final treatment in the "duster."

Flour Grades.

It is evident from the discussion of flour milling that the miller has at his disposal a number of flour streams of varying characteristics which may be mixed or blended in producing the grades of flour to be marketed. In case several grades are marketed, the highest, least fibrous and whitest grade is commonly called "patent" flour. A lower or "clear" grade is simultaneously produced, constituting most of the remaining streams after the patent has been separated; this has sometimes been known as "bakers" flour, but the latter term has been objected to by the commercial bakers, who actually use the higher patent grades or straight grades in bread making. In England the patent and clear grades are referred to as "whites" and "households" respectively, according to Hamill (1911). In Germany the terms "Auszngmehle," "Semmelmehle," "Brodmehle," "Schwarzmehle" have been used in describing flours of diminishing refinement (Note Dempwolf, 1869). A numerical system, beginning with No. 0 and increasing to No. 8, is also in vogue (Note Vedrödi, 1893; Cerkez, 1895; Wender, 1905; and Miller, 1909).

The streams entering into the patent and clear flour in the case of a certain Kansas mill were indicated by Swanson, Willard and Fitz (1915). In this instance the 70 per cent patent included the flour streams resulting from the reduction of the first, second, third, fourth and fifth middlings, and the first reductions of chunks and first sizings. The clear grade, representing 27-27½ per cent of the total flour, included all five break flours, the sixth and seventh middlings flours, and the flour from the second sizings, B middlings, and first and second tailings. In the event that the patent flour was extended in such a mill, the probabilities are that the miller would include the second and third break flours, these being commonly used in the longer patents. In compounding the patent the miller starts with the choicest stream and adds streams of diminishing degrees of refinement until the lowest limit of quality is reached which can be permitted in the grade or brand to be marketed. In the event that a very high standard of quality is set, and a relatively small percentage of the total flour is included in the patent, as from 40 per cent to less than 70 per cent, this would often be characterized as "short" or "fancy" patent. When the patent grade included from 70 to 90 per cent of the total flour, it has sometimes been known as a "long" patent. Obviously the longer the patent, the less highly refined will be the clear grade simultaneously produced. A mixture of the mill run of patent and clear grade has often been known as "straight" grade flour; and when the poorest streams amounting to 3 to 5 per cent are omitted, the flour is called "standard patent."

The distinction between "straight" and "standard patent" is not at all clear, and the two terms are used more or less interchangeably.

Wheats vary so from season to season, and in different localities during the same season; the milling equipment and practices of individual millers are so variable, that it becomes difficult to characterize flours on the basis of the methods employed in their production. It frequently happens that flours representing the same percentage of the total flour produced in different mills are distinctly unlike in composition and baking qualities. There are, accordingly, real difficulties in the way of promulgating flour standards which will so classify flours as to convey an adequate idea as to their characteristics and quality. Thus a standard stated in terms of the percentage of flour produced in milling, or the flour streams permitted in each grade, would be inadequate because of the variations encountered in wheats and milling practices. Definitions in terms of chemical composition would be objected to on the ground that no single and easily determinable constituent is uniformly and invariably correlated with the important baking properties.

It is of interest to note some of the outstanding differences in the various grades of flour that have been observed by various investigators. Certain of these have been discussed in the preceding section, particularly in the reference to the work of Vedrödi, Jacobs and Rask, Wender, Bailey et al., and others. Snyder (1901) stated that "The ash in the wheat kernel varies so little in different samples of the same wheat, and the ash content varies so regularly in the different grades of milling products that it is possible to determine the grade of flour by determining the amount of ash which it contains. In fact, the mixing of grades can be detected more readily by determining the percentage of ash than by the study of any other constituent." Data resulting from Snyder's analyses of flours of different grades show a regular increase in protein and acidity with increasing ash content in case of the edible flours.

Snyder (1904a) reported the ash content of the several flour grades milled from hard spring wheat, the ranges being given in Table 106.

TABLE 106

Ranges in Ash Content of the Several Flour Grades Suggested by Snyder (1904a)

Grade	Ash, Per Cent
First patent	.35– .40
Second patent	.40– .48
Straight grade	.48– .55
First clear	.60– .90
Second clear	.90–1.80

In a further discussion of flour grades Snyder (1905b) stressed the usefulness of the ash determination in establishing the commercial grade. The first and second grades of patent flour were found to contain less than .48 per cent of ash, and it was stated, "In case a flour contains 0.5 per cent of ash it would not be entitled to rank with the patent grades."

The commercial flours on sale in North Dakota were examined by Ladd (1908b), and the terms used in describing them were discussed. First patent represented 70-80 per cent of the total flour, patent constituted 80-85 per cent, and straight grade or second patent 92-95 per cent. First clear included the flour which is not incorporated in the patent, but excluding the most inferior streams which are combined in the second clear. Commercial flours offered for sale in North Dakota (about 1907-1908) were found to contain an average of 11.08 per cent of moisture, .458 per cent of ash, 1.22 per cent of fat, and 12.81 per cent of crude protein ($N \times 6.25$). Somewhat later a pamphlet was published by Ladd (undated) containing suggestions for discussion of flour standards by the Committee on Definitions and Standards in 1915, which included the limitations for the flours classified in the several classes and grades indicated in Table 107.

TABLE 107

Flour Standards Suggested by Ladd, with Limits Stated on the Basis of a Moisture Content of 11 Per Cent

	Class of Wheat			
	Hard Spring	Durum	Hard Winter	Soft Winter
Ash, per cent, not more than,				
Straight flour	.52	.65	.50	.44
Patent flour	.42	.55	.40	.37
First clear flour	.80	1.00	.80	.60
Fiber, per cent, not more than,				
Straight flour	.50	.50	.50	.50
Nitrogen, per cent, not less than,				
Straight flour	1.50	1.75	1.50	1.15

In the same pamphlet Ladd suggested the following terms to be used in describing the three principal grades of flour:

"Patent flour is the fine, clean, sound, unbleached product made from . . . wheat meal by bolting or by a process accomplishing the same result, produced by the reduction of the best of the purified middlings," etc.

"Straight flour is the fine, sound, unbleached product made from . . . wheat meal by bolting, or by a process accomplishing the same result," etc.

"First clear flour is a straight flour made from . . . wheat from which the patent flour or a portion of the purified middlings shall have been removed," etc.

In a footnote referring to standards for patent flour Ladd offers the opinion that the term "patent flour" should be limited to 70 per cent of the total flour.

Eleven series of flours were secured by Alway and Clark (1909) from Nebraska mills in 1906. The percentage of ash in the three grades examined was as follows:

	Ash, Per Cent		
	Maximum	Minimum	Average
Patent	.42	.30	.36
Straight	.47	.35	.42
Bakers	.71	.50	.58

The manner of producing different grades of flour was described by Teller (1909), who stated that the patent may constitute from 70 to 90 per cent of the entire flour produced. The composition and technical

TABLE 108

Composition and Technical Analysis of Average Hard Spring Wheat Flours, as Reported by Teller (1909)

	Composition		
	Patent	Straight	Clear
Moisture, per cent	11.20	10.86	10.30
Protein, per cent	12.60	12.80	14.20
Ash, per cent	.43	.49	.72
Fat, per cent	1.17	1.31	1.76
Crude fiber, per cent	.30	.34	.55
Carbohydrates, per cent	74.30	74.20	72.47
Technical Analysis			
Gluten, per cent	12.6	12.8	14.2
Absorption, per cent	63	63	65
Color	100	96	80
Loaves, per barrel	100	100	101.2
Size of loaf	100	100	95
Quality of loaf	100	99	95
Average value	100	98.7	92.8
Fermenting period	100	101	107.3
Quality of gluten	100	99	93.6

analysis of patent, straight and clear flour milled from hard spring wheat, as reported by Teller, are given in Table 108.

The present-day views of operative millers in regard to flour grades are indicated by the data published by the Association of Operative Millers (1923-24). The author has endeavored to summarize in Table 109 the extensive data included in the tabulations published by this association. Straight grade flours milled from hard winter, and soft winter wheat flours contained an average of 0.447, and 0.425 per cent of ash respectively. Patent flours milled from hard winter wheat, and representing 60 to 88 per cent of the extraction contained an average of 0.379 per cent of ash. The shorter patents milled from the soft winter wheats averaged 0.341 per cent of ash. The moisture content of these flours ranged around 13.0 per cent.

It was maintained by Smith (1913) that the production of middlings (or semolinas) by high grinding with *mill stones* resulted in the first production of "patent" flour. Kraft (1913) declared that patent flour originated with the purifier about 1870, and antedated the roller process in America, although only 25 to 30 per cent of the total flour was first separated as patent. Dedrick (1913) declined to recognize as "patent flour" anything other than flour produced from the middlings and not exceeding 80 per cent of the total flour. Any higher percentages of extraction would include break flour and tailings flours, and these were termed "fancy straight" when 80 to 90 per cent extraction, "standard straight" when 93 to 97 per cent extraction, and "full" or "long straight" when including all the flour milled. For the patents, "fancy patent" would represent 40 to 60 per cent, "regular patent" 60 to 70 per cent, and "standard patent" 70 to 80 per cent of the total flour extraction, the first two being purified middlings reduction, and the standard patent being all middlings and dunst reduction. Enumerating the particles of bran was suggested as a means of differentiating between the several grades of flour.

The usage of the term "patent" was traced by Dedrick (1921) from the time when middlings were produced by stone grinding and the purifier first developed, down to the present confusion of terminology. It was stated that no flour constituting in excess of 80 per cent of the total flour can justly be called a patent, and in a 75 per cent patent the ash content should not exceed .36 to .37 per cent of ash.[1] When so milled it may include the flours resulting from the reduction of prime (or first) down to fourth or fifth middlings together with that from the fine dust middlings.

The responses to a questionnaire sent to the millers of Kentucky by Allen (1910) indicate that in the majority of mills that replied the "patent" flour constituted 75 per cent or less of the total flour produced.

[1] Author's note. It is probable that this limit was intended for soft winter wheat patents.

In several instances it was insisted that only flour made by the reduction of middlings should be so classified. There seemed to be some confusion in the use of the term "straight grade flour," however, since apparently certain millers so designated the flour remaining after a short or fancy patent had been removed, while others applied the term to the total run, or the patent plus the clear.

At the hearings on flour standards held by the Committee on Food Definitions and Standards in 1915, several millers testified concerning their idea of the usage of various trade terms in describing flour. A portion of the report of this hearing was published by Snyder (1923b), from which the statements of Andrews, Plant and Meek are quoted. Andrews stated that the use of the term patent "Began, as nearly as I can recall, with the introduction of the middling purifiers, which came in at about the time the rollers did. We bought this middlings purifier and it was patented. The miller got to saying, 'This is my patent flour,' meaning he made it with the middlings purifier, and that is the way it started." Andrews also indicated that in his opinion it would be inadvisable to restrict the application of the term patent to the flour produced from middlings alone, since that might exclude satisfactory break flours. Plant maintained that patents should be milled up to 95 per cent of the mill run. Meek also indicated that "Our 'standard patent' is the 95 per cent flour."

A discussion of the terminology employed in describing flours has recently been presented by Snyder (1923b). The difficulties arising in the characterization of flour grades on a basis of chemical composition have been stressed. Accompanying Snyder's contribution is the report of a committee representing the Millers' National Federation, in which a system of nomenclature and definitions for flour is suggested. The committee considers, "That there are no chemical or physical constants that can be used to establish flour standards or definitions, because of the great difference in composition of wheats and the corresponding difference in composition that occurs when flours are milled from different wheats by various systems of milling. Also, wheat of the same characteristics will produce flour of varying chemical composition from day to day even in the same mill." The following definitions for the different grades of flour are suggested:

A. Flour is finely ground bolted wheat meal.

B. Straight flour (or 100% Flour) is all the bolted wheat meal recovered from the wheat after removal of the feeds.

C. Patent flour is the more refined portion of the wheat meal from which all or a portion of the clears have been removed.

D. Clear flour is the less refined bolted portion of the wheat meal recovered in the manufacture of Patent Flour (Millers, according to their processing or trade demand, divide this into First and/or Second Clears).

CHEMISTRY OF ROLLER MILLING

TABLE 109

MAXIMUM, MINIMUM AND AVERAGE MOISTURE, ASH AND PROTEIN CONTENT OF HARD WINTER, AND SOFT WINTER WHEAT FLOURS REPORTED BY THE ASSOCIATION OF OPERATIVE MILLERS

Week Ending		Moisture, Per Cent			Crude Protein, Per Cent			Ash, Per Cent		
		Maximum	Minimum	Average	Maximum	Minimum	Average	Maximum	Minimum	Average
Hard Winter Wheat, 100% Straight Grade Flours										
Aug. 11, 1923		14.00	11.45	12.93	13.60	10.50	11.18	0.560	0.403	0.451
Oct. 6, 1923		14.35	11.30	12.76	11.30	9.20	10.60	.500	.430	.454
Aug. 16, 1924		13.92	11.83	13.09	11.76	10.37	10.88	.492	.396	.438
Oct. 11, 1924		13.50	12.32	13.05	11.30	13.32	10.91	.466	.436	.444
Average				12.96			10.89			.447
Hard Winter Wheat, 95% Straight Grade Flours										
Aug. 11, 1923		14.00	10.85	13.09	11.88	9.96	10.99	0.490	0.396	0.425
Oct. 6, 1923		13.93	11.78	12.88	11.40	9.88	10.88	.480	.410	.450
Aug. 16, 1924		13.95	10.02	13.07	11.42	10.08	10.88	.467	.380	.417
Oct. 11, 1924		13.50	13.10	13.42	11.30	10.50	10.99	.440	.410	.426
Average				13.11			10.93			.429
Hard Winter Wheat, Patent Flour										
	Extraction, Per Cent									
Aug. 11, 1923	65–87	13.90	10.45	12.85	11.72	10.12	10.69	0.430	0.330	0.374
Oct. 6, 1923	65–88	14.46	11.20	13.06	11.40	9.00	10.48	.440	.350	.386
Aug. 16, 1924	60–87	13.90	11.54	13.03	11.00	9.85	10.49	.437	.343	.368
Oct. 11, 1924	70–85	13.88	13.05	13.42	11.25	10.30	10.83	.422	.365	.389
Average				13.09			10.62			.379
Hard Winter Wheat, Clear Flour										
Aug. 11, 1923	10–40	14.00	11.20	12.79	15.08	10.44	11.95	0.780	0.467	0.612
Oct. 6, 1923	15–35	13.92	11.50	12.67	12.60	11.08	11.83	.930	.550	.669
Aug. 16, 1924	10–37	14.00	10.51	12.94	13.64	10.78	11.77	.740	.500	.599
Oct. 11, 1924	10–26	13.50	11.42	12.90	12.50	12.00	12.27	.760	.520	.648
Average				12.82			11.95			.632
Soft Winter Wheat, 100% Straight Flour										
Aug. 11, 1923		13.53	12.00	12.90	12.20	8.78	10.32	0.398	0.393	0.423
Oct. 6, 1923		13.58	11.50	12.88	10.60	8.50	9.75	.520	.355	.431
Aug. 16, 1924		12.80	12.67	12.79	10.58	9.75	10.25	.440	.370	.410
Oct. 11, 1924		14.40	11.20	12.94	10.92	7.92	9.61	.520	.390	.436
Average				12.88			9.98			.425
Soft Winter Wheat, 95% Straight Flour										
Aug. 11, 1923		13.50	10.56	12.54	11.48	9.70	10.21	0.410	0.355	0.380
Oct. 6, 1923		13.30	11.80	12.91	10.52	8.00	9.57	.450	.350	.400
Aug. 16, 1924		13.62	12.64	13.05	10.54	8.60	9.83	.405	.330	.378
Oct. 11, 1924		14.01	12.54	13.23	10.30	7.34	9.16	.490	.365	.407
Average				12.93			9.74			.391
Soft Winter Wheat, Patent Flour										
	Extraction, Per Cent									
Aug. 11, 1923	40–78	14.06	11.23	12.93	12.28	8.68	9.90	0.370	0.290	0.333
Oct. 6, 1923	40–70	13.62	12.48	13.02	10.80	8.04	9.32	.390	.280	.340
Aug. 16, 1924	50–75	13.20	12.17	12.82	10.36	9.10	9.60	.380	.320	.344
Oct. 11, 1924	40–85	14.80	12.00	13.26	9.64	7.20	8.80	.420	.320	.346
Average				13.01			9.40			.341
Soft Winter Wheat, Clear Flour										
Aug. 11, 1923	5–60	13.39	10.68	12.43	13.48	9.98	11.63	0.960	0.400	0.501
Oct. 6, 1923	20–60	13.64	12.20	12.95	12.50	8.72	10.44	.740	.410	.506
Aug. 16, 1924	23–15	13.00	11.73	12.66	11.17	10.14	10.75	.485	.380	.438
Oct. 11, 1924	13–60	13.98	12.00	12.95	11.64	8.24	10.01	.616	.438	.467
Average				12.75			10.46			.478

E. First clear is the better portion of the Clear when separated into two parts.

F. Second clear is the remaining portion of the Clears when First Clear is removed.

These definitions are supplementary to the requirements mentioned in the Food and Drugs Act and the Mixed Flour Law.

The author suggests definitions for five terms that are used more or less loosely in milling terminology. "Yield" should refer to the proportion or percentage of the cleaned and untempered wheat that is converted into straight grade flour. It may be recorded either in terms of percentage of the wheat, or in terms of the quantity of wheat required to produce one barrel of 196 pounds of straight grade flour. "Extraction" should refer to the percentage of straight grade flour which appears in each grade marketed. Thus a patent of 70 per cent extraction should constitute 70 per cent of the choicest flour streams of the group of mill streams that, if combined, would constitute the straight grade flour. "Grade" should refer to the type of flour as determined by the process of milling. In ordinary terminology, American flour grades would be distinguished as Patent, Straight, First Clear, and Second Clear. Low grades and Red dog should be regarded as feed grades.

"Class" of flour should refer to the market class of wheat from which the flour was milled, such as hard spring, durum, etc.

Chapter 8.

The Changes in Flour Incidental to Aging.

Reference has already been made to the hygroscopic character of wheat, and its tendency to change in moisture content when the humidity of the atmosphere in contact with it is varied. Flour manifests a similar tendency, as has been shown by numerous experiments. It is true, however, that flour is probably more uniform in moisture content when it leaves the mill than is wheat arriving at the terminal markets. On the other hand, flour responds more readily to changes in the humidity of its environment than does bulk grain, due to its fine state of division and to the greater exposure resulting from handling it in sacks.

One of the earliest systematic studies of the hygroscopic properties of flour was conducted by Richardson (1884). Five samples of flour, containing varying percentages of moisture, were exposed in the laboratory from March 7 to March 14. The original moisture content of the flour, and the comparative weights of the samples are given in Table 110. These data show the drier samples to have gained appreciably in weight when exposed to an atmosphere ranging in relative humidity from 34 to 60 per cent. During a two day period, when the humidity was about 60 per cent, an increase in weight of about 3 per cent was registered. The sample containing 13.7 per cent of moisture at the outset lost about the same relative weight during the 7 day period that the dry samples gained during the same time.

TABLE 110

Changes in Weight of Five Samples of Flour, as Determined by Richardson (1884)

Moisture in Original Flour, Per Cent Mar. 7	Relative Weight on the Several Days of the Experiment					
	Mar. 8	Mar. 10	Mar. 11	Mar. 12	Mar. 13	Mar. 14
7.80	102.15	101.42	103.53	104.57	101.97	103.07
7.85	102.30	101.70	103.80	104.95	102.25	103.25
7.97	102.15	101.55	103.60	104.80	102.05	103.15
9.48	100.65	99.53	101.73	102.68	99.88	101.08
13.69	96.72	95.35	99.35	98.20	95.50	96.60
Rel. humidity	46.4	35.0	59.0	60.1	34.0	

Twenty-seven sacks of flour weighing about 43 pounds each were stored for a year and weighed at intervals, in an experiment reported by Willard (1911). The sacks were piled in 3 layers of 9 sacks each in an airy room heated to ordinary temperatures and screened for protection from rodents. The test was commenced in August, and during the ensuing year the average loss per sack was .52 pounds. The greatest loss in weight, .79 pounds per sack, was noted at the end of 8 months, showing that there was subsequently a slight gain in weight during the late spring and the summer months.

In one series of experiments conducted by Sanderson (1914a), 50 pound sacks of flour stored in a warm room from August 12 to December 16, 1912, lost an average of 3.27 per cent in weight. A subsequent gain in weight occurred, amounting to about 1 per cent, in 4 months, and 1.58 per cent up to July 28, 1913, when the average net loss in the period of nearly one year was 1.69 per cent. In another series a gain in weight of .33 per cent was observed. The average of all the observations made by Sanderson during this study represented a gain of .21 per cent in weight. The exact conditions of storage and changes in the moisture content were not reported. On continuing this work, Sanderson (1914b) concluded that flour with about 11 per cent of moisture will vary less in weight under the conditions of storage studied than if the moisture content be more or less.

The day to day variations in the moisture content of two samples of flour, one milled from strong Manitoba wheat, the other from weaker Australian wheat, were studied by Guthrie and Norris (1912b). The moisture content of the flour was plotted against the relative humidity at the time of day that the flour samples were drawn. The parallelism between humidity and moisture content was not exact, since the mean humidity for the daily period was not determined and recorded. The moisture content of the hard wheat flour varied between 8.1 and 13.4 per cent during the period of the experiment, and a scanning of the graph indicates that a moisture content of 11 per cent was about in equilibrium with an atmospheric humidity of 65 per cent.

The changes in weight of 48 pound sacks of flour stored in a heated room was determined by Swanson, Willard and Fitz (1915). The experiment commenced in August, and a steady loss in weight occurred until April, at which time the average loss per sack was .79 pounds, or 1.64 per cent of the original weight. A slight gain amounting to .27 pounds occurred during the following spring and summer months between April and August. In another series of studies involving observations of one kilo samples in cloth sacks, the losses were slightly larger than those recorded in case of the 48 pound samples. The moisture content of the flour at the outset was about 10.4 per cent.

Flour exposed to a saturated atmosphere by Stockham (1917) increased from its original moisture content of 11.97 per cent to a moisture content of 22.86 per cent in one day, and to 28.74 per cent at the end of 9 days, when the flour became mouldy. In a saturated atmosphere at 0°C. a moisture content of 34.78 per cent was reached in 17 days, which he states was not the maximum.

Flour containing 13.03 per cent of moisture and stored in a cotton sack during April, 1921, lost nearly 2 per cent of moisture in 18 days, in the experiment reported by Herman and Hall (1921). Flour was stored in 48, 24 and 12 pound sacks by Frank and Campbell (1922) on April 24, 1922, and the changes in moisture content noted for a period of 48 days. During the first 36 days the moisture content diminished steadily, the total decrease being about 1.5 to 1.7 per cent. During most of this period the relative humidity of the atmosphere ranged between 52 and 63 per cent, and the temperature between 74° and 86° F. Between the 36th and 48th days the humidity of the atmosphere increased, ranging between 54 and 72 per cent, and the moisture content of the flour likewise increased by about .3 to .4 per cent, so that the average loss during the 48 day period was about 1.25 per cent. Hard wheat flour was observed to retain more moisture in a dry atmosphere, and to gain more in a moist atmosphere than did soft wheat flour in another study conducted by Frank (1923). It was also noted that the moisture content of freshly milled flour did not affect its subsequent hygroscopicity.

The significance of the size of the flour package in determining the rate of decrease in weight was stressed by Arpin and Pecaud (1923b), who found that flour in 0.5 kilogram cartons lost 4.60 per cent in weight, while 100 kilogram bags lost 2.05 per cent.

These several experiments demonstrate that flour is hygroscopic, and that its moisture content consequently varies with the humidity of the atmosphere in contact with it. Since no effort was made to control the atmospheric humidity in these experiments, however, there is little to indicate what the moisture content of the flours studied may have been when in hygroscopic equilibrium with the atmosphere. In certain studies, notably those of Guthrie and Norris (1912b), there was evidence that hard wheat flour containing 11 per cent of moisture was in hygroscopic equilibrium with an atmospheric humidity of about 65 per cent. This lack of precise data is in consequence of the mechanical difficulties involved in maintaining a constant humidity in a large chamber over an extended period of time. Small scale studies were conducted by Bailey (1920), which yielded data of value in measuring this important property of flour. Small portions of a sample of patent and of clear flour milled from hard spring wheat were exposed to atmospheres of different relative humidity. The humidity was controlled by contact of the atmosphere with a comparatively large volume of water solution of sulfuric acid having the appropriate aqueous vapor pressure. Four degrees of humidity were employed, namely 30, 50, 70 and 80 per cent at 25° C. When the flour varied appreciably from equilibrium with the vapor pressure of the atmosphere to which it was exposed, about 3 days were required to reach equilibrium. There was comparatively little difference in the moisture content of the patent and the clear flours on exposure to atmosphere of the same humidity. This is shown by the data in Table III, which are expressed graphically in Figure 16. The moisture content of the flour (vacuum oven, official

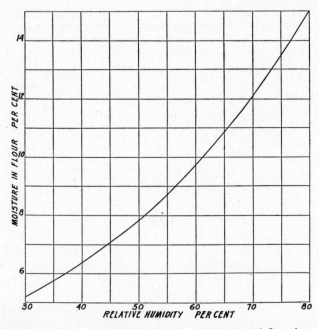

Fig. 16.—Graph showing the hygroscopic moisture content of flour in equilibrium with atmospheres of different relative humidities. (Bailey, 1920.)

A. O. A. C. method) was found to vary from a little more than 5 per cent at 30 per cent relative humidity, to 15 per cent at 80 per cent relative humidity. The curves showing this correlation have the shape of a simple parabola, indicating that hygroscopic moisture is not a linear function of the vapor pressure of the atmosphere.

The hygroscopicity of flour attracts the attention of food control officials largely because of its bearing upon net weight laws. Losses or gains in moisture effect corresponding changes in the weight of flour packages, since there is an insignificant change in the total weight of dry matter except in the event of actual leakage of flour particles from the package. The rational enforcement of net weight laws, when the laws and the regulations are properly drawn, involves a recognition of the possibility of change in weight of the contents of flour packages in consequence of changes in moisture content. About 20 years ago a committee commissioned by the Secretary of Agriculture recommended that flour containing in excess of 13.5 per cent of moisture should be considered adulterated by the addition of water. According to Wiley (1921) the committee was unanimous in its opinion that no flour could be considered sound and safe if its moisture content exceeded 13.5 per cent . . . and that this per cent of moisture represented the absolute maximum that should be obtained in the milling process by any justifiable milling operations. The experimental data considered in arriving

TABLE III

Hygroscopic Moisture of Patent and Clear Flour in Equilibrium with Atmospheres of Different Relative Humidity (Bailey, 1920)

Relative Humidity of Atmosphere at 25° C., Per Cent	Moisture Content of Flour	
	Patent, Per Cent	Clear, Per Cent
80.0	15.00	15.00
69.8	12.05	11.65
50.3	7.93	7.81
29.4	5.18	5.11

at this conclusion are not known to the author, nor is there in the literature, so far as he is aware, any extensive data indicating the exact moisture limits beyond which spoilage will result under specific conditions of storage. It is indeed strange, with the interest manifested in flour storage, that no one has conducted and published the results of a critical and comprehensive study of the relation of moisture content to keeping qualities at different temperatures.

At the time the moisture limit of 13.5 per cent was incorporated in the Federal standards of purity of flour, the percentage of moisture was determined by drying representative aliquots of the flour to constant weight in a water-jacketed oven, at the temperature of boiling water. Since that time the vacuum oven method has been made official for the determination of moisture in flour. This method gives somewhat higher results than the water oven method, and it has accordingly been contended by Snyder (1923a) and others that flours should not be considered as adulterated if they contain in excess of 13.5 per cent of moisture, when the determination is made by drying in vacuum. Snyder agrees with the committee referred to above that "the 13.5 per cent flour moisture standard applied to freshly milled and packed flour is a reasonable and just standard when based on the water oven method. . . ."

While the comprehensive study of Bailey and Thom (1920) of the changes occurring in stored corn meal may not afford a criterion of the safe limits in the moisture content of wheat flour, they are of interest in this connection. Corn meal containing in excess of 13 per cent of moisture supported the growth and propagation of microorganisms which caused it to deteriorate. The damage progressively increased with the moisture content.

In addition to the changes in hygroscopic moisture during storage of flour, other modifications of its composition and physical properties have been observed. Mention has already been made of the spontaneous oxidation of the carotinoid pigments of flour with the lapse of time, resulting in a progressive bleaching of the flour. The data reported by Winton (1911) and recorded in Table 122 afford a quantitative measure of the extent and rate of this change. The color of crumb of unbleached flours stored for a protracted period under different conditions of storage and baked at intervals, was recorded by Swanson, Willard and Fitz (1915). The color score of the freshly milled flour was 94, and after 305 days a score of 100 was assigned each of the lots, whether stored in a heated or unheated room. A similar improvement in color was noted by Saunders, Nichols and Cowan (1921). Flours were milled from different varieties of wheat in September, 1907, and baked at intervals. The color score assigned reached a maximum within about three years, and maintained the same level for about ten years thereafter.

Attention has been attracted to the relative acidity of stored flour, since acidity has been assumed by many flour specialists to afford a measure of the soundness of this product. Since the methods employed in the determination of acidity are purely empirical, and the results are reported in various terms, there is some difficulty in the way of comparing the findings of different analysts. Each set of data must be studied separately, therefore, and these data regarded as merely comparative in each instance.

Acidity in stored flour increased in consequence of, and was not the cause of gluten dispersion, in the opinion of Balland (1883b), who reported one of the earliest investigations of changes occurring in old flour. Balland (1883c) observed that soft wheat flour changes in acidity more rapidly than flour milled from the hard wheats. Farinaceous parts near the bran of the wheat kernel altered more rapidly than the central portion of the endosperm.

Hard spring wheat and durum wheat flours were stored by White (1909) in an unheated grain elevator and in a cellar, and no appreciable difference was noted in the relative increase in acidity in the two locations. Data indicating the initial acidity of the flours were not reported, but between the 4th and the 8th month of storage the relative increase in acidity of the hard wheat and the durum wheat flours was 8 per cent and 12 per cent of the original acidity respectively. On continuing these experiments for a total of 23 months, White (1912) observed little further change in the acidity of these flours. The acidity of flours stored under different sets of conditions by Swanson, Willard

THE CHANGES IN FLOUR INCIDENTAL TO AGING 183

and Fitz (1915) evidenced little change during the year that they were under observation. A slight decrease in the relative acidity was observed between the 125th and the 366th day in the unbleached flours sealed in jars and in cloth sacks. This is shown by the data in Table 118.

Patent and clear flours were stored by Bell (1909) under four sets of conditions with respect to combinations of temperature and humidity, and in all instances, as shown by the data in Table 112, there was a substantial increase in acidity (expressed as lactic acid) over a period of about 10 months. The clear flour increased more in acidity than the patent grade, the average increase under the four conditions of storage being 0.362 per cent and 0.085 per cent for the clear and patent flours respectively.

TABLE 112

Acidity (as Lactic Acid) in Samples of Stored Flour, as Reported by Bell (1909)

Conditions of Storage	Months of Storage			
	July, 1907	Oct., 1907	Jan., 1908	May, 1908
Patent Flour				
High temperature, low humidity	.333	.315	.378	.450
High temperature, high humidity	.369	.315	.369	.459
Low temperature, high humidity	.414	.324	.387	.468
Low temperature, low humidity	.351	.306	.374	.432
Clear Flour				
High temperature, low humidity	.522	.549	.720	.847
High temperature, high humidity	.477	.486	.801	.855
Low temperature, high humidity	.576	.549	.774	.900
Low temperature, low humidity	.452	.490	.642	.873

An increase in the acidity of flours stored by Marion (1909) was observed during the first year, followed by a slow but regular decrease during the second year of storage. This decrease in acidity was deemed by Marion to have been probably due to the formation of ammonia in consequence of the progressive decomposition of the flour proteins. Zak (1924) employed the determination of acidity, in connection with other chemical determinations in following the deterioration of flour over short periods of time. Vuaflart (1908) counselled that an opinion as to the age of flour must be formed cautiously from such chemical data as result from the determination of acidity, since new flours had been examined which had the characteristics of old flours in this particular.

A series of determinations of the hydrogen ion concentration of flours stored for protracted periods was recently reported by Bailey and Johnson (1924a). A commercial patent flour was sealed in a mason jar on August 8, 1921, and aliquots withdrawn at intervals until April 9, 1924. The hydrogen ion concentration increased fairly regularly during this period, the initial pH being 6.07, while on April 9,

1924, it had reached 5.53. In a second series, five samples of flour, representing as many commercial grades, were obtained in a freshly milled condition on March 13, 1922, and stored for somewhat more than two years. During this period their hydrogen ion concentration was determined at intervals. It was found that the lower or clear grades changed in degree of acidity at a slightly greater rate than the more refined or short patent and patent grades, in spite of the higher buffer action of the former. This is shown by the data in Table 113. While no determinations of titratable acidity were made in these cases, it is probable that such determinations would have shown a substantially greater increase in the acidity of the clear grades than in the fancy patents. This supports the observations of Bell (1909) noted in Table 112.

TABLE 113

Hydrogen Ion Concentration (as pH) of Several Grades of Flour Milled on March 10, 1922, and Stored for 25 Months (Bailey and Johnson, 1924 a)

Flour Grade	Ash Content, Per Cent	Hydrogen Ion Concentration as pH on					
		Mar. 13, 1922	Oct. 13, 1922	Dec. 29, 1922	June 1, 1923	Dec. 28, 1923	Apr. 9, 1924
Short patent ..	0.33	5.78	5.70	...	5.65	5.56	5.51
Patent	0.40	5.92	5.79	5.65	5.61	5.58	5.58
Straight	0.64	6.03	5.83	5.67	5.56	5.60	5.56
1st clear	0.82	6.24	6.09	5.93	5.93	5.87	5.85
2nd clear	1.37	6.38	6.17	6.04	6.02	6.00	5.97

Incidental to these studies it was found that there was no change in the electrolytic resistance of water extracts of flours with the lapse of time in storage.

Chlorine bleached flours changed less in hydrogen ion concentration on storage than the natural or untreated flours. This is shown by the data recorded in Table 114, which were accumulated in consequence of observations made on straight flour, portions of which were treated with varying dosages of chlorine and stored for 29 months. The flour samples were placed in mason jars immediately after milling and bleaching them, so that there was no significant change in their moisture content during the period of the experiment. While the chlorine-treated flours decreased somewhat in pH, the extent of change diminished with increased dosages of chlorine. There appears to be a tendency in the direction of arriving ultimately at the same hydrogen ion concentration, this condition having been practically reached during the second year of storage.

A large sample of flour which had an initial pH of 6.07 on August 6, 1924, when freshly milled, was subsequently extracted with ether, the ether vaporized at about its boiling point, and the extracted flour stored

THE CHANGES IN FLOUR INCIDENTAL TO AGING

in a tight jar. This flour after extraction contained 12.63 per cent of moisture, and was devoid of any perceptible odor of ether. The hydrogen ion concentration of this extracted flour was determined at intervals between May 23, 1922, and April 9, 1924, and no change could be detected during this period. The results of all these determinations ranged between pH 5.95 and 5.98, which are within the limits of experimental error. It is not entirely clear from this study whether (a) the ether removed from the flour those substances which undergo changes and give rise to acid reacting material, or (b) the ether inactivated the oganisms or enzymes instrumental in modifying the hydrogen ion concentration. It seems probable that the first alternative may be accepted as the explanation of the small change in pH of the flour thus extracted

TABLE 114

Hydrogen Ion Concentration (as pH) of Straight Flour Treated with Varying Dosages of Chlorine and Stored for 29 Months (Bailey and Johnson, 1924a)

Date of Test	Chlorine per Barrel of Flour, Ounces						
	0.00 pH	0.42 pH	0.50 pH	0.58 pH	0.67 pH	0.75 pH	0.83 pH
Nov. 9, 1921	6.04	5.93	5.90	5.88	5.87	5.83	5.80
Jan. 13, 1922	6.03	5.97	5.95	5.93	5.92	5.92	5.87
Feb. 12, 1922	6.04	5.95	5.93	5.91	5.90	5.89	5.87
May 26, 1922	5.95	5.90	5.89	5.90	5.90	5.83	5.81
Oct. 18, 1922	5.95	5.87	5.86	5.86	5.86	5.82	5.81
Dec. 29, 1922	5.87	5.78	5.78	5.73	5.71	5.71	5.70
June 1, 1923	5.80	5.75	5.73	5.71	5.73	5.71	5.70
Dec. 28, 1923	5.72	5.75	5.78	5.73	5.73	5.73	5.71
April 9, 1924	5.75	5.73	5.75	5.73	5.75	5.71	5.65

with ether. This view harmonizes with that previously expressed by Balland (1904a), who found an increase in the organic acid content of aged flour at the expense of the fats.

The effect of several variables, including temperature, moisture content and soundness, on the rate of change in hydrogen ion concentration of stored flours, is developed in the studies reported by Sharp (1924). A long patent flour containing 12.2 per cent of moisture was divided into three portions, and two of these were so manipulated that the moisture content of one was decreased to 5.3 per cent and that of the other was increased to 14.4 per cent, while the moisture content of the third was not modified. Each lot was then divided into three portions, which were stored under as many sets of conditions. Thus one portion of each flour sample was stored out of doors, one was stored in the laboratory at 22° C., while the third was stored in a thermostat at 35° C. The experiment began on December 20, 1923, and extended through one year. All the flour samples were stored in glass bottles closed with paraffined cork stoppers to prevent any changes of moisture during the period of the experiment. At intervals the stoppers were momentarily removed to permit the withdrawal of a 5 gram portion for the hydrogen ion determination.

The data resulting from this study are given in Table 115, and show that moisture content had an important bearing on keeping qualities as evidenced by the changes in degree of acidity. Thus the flours containing 5.3 and 12.2 per cent of moisture remained sound under all three conditions of storage. The sample containing 14.4 per cent of moisture changed only slightly in degree of acidity when stored out of doors during the cool winter months. The following summer, following July 5th, it increased in acidity rapidly, and went out of condition as shown by its odor and appearance. The flour containing 14.4 per cent of moisture stored in the laboratory changed in pH fairly rapidly during the winter months, and went out of condition following May 9th. Flour of the same moisture content stored in a thermostat at 35° changed in pH still more rapidly, and went out of condition following March 14th. From these observations it appears that flour containing 14.4 per cent of moisture cannot be expected to remain sound in storage unless maintained at an abnormally low temperature. It might not spoil in an unheated warehouse during the winter months, but will support the growth and propagation of microörganisms when the temperature is high enough for the development of fungi.

Flour milled from frosted, immature wheat was subjected to a similar study by Sharp. Without reporting the data relating to these samples in detail, it appears that it evidenced a tendency to go out of condition more readily than did flour milled from normal wheat.

TABLE 115

Changes in Hydrogen Ion Concentration of Long Patent Flour Containing Different Percentages of Moisture, and Stored for One Year Under Different Conditions, as Reported by Sharp (1924)

Moisture	Dec. 20	Jan. 17	Feb. 14	Mar. 14	Apr. 11	May 9	July 5	Dec. 20
Outside Temperature								
5.3	6.14	6.12	6.12	6.07	6.25	6.06	6.13	6.06
12.2	6.14	6.13	6.08	5.98	6.26	5.89	6.06	5.99
14.4	6.16	6.07	5.99	6.00	6.14	5.98	5.94	4.72[1]
Room Temperature, About 22° C.								
5.3	6.14	6.10	6.04	5.99	6.05	6.05	6.08	6.04
12.2	6.14	6.01	6.15	6.01	6.05	6.01	6.01	5.98
14.4	6.16	6.01	5.94	5.90	5.82	5.63	4.82[1]	4.82[1]
Thermostat at 35° C.								
5.3	6.14	6.08	5.98	6.05	6.03	6.03	6.09	6.01
12.2	6.14	6.01	5.90	5.91	5.91	5.81	5.92	5.81
14.4	6.16	5.81	5.61	5.48	5.23[1]	4.79[1]	4.86[1]	...

[1] Indicates growth of microörganisms.

THE CHANGES IN FLOUR INCIDENTAL TO AGING

The importance of storing flours at a comparatively low temperature was stressed by Mangels (1924), whose experience indicated that flour stored in an unheated warehouse improved in quality, while that held in the heated laboratory was impaired in baking strength during the same period from December, 1922, to November, 1923.

If the moisture content of flour did not exceed 15 per cent, it could be stored at 18° C. without spoiling, in the opinion of Reimund (1916). Measuring changes in terms of respired carbon dioxid, the respiratory activity of flour containing varying percentages of moisture was as follows:

Moisture, Per Cent	Respired CO_2
15	1 unit
19	100 "
20	200 "
30	1000 "

These data indicate the substantial acceleration of respiration resulting from relatively small increases in moisture content between 15 and 20 per cent.

The inclusion of 2 or 3 per cent of sprouted kernels in the wheat mixture did not appear to unfavorably affect the keeping qualities of the flour so far as is indicated by the determinations of hydrogen ion concentration, titratable acidity, and baking tests reported by Sherwood (1925). This is indicated by certain of Sherwood's data recorded in Table 116.

TABLE 116

CHANGES IN ACIDITY OF FLOURS MILLED FROM WHEAT MIXTURES CONTAINING SPROUTED KERNELS, AS REPORTED BY SHERWOOD

Sprouted Kernels in Wheat, Per Cent	H-ion Concentration as pH			Titratable Acidity as Lactic Acid		
		After 11 Months Storage			After 11 Months Storage	
	When Milled, pH	Stored in Mill, pH	Stored in Refrigerator, pH	When Milled, Per Cent	Stored in Mill, Per Cent	Stored in Refrigerator, Per Cent
0	6.12	5.95	6.05	0.13	0.14	0.14
2	6.13	5.98	6.04	0.13	0.14	0.13
3	6.14	5.98	6.06	0.14	0.14	0.14

The liability of bacteria and molds to propagate on moist flour was emphasized by Bell (1909) and Kuhl (1911). The observations made by Sharp, and referred to in the foregoing paragraphs, indicate the coincident appearance of the characteristic "musty" odor with the sudden changes in pH of the flour. Because of the rôle played by fungi in flour spoilage, efforts have been made to so treat the flour as to inhibit the action of such organisms. Thus in several patents reference is made to methods of treating the flour with ozone, oxides of chlorine, chlorine, bromine and analagous substances (British patent 211, 505, Feb. 16, 1924, granted Soc.Anon.des Rizeries Francaises), and of replacing the air in sealed packages of flour with inert gases, such as nitrogen or carbon dioxid (British patent 203,661, July 5, 1923, granted M. Schoen). In Chapter 9 appears a reference to Dunlap's contention that chlorine treatment inhibits the activity of enzymes which are responsible for the principal changes occurring in stored flour. Cold storage of flour might likewise prove effective in preventing fungi from damaging flour. This means of preserving flour has been stressed by Brandeis (1908), and Balland (1904a).

The comparative changes in moisture content, and acidity resulting from prolonged storage of flour at ordinary temperatures, and in a refrigerator, as traced by Balland, are shown in Table 117. Development of a rancid condition was retarded by refrigeration of the flour, and no increase in acidity occurred. The moisture content of the refrigerated flour increased, in consequence of the high aqueous vapor pressure of the atmosphere to which it was exposed.

TABLE 117

Moisture Content and Acidity of Flour Stored for Three Years (Balland, 1904 a)

Fine Flour Samples	Moisture, Per Cent	Acidity
Original flour	12.83	0.033
Flour stored for 3 years at ordinary temperature	11.70	0.098
Flour held in refrigerator for 3 years	17.30	0.033

The effect of mold upon the oil of corn was traced by Rabak (1920), and his observations are of interest in indicating the possible changes thus effected in the oil of moldy flour. The percentage of fat was found to decrease progressively with mold decomposition. An increase of free fatty acids, soluble acids, hydroxylated acids (acetyl value) and unsaponifiable constituents, was traced, and a decrease in the percentage of volatile acids, insoluble acids and unsaturated acids.

Decomposition in old flour was followed by Norzi (1915), who found that the increase in amino acids was a suitable criterion of spoilage. On dialysing 0.5 grams of flour in sterile water and adding to 5 c.c. of dialysate, 5 drops of 1 per cent ninhydrin solution, a violet color developed on heating this preparation for 2 or 3 minutes. The depth of color was regarded as an index of the degree of decomposition.

The comparatively low moisture content of normal cereal products is responsible for practical inactivity of the microörganisms that are present, in the opinion of Thom and Hunter (1924). The outer, or branny coat of the wheat kernel is always contaminated and frequently infected, the germ is often infected with organisms of several species, but the endosperm is much less commonly infected. In milling the grain, the tendency is for the bulk of the original micro-flora to go with the branny by-products. The products that include chiefly endosperm particles are freer from viable microörganisms than the other mill products. Acidity of cereal products may increase without any multiplication of the microbial flora, in consequence of the activity of flour enzymes. The development of mustiness in cereal products is commonly

TABLE 118

ACIDITY AND BAKING QUALITY OF FLOURS STORED BY SWANSON, WILLARD AND FITZ (1915) UNDER VARIOUS SETS OF CONDITIONS

Condition of Storage	Days Stored	Acidity Dry Basis	Loaf Volume, cc.	Gluten Quality Factor	Baking Tests	
					Texture of Crumb	Color of Crumb
Unbleached Flour						
Sealed jars in heated room	0	.184	1,635	1,918	95	94
	125	.185	2,040	3,827	93	96
	366	.170	1,915	2,783	95	99
Cloth sacks in attic...	0	.184	1,625	1,918	95	94
	125	.184	1,990	3,516	94	96
	366	.168	1,945	2,827	95	100
Cloth sacks in steam-heated room	0	.184	1,635	1,918	95	94
	125	.178	1,990	3,622	94	96
	366	.165	1,930	2,441	96	99
Moderately Bleached Flour						
Cloth sacks in steam-heated room	0	.172	1,610	1,983	95	94
	125	.173	1,980	3,493	96	97
	366	.171	1,875	2,251	96	98
Over-bleached Flour						
Cloth sacks in steam-heated room	0	.176	1,653	2,036	94.5	95
	125	.177	2,020	3,632	95	98
	366	.165	1,865	2,175	96	97

due to the penetrating odor of certain fungi of the Actinomyces group. Moldy odors are the result of the activity of mycelial forms.

Diastatic activity increased during storage in 5 of the 7 flours studied by Baker and Hulton (1908). The proportional increase in these 5 cases was about 50 per cent when diastatic activity was expressed in terms of Lintner degrees.

It appears to be generally conceded that sound flours improve in baking qualities on aging when the results of baking tests are expressed in terms of loaf volume, texture and other units which denote comparative baking strength. Flour was found to improve more rapidly in this respect by Fitz (1910) than did the wheat from which it was milled.

Every flour but one of 15 samples studied by Harcourt (1910) that were stored for 6 months baked into bread of larger size, or with a higher quality rating than at the beginning of the period. The mean increase in "quality of loaf" rating for the 15 samples during 6 months' storage was 1.5 units, on the basis of about 100 units for good bread.

Several experiments were conducted by Swanson, Willard and Fitz (1915) which involved the changes in baking qualities on storage. Typical data resulting from these studies are given in Table 118. Different portions of the same lot of flour, containing 12.98 per cent of moisture (vacuum desiccator method) were stored in sealed jars, and in cloth sacks. Certain of the sacks were placed in an unheated attic, and others in a steam heated room. Baking tests were made at intervals for one year. The results of only three of these baking tests are recorded in the table in the interests of brevity, but these show the general tendency. The loaf volume and "gluten quality factor" increased fairly regularly up to 125 days or more, and then diminished up to 366 days. The baking tests indicated superior "strength" at the end of the year than at the beginning of the period, however.

Flour milled from sound wheat was found by Whitcomb, Day and Blish (1921) to improve in baking quality when stored from February, 1920, until June, 1921. Flour milled from frosted wheat and stored for a like period was impaired in baking quality. These observations are of interest in the light of Sharp's (1924) later findings which showed a tendency toward a more rapid increase in hydrogen ion concentration of the flour milled from frosted wheat than was encountered in sound wheat flour.

Observations made by Saunders (1907, 1910) were extended over a period of nearly 14 years in the instance of certain samples, the data being reported by Saunders, Nichols and Cowan (1921). A more rapid improvement in color and strength was observed when material was kept over in the form of flour for limited periods than when stored as wheat. Samples of flour ground in September, 1907, and baked at intervals until January, 1921, evidenced a general increase in loaf volume and baking strength for 2 to 3 years. This was followed by a decrease in these values, although so far as the data recorded indicate, the baking strength at the end of nearly 12 years' storage was superior to that at the outset. Deterioration was found to likewise occur in flour sooner than in wheat, but under good storage conditions flour could be kept for 10 years. Certain typical data taken from the extensive tables published by Saunders, Nichols and Cowan are recorded in Table 119. It was stated that the dough made from very old samples of flour was quite short, and lacked the stickiness necessary to reach a maximum volume of loaf. Important changes had apparently occurred in the gluten which were responsible for this condition.

Flour was stored at three different temperatures by Neumann

(1909), 6°-17.0°, 10.5°-18.5°, and 17.5°-22.7° C. All improved in baking quality for a period of 12 weeks, the greatest improvement being observed in the flour stored at the highest temperature. Additional observations on aged flour were later reported by Neumann (1911).

TABLE 119

Effects of Storage on the Baking Properties of Flour, Summarized from the Data of Saunders, Nichols and Cowan (1921)

	Date of Test		
	Sept., 1907	Jan., 1910	Jan., 1921
Water absorbed			
Huron wheat flour	58.3	62.8	74.0
Red Fife wheat flour	58.1	63.4	75.1
Yellow Cross wheat flour	56.9	60.9	72.0
Loaf volume, c.c.			
Huron	425	509	377
Red Fife	522	532	417
Yellow Cross	394	530	443
Baking strength			
Huron	86	107	87
Red Fife	98	109	95
Yellow Cross	74	109	100
Color of crumb			
Huron	88	102	102
Red Fife	100	104	102
Yellow Cross	90	104	103

Absorption of flour, or the quantity of water required to produce a dough of standard consistency, increased on aging the flour. In these experiments of Saunders, Nichols and Cowan, recorded in Table 119, the water absorption increased enormously before the end of the experiment. Since the absorption was calculated and recorded on a basis of 10 per cent of moisture in the flour, this increase on aging cannot be attributed to a simple drying out of the flour. Neither is it probable that increasing acidity was entirely responsible for the increased absorption. Some modification of the colloidal condition of the gluten must have occurred with the lapse of time which effected this increased imbibitional capacity. The easiest way to dispose of this hypothesis is to assert that it may be attributed to the action of enzymes in the flour, but even though this be true, the nature of these enzymes and the changes which they effect still remains undetermined.

Should support for the data recorded in Table 119 be required, it will be found in the results of Stockham's (1917) experiments. In these experiments it was observed that fresh flour which absorbed 55.63 per cent of moisture, absorbed 56.85 per cent after 8 months' storage, with the result calculated to a basis of 11 per cent of moisture in both instances. In a comparison of flour milled from normal wheat with flour from sprouted wheat, the former increased in absorption to the extent of 2.9 per cent, while the latter increased only 1.2 per cent when both were stored for the same length of time.

An effort to establish a chemical basis for the improvement in flour on aging was made by Shutt (1911c), but with rather inconclusive results. A slight increase in the nitrogenous compounds soluble in alcohol (gliadin) was noted with the lapse of time, but the other constituents of the flour that were determined remained fairly constant. Marion (1909) noted that the percentage of fat in flour diminished on aging, while the gliadin content showed two maxima, the first after 3 months, and the second after 15 months, following which there was a decrease. The content of crude gluten showed only slight variations up to the 19th month, after which it was found to be difficult to recover the gluten as a coherent mass. In these experiments the flour must have deteriorated much more rapidly than has been the case in similar storage experiments conducted in America, since Marion found that the flour which retained its baking qualities for three months was no longer suitable for baking after 6 to 7 months.

A fantastic but interesting hypothesis to account for the improvement in baking strength of flour on aging was offered by Cobb (1908). It was suggested that the change which flour undergoes after milling is nothing less than the gradual dying of the protoplasmic matter contained in the flour. One of the reasons that fresh flour is more difficult to convert into bread than ripened flour, according to this hypothesis, may be because the yeast cells do not act so readily upon the fresh flour. In other words, the yeast acts upon living flour less readily than upon dead flour, owing to the better defenses of the former. Cobb thus makes of aging an increase in the fermentability; but it seems reasonable to believe that the changes go farther than this and involve modifications in the gluten which may increase the gas retaining power of the dough as well.

An exponent of the hypothesis that increasing hydrogen ion concentration is responsible for progressive improvement in the baking strength of flour during aging is found in Jessen-Hansen (1911). His experiments indicated to him (although no extended data are reported) that flour becomes acid when stored, and that the concentration of fats diminishes simultaneously. Moreover, it was found that old flours, which presumably had increased in hydrogen ion concentration, responded less in terms of baking properties on the addition of acid than did fresh flour. In his discussion of these studies Jessen-Hansen concluded that additional and extended researches would be necessary to dispose of this problem.

That the observed progressive increase in hydrogen ion concentration is the only factor responsible for the improvement in the baking strength of flour on aging appears to the author to be improbable.

If acidity were the cause of improvement it would indeed be fortunate, since in that event all the advantages of natural aging, with perhaps the exception of bleaching, could be achieved by a regulated addition of hydrogen ions. Extended efforts to accomplish this effect by means of graduated additions of various acids have failed to yield the results of natural aging. Other changes seem to be in progress in stored flour, which, however, have thus far eluded detection and measurement. That these may involve diastatic activity is suggested by the work of Baker and Hulton; that a modification of the colloidal condition of the gluten is likewise involved appears probable, but has not yet been established. With the present state of our knowledge it seems unwise to declare that the problem is solved, and further and more detailed studies must be conducted to dispose of it effectively.

Kent-Jones (1924) referred to the changes induced in the flour mill streams on aging them for 5 months. In general the crude gluten tended to diminish; the "maltose" figure was unaltered in the higher grades, but increased slightly in the lower grades; the hydrogen ion concentration increased, and the buffer action increased (note pp. 160-164).

The effect of elevated temperatures on the baking properties of flour is indicated by the experiments of Snyder (1901), who held flour for 4 hours at temperatures of 50°, 70°, and 100° C. On baking these heated flours in comparison with the control or unheated flour, the flour heated at 50° appeared normal, that heated at 70° baked into bread that was nearly normal, while the bread baked from flour heated to 100° was inferior, having a smaller volume and darker color.

Chapter 9.

The Color of Flour and Flour Bleaching.

The color of flour must be considered in connection with the discussion of flour bleaching, since it is this property of flour which bleaching is designed to modify. Dry granular flour owes its color or visual appearance to the joint operation of several factors. First, there is the influence of the granulation, or size of flour particles. Other things being equal, large granules have a darker appearance than small granules. Thus, if relatively coarse middlings be ground or pulverized to varying degrees of fineness, the finest product appears much whiter in the dry condition than the coarse flours. This was demonstrated by Shollenberger, Marshall and Hayes (1912).

This effect of granulation is due to the purely physical phenomenon of the difference in the refraction of light in the dense endosperm structure of which the granule is composed, and in the air which surrounds the particle. Large granules also cast shadows, registering as a gray hue, unless unusual care is taken in insuring that white light rays impinge on the flour surface from several angles. The difference in visual appearance of large and small granules, otherwise identical, disappears when they are kneaded with water into a smooth homogeneous dough. Hence some attention must be given to the effect of granulation upon flour "color" when the examination is confined to dry flour; but granulation is of less significance in examining a flour dough.

Secondly, there are the changes in color which result when such a dough as was mentioned in the foregoing paragraph is allowed to stand in a moist condition in contact with air, and particularly at an elevated temperature. According to Bertrand and Mutermilch (1907a), the joint action of a glutenase and tyrosinase results in the progressive production of a brown pigment in the surface layer of such a dough. These enzymes are chiefly present in the pericarp, or outer layers of the branny coat of the wheat berry, and consequently those doughs which contain the smallest proportions of bran fragments undergo the least change in color. It has been suggested that the measurement of the rate of change in color of wetted flour would serve as an index of grade or refinement. The Pékar[1] test, as commonly employed by flour millers, makes use of the same principle applied in a crude manner.

[1] It is not known when Pékar devised this test. His book, "Weizen und Mehl und Unserer Erde," does not enlighten us (see Maurizio [1917] Vol. I, p. 162).

In this test the flour, smoothly flattened on a slab of glass, metal, or wood, is dipped in water, and the moistened material incubated for 30 to 60 minutes at a temperature of about 50° C. Highly refined flours acquire a creamy tint on such treatment, while the lower grades become distinctly brownish in hue. It is obvious that the color of either a dough or the wetted Pékar or "slick" test, is not the color of the flour, because of these changes in hue which occur in contact with oxygen in the air. A quantitative measurement of the color of such preparations, while of value in classifying flours on the basis of refinement or freedom from branny particles, would not constitute a true measurement of the flour color, nor of the bread which might be prepared therefrom.

The third factor which contributes to the color of both flour and dough is the reddish-brown pigment present in the bran particles derived from "red" wheats. The chemical constitution of this pigment has apparently not been determined as yet. It is found chiefly in the thin residue of the testa of the typical red wheats. Many white wheats are practically devoid of this pigment, and the quantity in the testa of amber durum wheat kernels is comparatively small. Red durum wheats, particularly of the variety known as Pentad, are heavily pigmented. In milling red wheats, the reddish-brown particles of bran effect a tinting of the flour in the proportion that they are present. In highly refined flours, as already developed in the preceding section in which criteria of flour refinement or grade are discussed, the numbers of such fragments are comparatively small. Straight and clear grade flours contain about two and five times as many such particles respectively as the patent grades, as shown by Keenan and Lyons (1920). The lower grades owe their darker color and specky appearance in the dry state in large part to the presence of relatively large numbers of pigmented bran particles.

Dirt and foreign matter, such as soil, fragments of weed seeds, and smut spores, constitute a fourth contribution to the color of flour. Such material is more or less completely removed from the wheat before milling, depending upon the cleaning, scouring and washing system employed. It only requires a trace of the spores of bunt or stinking smut to paint the flour a dark hue. In milling smutty wheat the greatest concentration of spores will be encountered in the first break stream, which should in such cases be incorporated with the offal known as red dog. Unfortunately these spores are not confined to the first break flour, however, and scouring with lime or thorough washing with water should be resorted to in cleaning wheat in which they are present. Soil particles may be removed in a similar manner, and the color of the flour thereby improved. Certain weed seeds have already been mentioned as a source of difficulty in the production of white flour. The seed coat of corn cockle is heavily pigmented, and when mixed with the wheat may become pulverized and contribute black specks to the flour. The same is true, to a lesser degree, in case of the seeds of wild vetch. The latter also contains a yellow pigment which may contribute to the yellowness of flour produced from wheat when mixed with such seeds. The seeds of giant ragweed, known to the grain trade as "kingheads," are a source of difficulty when milled with wheat. As shown by Miller (1915), kingheads mixed with wheat at the time of milling result in a greater lowering of the color score of the flour than equal amounts of seeds of corn cockle or vetch.

Fifth and last of this group of factors contributing to the color of flour are the carotinoid pigments. Monier-Williams (1912) concluded that the yellow pigment of flour was largely carotin. This pigment has been extensively studied by Willstätter and Mieg (1907), who reported its formula to be $C_{40}H_{56}$. Its chemical constitution has not yet been determined. The crystals of carotin melt at 167.5-168° C. These are soluble in ether, petroleum ether, chloroform, carbon disulfid, carbon tetrachloride, ethereal and fatty oils, and oleic acid. In carbon disulfid the solution appears red-orange to blood red in color; in other solvents its solutions are yellow to golden yellow. The crystals are insoluble in absolute alcohol. Carotin is not adsorbed by calcium carbonate, inulin, or powdered sucrose from petroleum ether or carbon disulfid solutions.

Solutions of carotin in oil give a green color reaction with a small crystal of ferric chloride when warmed. The crystals are bleached by oxidation, one-third their weight of oxygen being required to completely decolorize them. A di-iodo addition product of carotin has been described, $C_{40}H_{56}I_2$ (Willstätter and Mieg, 1907), while a colorless bromine substitution derivative $C_{40}H_{36}Br_{22}$ was described by Willstätter and Escher (1910).

Carotin is distinguished by a characteristic absorption spectrum, with three absorption bands in the green and blue parts of the spectrum when projected through a solution of carotin in the ordinary fat solvents.

In the quantitative estimation of carotin by a colorimetric precedure, Willstätter and Stoll (1913) used a 0.25 per cent solution of alizarin in chloroform, or a 0.2 per cent solution of potassium dichromate in water as color standards. The latter was found suitable for ordinary work, and using 5×10^{-5} molar solutions of carotin, equivalent to 0.0268 per cent, the following relations in the depth of color of the two solutions were observed.

```
100 mm. carotin solution = 101 mm. K₂Cr₂O₇ solution
 50 mm.    "       "     =  41 mm.    "       "
 25 mm.    "       "     =  19 mm.    "       "
```

Palmer (1922) suggests that the most accurate estimates of the carotin content of a solution can be made by matching the unknown with standard carotin solutions, 100, 50 and 25 mm. deep, and averaging the readings.

THE COLOR OF FLOUR AND FLOUR BLEACHING

The addition compounds which carotin forms with oxygen and chlorine lack the deep yellow color of the natural pigment. The formation of these addition products in flour accordingly results in bleaching the latter.

The flours examined by Monier-Williams (1912) were found to contain 1.3 to 2.0 parts per million of carotin. That varying quantities of the carotinoid pigments are found in flours milled from different types of wheat was shown by the work of Winton (1911), who determined the relative content by measuring the depth of color of gasoline extracts, using potassium chromate solution (0.005 per cent) as a standard. The relative carotinoid pigment content of the Nebraska hard winter wheat flour that Winton examined was nearly twice that of Missouri and Michigan soft winter wheat flours. Minnesota hard spring wheat flour was intermediate in this particular. This is shown by the data in Table 120. There is less difference in the carotinoid pigment content of patent and clear flours milled from the same wheat than might be anticipated in view of the higher fat content of the latter. This indicates that while carotin may be soluble in fat and in the fat solvents, it is not of necessity associated with the fat in the wheat kernel, nor quantitatively correlated with the percentage of fat.

TABLE 120

GASOLINE COLOR VALUE OF FRESHLY MILLED, UNBLEACHED PATENT AND CLEAR FLOURS PRODUCED FROM FOUR TYPES OF WHEAT (WINTON, 1911)

Wheat Type	Gasoline Color Value		Fat, Per Cent	
	Patent Flour	Clear Flour	Patent Flour	Clear Flour
Minnesota hard spring	2.00	2.00	1.09	1.98
Nebraska hard winter	2.63	2.50	.85	1.32
Michigan soft winter	1.43	1.61	1.11	1.77
Missouri soft winter	1.47	1.60	.87	1.15

The gasoline color values of straight flours milled from the several market classes of wheat have recently been determined by D. A. Coleman, of the U. S. Department of Agriculture, Bureau of Agricultural Economics, who has kindly permitted the use of his data. These are recorded in Table 121, and show that the flour milled from the hard red winter wheats had the highest gasoline color value, followed closely by the soft red winter wheat flour. The soft white wheat flour contained the least yellow pigment of the several classes examined. The gasoline color value of durum wheats, as shown by Clark (1924) ranged around 1.5 to 1.6 in the best varieties.

The effect of removing the carotinoid pigments with fat solvents upon the color of flour was shown by Stockham (1920) who reported a color score of 96 for the crumb of bread baked from normal flour, and

104 for the bread baked from the same flour after it had been extracted with ether.

It was stated by Coward (1924) that no detectable lipochrome (including carotin and xanthophylls) could be found in the grain of a variety of English wheat known as Carter's "Red Stand-up," even on saponifying and extracting 100 grams. This suggests that some varieties of wheat may be devoid of yellow pigments, and would accordingly yield a dead-white flour.

TABLE 121

GASOLINE COLOR VALUE OF STRAIGHT GRADE FLOUR AS DETERMINED BY COLEMAN

Class of Wheat	Number of Samples	Gasoline Color Value
Hard red spring	18	1.39
Hard red winter	22	1.69
Soft red winter	10	1.67
Hard white	8	1.41
Soft white	5	1.13

Palmer (1922) identified xanthophyll in wheat bran, and in a private communication stated that in the specimen of bran in question xanthophyll constituted the major and carotin the minor fraction of the yellow pigments.

Flour is in such finely divided condition, and the bulk includes such a large proportion (about one-half) of air, that conditions are favorable for the spontaneous oxidation of carotin in the flour, resulting in natural bleaching. The progress of this change was traced by Winton (1911) in the case of the eight flours referred to in the preceding paragraph, and his data are recorded in Table 123. These data show that after 20 weeks of storage the gasoline color value of the unbleached or natural flours diminished to about one-half that of the freshly milled flours.

TABLE 122

COLORING MATERIAL OF NATURALLY AGED FLOUR, AS REPORTED BY MONIER-WILLIAMS (1912)

Flour Treatment	Coloring Matter in Parts per Million
Freshly milled	2.00
Stored in closed tin for 2 months	1.40
Exposed in office " " "	1.12
Exposed on roof " " "	1.12

Carotin content of flour was found by Monier-Williams (1912) to diminish when stored for two months. The flour exposed to the air changed somewhat more than the flour that was stored in a closed tin. This is evident from the data recorded in Table 122.

That natural bleaching of stored flour is due to oxidation of the carotin is suggested by an experiment conducted by Kent-Jones (1924, note pp. 175-176). Flour stored in a bag for two months bleached in the usual manner, while flour kept in a vacuum did not bleach at all, and flour exposed to hydrogen (in the absence of air) was scarcely changed.

The rate of natural bleaching was hastened by exposing unbleached flour to sunlight, in the experiments conducted by Shutt (1911b). Layers one-fourth inch thick placed between glass plates and exposed to direct sunlight showed a bleached appearance at the end of one hour, and still more at the end of three hours. Flour exposed to the atmosphere in the dark showed a light bleach; that exposed to both air and sunlight bleached still more. This effect of direct sunlight was also noted by Avery (1907).

A similar altering of the color of flour pigments may be effected by chemical treatment or bleaching. As early as 1879 Beans was granted British patent No. 2502 covering the use of chlorine as a bleaching agent. Commercial application of bleaching practice apparently dates from the granting of patents to Frichot (French patent No. 277,751, and British patent No. 21,971) in 1898. Frichot referred to the use of the flaming arc, and the production of ozone by this means. In this and other "ozone" generators it is probable that the active agent was nitrogen peroxide. The Andrews patents, to which reference will later be made, involve the use of nitrogen peroxide resulting from a chemical reaction.

Small scale bleaching experiments were conducted by Snyder (1904a) before commercial bleaching became common in America. Oxygen generated from potassium chlorate appeared to whiten the flour; later Snyder suggested that the bleaching action observed in this experiment may have been due to chlorine present as an impurity in the unwashed oxygen. What must have been an extensive study of a variety of bleaching agents was reported by Avery (1907). Ozone was believed to have been without effect except when mixed with nitrogen peroxide as an impurity. Carbon dioxide was without effect on color, as was oxygen, even at a temperature of 98° C. Sulfur dioxide bleached slowly, and a large excess was required, which conferred a pronounced odor.[1] Hydrogen peroxide solution shaken with a benzine extract of flour did not bleach the flour pigments dissolved in the latter, although such extracts were bleached with solutions of nitrogen peroxide, chlorine, bromine and sulfur dioxide. Bromine (gas) bleached effectively, 4 c.c. at 25° C. producing a noticeable bleach, while 150 c.c.

[1] Author's note. It is also probable that sulfur dioxide would increase the hydrogen ion concentration of flour treated with it, to the extent that the gluten would be dispersed in the dough, and the resulting bread would be of poor quality. Activity of yeast in such a dough would also be unfavorably affected, owing to the toxicity of sulfurous acid.

TABLE 123

Effect of Variety, Grade, Aging and Bleaching on the Gasoline Color Value (Winton, 1911)

Description of Samples	Minnesota Hard Spring		Nebraska Hard Winter		Michigan Soft Winter		Missouri Soft Winter	
	78 Per Cent Patent	22 Per Cent Clear	80 Per Cent Patent	20 Per Cent Clear	80 Per Cent Patent	20 Per Cent Clear	40 Per Cent Patent	60 Per Cent Clear
Unbleached:								
New (February)	2.00	2.00	2.63	2.50	1.43	1.61	1.47	1.60
Aged 10 weeks	1.78	1.82	2.12	2.17	1.22	1.49	1.22	1.33
Aged 20 weeks	1.20	1.34	1.36	1.68	.80	1.20	.68	.88
Aged 30 weeks	.72	.88	.70	.82	.56	.72	.48	.52
Bleached:[1]								
New (February)	.60	.66	.80	.80	.40	.50	.32	.40
Aged 10 weeks	.44	.54	.46	.48	.20	.38	.22	.26
Aged 20 weeks	.30	.50	.34	.40	.20	.36	.18	.24
Aged 30 weeks	.30	.50	.24	.36	.18	.40	.14	.16

[1] All the samples contained, when freshly bleached, approximately 2 parts of nitrous nitrogen per million. After aging 30 weeks nearly all the nitrous nitrogen had disappeared.

per kilo of flour produced the maximum effect. Chlorine acted much like bromine. A unit quantity of nitrogen peroxide bleached more flour than any other reagent studied.

Nitrogen peroxide, nitrosyl chloride, chlorine and mixtures of nitrosyl chloride and chlorine, nitrogen trichloride, and certain organic peroxides, notably benzoyl peroxide, have been extensively used in commercial flour bleaching. These have been enumerated in about the chronological sequence of their industrial application, and will be discussed in the same order.

Nitrogen peroxide was the first agent used extensively in the bleaching of flour in America. Introduced in England early in the century under the name of the Andrews' process, it soon made its appearance in the United States, where it became known as the Alsop process. The patent granted John Andrews and Sidney Andrews (U. S. Patent No. 693,207) covered the production of nitrogen peroxide by the reaction between nitric acid and ferrous sulfate. In the patents granted James N. Alsop (U. S. Patents Nos. 758,883 and 758,884, dated May 3, 1904, and No. 759,651, dated May 10, 1904) reference is made to the flaming electric arc as a means of producing the bleaching agent. Various types of generators have been used in producing nitrogen peroxide by means of the flaming arc. In the simplest and earliest form the two electrodes were housed in a metal tube, and by means of an eccentric one electrode was caused to wipe against the face of a stationary electrode, and as they were drawn apart an electric current of high potential from a step-up transformer or coil was caused to flow. This occasioned a flaming arc to play between the electrodes until the arc was extinguished because it could no longer jump the gap of increasing width. At the high temperature of the arc atmospheric nitrogen was oxidized to NO, and this in turn to NO_2 and N_2O_4, and these in turn were removed in the current of air which was swept through the generator tube.

Hulett (Notice of judgment No. 722, p. 45) sampled the gas produced by a 500-watt direct-current generator of this type, and at a time when it was drawing about 4 amperes the discharge mixture contained 300 parts of nitrogen peroxide in a million parts of air. Snyder (1908a) analyzed the gas from a similar generator and found an average of 0.205 milligrams of nitrogen as nitrite per liter. This is equivalent to 574 parts of nitrogen peroxide per million of air. It is obvious that the concentration of NO_2 in the air leaving the generator must have been variable. To reduce the bleaching gas mixture to a uniform concentration of its active constituent, as well as to cool the mixture, it was a common practice to pump the mixture of nitrogen peroxide and air into a large tank or reservoir, from which it was blown into the agitator and mixed with the flour. The temperature of the mixture was found to be of importance, since the temperature equilibrium of $NO_2 \rightleftarrows N_2O_4$ shifts to the right with reducing temperature,[2] and the N_2O_4 appears to be more active in bleaching carotin than is NO_2. Cooling the gas mixture is accordingly advantageous in securing the maximum effect from the oxides of nitrogen produced by this method.

Another type of generator was developed later, in which a number of stationary electrodes were mounted on the periphery of a circle, and several rotating electrodes made contact and "drew" arcs in such rapid succession as to give the appearance of a practically continuous circle of flaming arc. These electrodes were mounted in a housing, through which air was circulated and the resulting mixture of air and nitrogen peroxide was cooled, and then blown into the flour agitator. This type of generator was doubtless more efficient than the earlier type, in so far as current consumption was concerned, and also tended to produce a more uniform concentration of nitrogen peroxide in the gaseous discharge.

Still more recently a third form of electrical generator has been developed. This involves a step-up transformer which delivers electric current at a potential of about 10,000 volts. The current consumption or amperage is controlled by modifying the interval in the gap of the

---

[2] Temperature equilibrium of $NO_2 \rightleftarrows N_2O_4$, according to Bodenstein (1922), follows the equation:

$$\log \frac{p^2 (NO_2)}{p^2 (N_2O_4)} = \log Kp = \frac{-2692}{T} + 1.75 \log T + 0.00483\ T\quad 7.144 \times 10^{-6}\ T^2 + 3.062.$$

In this equilibrium equation p is measured in atmospheres.

movable core of a choke coil. At this high potential, and with a suitable current or amperage, a flaming arc can be produced between two stationary wire electrodes, the gap between which is approximately ¼ inch. These electrodes are mounted inside of an aluminum dome,

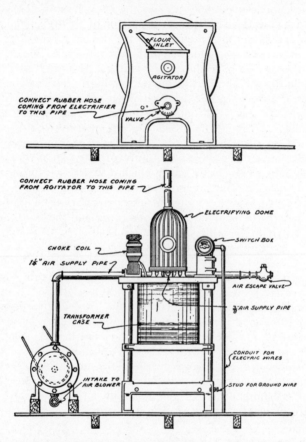

Fig. 17.—Alsop continuous electrifier or nitrogen peroxide generator. (Dedrick, "Practical Milling.")

Reproduced by permission of the National Miller, Chicago.

closed at the bottom except for the air inlet, and discharging at the top into the pipe leading to the flour agitator. A pump provides a constant movement of air through this dome which sweeps out the nitrogen peroxide, and by extending the length of the discharge pipe to 20 feet or more, the temperature of the gas mixture can be reduced to approximately that of the atmosphere of the mill before blowing the gases into the flour agitator. An illustration of such a generator will be found in Figure 17.

In addition to the flaming arc generator, a number of other methods have been used in producing nitrogen peroxide for flour bleaching. Mention has already been made of the original Andrews patent covering the reaction between nitric acid and ferrous sulfate. The reaction between nitric acid and certain metals, notably iron and copper, has been used in industrial practice. U. S. Patent No. 832,372, granted Gerrard and Naylor (Oct. 2, 1906), covered a machine for feeding the metal in form of wire into the acid container, and removing the oxides of nitrogen thus produced. Recently a simpler form of generator has appeared, in which nitric acid is caused to drip on metallic copper, and the nitrogen peroxide formed in an excess of air is blown out of the generator and into the flour agitator. Electrolysis of nitric acid has also been employed to a limited extent. In U. S. Patent No. 769,522, Williams (Sept. 6, 1904) describes an electrolytic cell for use in producing flour bleaching gases, although nitric acid or oxides of nitrogen are not referred to in the patent claims. Later, in U. S. Patent No. 963,970, Williams (July 12, 1910) refers particularly to the use of nitric acid, 36° Baumé, in an improved form of cell. Oxidation of anhydrous ammonia constituted another method of producing nitrogen peroxide. This process is referred to in U. S. Patent No. 812,777, granted to E. E. Werner (Feb. 13, 1906), but has not been used extensively in commercial operations. Of all these processes, the flaming arc and the decomposition of nitric acid with a metal have been most widely used in American flour mills.

Several reactions may be made use of in the laboratory in producing small quantities of nitrogen peroxide for experimental studies. In a closed vessel, free from oxygen or atmospheric air, these yield NO, in which form the gas is measured for addition to the flour. When the NO is passed into the vessel containing the flour and a large volume of air, the NO is oxidized to nitrogen peroxide. Hale (1910) produced NO through the reaction of double normal solutions of sodium nitrite, sulfuric acid, and potassium iodide, the reaction proceeding as follows:

(a) $2\ NaNO_2 + H_2SO_4 = Na_2SO_4 + 2\ HNO_2$
(b) $2KI + H_2SO_4 = K_2SO_4 + 2\ HI$
(c) $2HNO_2 + 2HI = 2\ H_2O + 2I + 2NO$

A method employed by A. S. Mitchell (privately communicated) involved the following reaction:

$K_8Fe_2(CN)_{12} + 2NaNO_3 + 4CH_3COOH = K_6Fe_2(CN)_{12} + 2\ CH_3COOK + 2\ CH_3COONa + 2NO_2 + 2H_2O$. A little more than the molecular proportion of $K_8Fe_2(CN)_{12}$ was placed in the Kipp generator, and 20 per cent solution of acetic acid was used.

Monier-Williams (1911) produced NO for flour bleaching by heating a solution of ferrous sulfate which had previously been saturated with the gas evolved by treating nitric acid with copper. The gas thus liberated was found to contain 99 per cent of NO. The latter, on dilution with air, spontaneously oxidized to nitrogen peroxide.

In applying the nitrogen peroxide, or any gaseous bleaching agent, to the flour, an agitator of some sort must be available. Various ingenious mechanical devices have been invented for this purpose. It is beyond the scope of this treatment of the subject to discuss the mechanical principles involved in the construction of these devices. It should be indicated, however, that the purpose of each is to throw the flour into a cloud in the agitator, or spread it out in a thin sheet. In this condition the bleaching gas is brought into contact with each flour particle, and the action of the gas on the carotin is immediately evident.

To secure the bleaching effect of nitrogen peroxide upon the carotin of the flour in a dough, a process was patented by Lunt (U. S. Patent No. 1,143,413, June 15, 1915), in which a mixture of nitrogen peroxide and air in suitable proportions is to be blown through the dough in the process of mixing. While the dough may be given this treatment during the initial mixing process, it is stated that the "Permeation is preferably caused to take place at a period subsequent to the formation of the dough in the mixer, and preferably after the lapse of a part of the fermentation. . . ." Apparatus is described by means of which this "permeation" may be accomplished.

The results of numerous investigations indicate that in the ordinary bleaching of dry flour a large portion of the nitrogen peroxide passed into the flour agitator is absorbed by the flour. Only a part of that absorbed could be expected to appear in the form of nitrite reacting material, however, since it is assumed that, apart from any other reactions, half of the nitrogen of the nitrogen peroxide would produce nitrous acid or nitrites in solution in the moisture of the flour, while the other half would yield nitric acid or nitrates. Alway (1907a) found that when varying quantities of nitrogen peroxide were added to flour, the quantity of nitrite found in the freshly treated flour was approximately proportional to the quantity of nitrogen peroxide used, within the limits of ordinary treatment. It was computed by Alway that an average of about 5 cc. of nitrogen peroxide per kilo of flour had been used in bleaching the Nebraska flours examined by him. In the bleached samples of all grades of flours that he examined the average quantity of nitrite, expressed as sodium nitrite, was 6.3 parts per million. The average by grades was: patent, 5.8; straight, 5.1; and baker's, 8.4 p.p.m.

It was observed by Snyder (1908a) that about 96 per cent of the nitrogen peroxide added to the atmosphere in contact with flour could be accounted for, either in the flour (86-88 per cent) or in the air over the flour (8-10 per cent). Flour was effectively bleached with 5 cc. of nitric oxide (which spontaneously oxidized to nitrogen peroxide when mixed with atmospheric air) per kilo of flour. Increasing the dosage of nitrogen peroxide resulted in a regular increase in the nitrite nitrogen of the flour up to 50 cc. of gas per kilo. When more than the latter amount of gas was used the nitrite nitrogen of the flour was not found to increase at the same proportional rate.

The testimony of several Government witnesses on the Kansas City (1910) bleached flour case (U. S. Dept. of Agr., Notice of Judgment No. 722) is of interest in indicating the conclusions which they had reached relative to the ratio between the quantity of nitrogen peroxide used in treating flours and the quantity of nitrites appearing in the flour. Winton testified that about one-fifth of the nitrogen peroxide used in bleaching flour appears in the form of recoverable nitrites. Mitchell computed that in bleaching the flour involved in the case, containing 2.3 parts per million of nitrogen as nitrites (equivalent to 7.5 p.p.m. as NO_2), there was required 30 cc. of nitrogen peroxide per kilo of flour. Shepard testified that he had been able to recover from flour as nitrite nitrogen only 10 to 14 per cent of that used in the bleaching reagent. It was suggested that the nitrites oxidized spontaneously to nitrates, thus causing a gradual disappearance of the former. Hulett presented the results of treating flour with varying concentrations of nitrogen peroxide in mixture with air, and found the proportion of the nitrogen peroxide in the residual air in contact with the treated flour increased as the concentration of nitrogen peroxide was increased. This is shown by the data in Table 124.

TABLE 124

NITROGEN PEROXIDE IN PARTS PER MILLION IN THE MIXTURES WITH AIR BEFORE AND AFTER CONTACT WITH FLOUR (HULETT'S TESTIMONY)

Concentration in the Mixture with Air	
Before Treating the Flour	After Contact with the Flour
18.6	0.6
28.4	1.4
74.0	5.0

Hulett also testified (N. J. 722) that when 500 grams of the bleached flour involved in the controversy was placed in a flask, the latter evacuated, and the vapors which distilled off at room temperature were condensed, the distillate was found to contain nitrous acid. Bleached flour was also observed to give up nitrogen peroxide to the atmosphere in contact with it. These observations suggest that a part at least of the nitrite reacting material which can be detected in flour exists either in aqueous solution as nitrous acid, or as gaseous nitrogen peroxide in the atmosphere held between the flour particles.

Hamill (1911) decided, however, that the quantity of nitrite-reacting material in flour is not a criterion of the quantity of nitrogen peroxide used in bleaching the flour.

Varying quantities of nitrogen peroxide were added to unit quantities of flour by Monier-Williams (1911), and the proportion absorbed was determined. As shown by the data in Table 125, a larger percentage of the added nitrogen peroxide was absorbed by the flour when large dosages were employed than with the light treatments such as are applied in commercial practice. Thus only 74 per cent of the 10 cc. per kilo treatment was absorbed, while 92 per cent was absorbed when 100 cc. of nitrogen peroxide was applied.

TABLE 125

NITROGEN PEROXIDE ABSORBED BY FLOUR IN CONTACT WITH VARYING QUANTITIES OF THE GAS (MONIER-WILLIAMS, 1911)

Nitrogen Peroxide Introduced per Kilo of Flour,		Nitrogen Peroxide Absorbed per Kilo of Flour,
cc.	Grams	Grams
5	.0027	.0011
10	.0054	.0040
15	.0082	.0064
20	.0109	.0083
30	.0163	.0133
40	.0218	.0192
60	.0327	.0277
100	.0544	.0502
160	.0871	.0841
230	.1252	.1182
300	.1638	.1579

About 1 milligram of nitrite (as sodium nitrite) will be added to flour by treating it with 1 cc. of nitrogen peroxide, according to Sheringa (1916). Unbleached flour was found to contain ordinarily less than 1 milligram of nitrite per kilo.

It is generally agreed that the nitrite-reacting material in flour tends to gradually disappear when the flour is stored in an atmosphere which is free from nitrogen peroxide. This was mentioned by Winton (1911), who found that flours containing 2 parts per million of nitrous nitrogen when freshly bleached failed to react when tested for nitrites after being stored for 30 weeks. The flours treated with varying proportions of nitrogen peroxide by Monier-Williams (1911) were held in storage, and it was found that with flours treated with as much as 20 cc. of nitrogen peroxide per kilo there was no appreciable decrease in their nitrite content after 20 days' storage. After 62 days' storage all the flours treated with 15 cc. or more of nitrogen peroxide per kilo lost a substantial fraction of the nitrite-reacting material which they contained, the fraction increasing with the dosage of bleaching gas. This is shown by Monier-Williams data recorded in Table 126.

TABLE 126

Change in Nitrite Content of Flour When Stored (Monier-Williams, 1911)

Nitrogen Peroxide Used in Bleaching Each Kilo of Flour, cc.	Nitrogen Peroxide Absorbed per Kilo of Flour, Grams	Nitrites (as $NaNO_2$ Parts per Million of Flour)		
		1 Hour After Bleaching	20 Days After Bleaching	62 Days After Bleaching
0	.0	0	0	0
5	.0011	4	3	3
10	.0040	6	5	5
15	.0064	10	7	6
20	.0083	12.5	12	8
30	.0133	20	20	11
40	.0192	30	30	15
60	.0277	45	42	17.5
100	.0502	90	63	25
160	.0841	150	92	25
230	.1182	225	100	20
300	.1579	330	103	17.5

The reduction in the nitrite content of moderately bleached and heavily bleached flours was determined by Swanson, Willard and Fitz (1915). After storing these flours for a year an appreciable reduction in nitrites was observed, as shown in Table 127.

TABLE 127

Nitrites in Bleached Flour When Freshly Bleached, and After Being Stored for 366 Days (Swanson, Willard and Fitz, 1915)

	Nitrites in Parts per Million	
	Moderately Bleached	Over-Bleached
Fresh flour	2.0	4.0
After 366 days	1.5	2.0

When flour freshly treated with nitrogen peroxide was heated to and maintained at a temperature of 85-95° C. for 4 hours by Snyder (1908a), it was found that 66 per cent of the nitrite nitrogen was lost. The nitrite-reacting material was found by Wesener and Teller (1907) to disappear over night when bleached flour was maintained at a temperature in excess of 100° C.

No nitrite was detected by Alway (1907b) in any fresh, unbleached flour, nor did any such flour subsequently acquire a nitrite-reacting material when exposed to the gases or vapors evolved from bleached flour. It was accordingly concluded that there was no nitrous acid in bleached flour. When unbleached flour was exposed for a time to the atmosphere of the laboratory it was found to react positively when subjected to the test for nitrites.

Data was presented by Wesener and Teller (1907) showing small quantities of "nitrous anhydride" in many unbleached flours, the quantity being less than was found in certain bleached flours. Flour exposed to ordinary air was observed by Wesener and Teller (1911) to take up the oxides of nitrogen. As high as 4 parts of nitrite nitrogen per million of flour was found in flour thus treated. Flours which gave no reaction for nitrites were exposed to the atmosphere in various ways and were analyzed by McGill (1910) and found to acquire a nitrite reaction. This is shown by the data in Table 128, which also indicates the treatment accorded each sample.

TABLE 128

Nitrites Found by McGill (1910) in Unbleached Flour After Exposure to the Atmosphere in Various Ways

Flour Treatment	Nitrite Content in Parts per Million
Original flour	0
After exposure for 10 days on roof of building	1.06
After exposure for 25 days on roof of building	3.00
After exposure for 4 days in laboratory	5.20
Under bell jar with bleached flour for 12 days	0.50

Two unbleached flour samples were found by Shutt (1911b) to contain .09, and .40 parts per million respectively of nitrite nitrogen, while in certain other samples no nitrite was detected. Nitrites in a pound of unbleached flour equivalent to the quantity of nitrogen peroxide in 5 cubic feet of air was detected by Thomson (1914). The range in nitrite content of unbleached flour samples which he examined was from .10 to .62 parts per million (as sodium nitrite). A sample of such flour, containing .34 p.p.m. of nitrites was found, after 34 days, to contain 3.12 p.p.m. in the surface layer, and an average of .83 p.p.m. in the contents of the entire bag.

Thus it appears that not only may flour treated with nitrogen peroxide actually fail to react when tested for nitrites, or at least diminish in the intensity of reaction after a period of storage, but, on the other hand, unbleached flour, may react positively when subjected to the test after prolonged exposure to an atmosphere containing nitrogen peroxide.

It appears reasonable to conclude that the concentration of nitrite reacting material in an untreated flour will be greatest in the surface layers, and that the center of large packages will give a negative reaction even after exposure to an atmosphere containing nitrogen peroxide. This complicates the identification of bleached flours, and necessitates the careful handling of all samples to be tested, as well as making

THE COLOR OF FLOUR AND FLOUR BLEACHING

it imperative that such tests be applied within a few days after treatment.

It is also evident that the nitrite-reacting material in a bleached flour largely disappears when the flour is converted into bread. Bread baked from the flours with which Alway (1907b) experimented was found to react positively, when tested for nitrites, in the case of only about half the samples studied. In these instances the quantity found in the bread was not correlated with the quantity in the flour. Alway (1907a) says that the quantity of nitrites in the bread was invariably less than that in the flour, the ratio apparently depending upon the method employed in making the bread. Bread was found by Ladd (1908a) to contain about one-third as much nitrite-reacting material as did the flour from which it was baked, an observation confirmed by Ladd and Bassett (1909).

Bread baked from flour containing .4 parts per million of nitrite nitrogen was found by Snyder (1908a) to give no nitrite reaction when the baking was done away from contact with flue gases. Harcourt (1911) likewise failed to find nitrites in bread baked from flours that had been bleached with nitrogen peroxide. A sample of flour which had been analyzed by Shutt (1911b) and found to contain .90 parts per million of nitrite nitrogen was subjected to two series of baking trials. In one instance the resulting bread contained .109, and in the other .15 p.p.m. of nitrite nitrogen. Other flours containing less nitrite nitrogen were converted into bread which was free from nitrites.

Flours which had been bleached by Monier-Williams (1912) with varying quantities of nitrogen peroxide, were converted into bread. An unbleached or control flour, which reacted faintly when tested for nitrites, being reported as containing .5 parts per million of sodium nitrite, yielded a bread containing more nitrite than the flour, namely 1.2 p.p.m. In the case of 4 bleached flours, the loaves baked contained less nitrite than the raw flour. The data resulting from this study will be found in Table 129.

TABLE 129

Nitrites Found in Flour and in Bread Baked Therefrom, as Reported by Monier-Williams (1912)

Nitrogen Peroxide Used in Bleaching Each Kilo of Flour	Sodium Nitrite in Parts per Million	
	In Flour	In Bread
0	0.5	1.2
10	11.0	8.0
40	28.0	17.5
100	68.0	42.0
300	158.0	75.0

Purified carotin readily absorbed a quantity of nitrogen peroxide equivalent to the weight of the carotin, according to Monier-Williams (1912). Continuing the reaction period for several hours resulted in further increases in the weight of nitrogen peroxide absorbed, 0.2 gram of carotin absorbing 0.268 gram of nitrogen peroxide. The treated carotin, on analysis by the Kjeldahl method, was found to contain about 10 per cent of nitrogen. Carotin in solution in petroleum ether was immediately decolorized by nitrogen peroxide, yielding a derivative insoluble in petroleum ether. The resulting substance reacted positively with the Griess-Ilosvay reagent. It would thus appear that the nitrogen peroxide bleached by forming an addition product, which reacts as a nitrite. Monier-Williams calculated that about one-fourth of the nitrite-reacting material in bleached flour is in this combination. It had previously been contended by Wesener and Teller that all the nitrite-reacting material of flour was in the form of this nitro-carotin derivative, but this view is hardly tenable in the light of Hulett's findings concerning the distillation of nitrous acid from bleached flour at room temperature.

A decrease in the iodine value of flour fat was effected by treatment of the flour with nitrogen peroxide, according to Fleurent (1906). The iodine value of fat extracted from bleached flour after 9 months' storage was found by Ladd (1908a) to be 84.1, while that of the fat of unbleached flour stored for a like period was 101.2. Qualitative evidence of combined nitrogen in the fat of these bleached flours was obtained. These findings were further emphasized by Ladd and Bassett (1909). Flour was bleached with 40 cc. of nitrogen peroxide per kilo by Gill (Notice of Judgment No. 382), and the fat was then extracted. It was found that the olein of the fat had been converted into a solid isomer, elaidin, from which elaidic acid was isolated and identified. Free fatty acids (calculated as oleic) of fat from unbleached flour was 5 per cent, and that of fat from bleached flour was 8.5 per cent. The iodine number of unbleached flour fat was 101.5, and of the bleached flour fat 98. The nitrite nitrogen content of the bleached flour fat was found to be equivalent to 1.2 milligrams per kilo of fat. This nitrite nitrogen could be removed from the fat by washing it with a dilute solution of potassium hydroxide. Avery (1907) also found that the crude fat extracted from flour with benzine reacted positively with the Griess reagent.

Fats extracted from bleached flour failed to give a positive test for nitrites when examined by Acree (Notice of Judgment No. 722), however. No modification of the iodine number of flour fat on commercially bleaching the flour could be detected by Snyder (1908a), nor could he find that the nitrogen content of the fat was altered by such treatment. When flour was bleached by Moore and Wilson (1914) with varying dosages of nitrogen peroxide, modifications of the fat were effected. With increasing dosages of nitrogen peroxide, as shown in Table 130, the percentage of nitrogen in the fat increased, the iodine value decreased, and the saponification value decreased at first and then increased because of the nitrous acid splitting off and combining with the alkali used in saponifying the fat.

A slight increase in the nitrogen content of fat from bleached flour was detected by Monier-Williams (1911). No substantial changes in the other fat constants, with the exception of color, was found except when the dosage of nitrogen peroxide was very heavy.

TABLE 130

Constants of Fat Extracted from Flour Bleached With Nitrogen Peroxide (Moore and Wilson, 1914)

Nitrogen Peroxide Used per Kilo of Flour, cc.	Nitrogen in Flour Fat, Per Cent	Iodine Value	Saponification Value	Free Fatty Acids, Per Cent
0	.68	111.4	200	10.8
20	.74	107.4	174	9.9
30	.82	96.5	155	8.5
100	.97	85.8	162	7.6
4500	1.26	82.5	245	28.5

The seeming lack of agreement in these observations on the changes in the characteristics of the fat of flour effected by treatment of the flour with nitrogen peroxide can probably be explained by the varying dosages of the bleaching reagent employed. In those instances in which the dosage was heavy, a substantial change in the fat constants and properties could be detected. When dosages similar to those employed in commercial practice were used, the changes in fat properties were probably within the range of the experimental error of the methods followed, and negative findings resulted. It may be assumed that if a heavy treatment substantially modifies the fat, a lighter treatment would occasion similar changes but of lesser degree. In the latter case, the changes would probably be too small to prove of significance in altering the general properties of the fat present in the flour so treated.

Owing to the complex nature of gluten, it has not been possible to trace quantitatively any modifications in its composition or properties which may result from treating flour with nitrogen peroxide. This has resulted in no little speculation concerning the manner in which gluten reacts with this bleaching reagent. The situation is further complicated by the fact that such small quantities of the reagent are used in treating flour.

No change in the gluten could be detected by Balland (1904b) when it was separated from flour treated with "electrified air," except that it appeared lighter in color. Crude gluten separated from bleached flour by Ladd and Stallings (1906) was tested with the Foster gluten tester and found to be inferior to that separated from unbleached flour. It was later suggested by Ladd (1908a) that a diazo or similar reaction occurred when nitrogen peroxide and gluten were in contact, as shown by the liberation of elemental nitrogen on treating bleached flour with an acid, an observation which was again referred to by Ladd and Bassett (1909). No modification of the gluten by ordinary bleaching could be detected by Snyder (1908a), and it was contended that a concentration of nitrogen peroxide at least 200 times that employed in commercial bleaching would be required to form nitro-substitution products of flour protein. Mann (Notice of Judgment No. 722) testified that the physical properties of gluten were altered when flour was bleached, and Moore and Wilson (1914) arrived at the same conclusion, although the effects could not be detected with small dosages of nitrogen peroxide. Folin (Notice of Judgment No. 722) ventured the opinion that it is inevitable that nitro bodies will be formed when flour is treated with nitrogen peroxide.

No evolution of free nitrogen from flour treated with nitrogen peroxide and then acidulated could be detected by Monier-Williams (1911), and a negative test for diazo compounds was given with an alkaline solution of beta naphthol and salicylic acid. A slight change in the percentage of soluble nitrogen in extracts of flours bleached with commercial dosages of nitrogen peroxide was observed, while with large dosages the soluble nitrogen was increased.

Flour has been removed from old style agitators in which it has remained undisturbed and subjected to the long continued action of nitrogen peroxide, which had all the appearances of flour sprayed with nitric acid. It was yellow in color, and would be characterized as having responded positively to the xantho proteic reaction. No such condition results from the ordinary momentary bleaching treatment, however, and the evidence relating to the baking qualities of bleached flour presented in one of the following paragraphs points to the conclusion that the gluten has been injured little if at all by the commercial treatment with nitrogen peroxide.

No substantial modification of the starch of flour in consequence of treating flour with nitrogen peroxide has been detected. Mann (Notice of Judgment No. 722) testified that the effect on starch was small, while Ladd and Bassett (1909) stated that no effect was observed.

It might be assumed that the use of any reagent which normally yields acids in aqueous solution should tend to increase the acidity of flour which has been treated with it. Yet the dosage of nitrogen peroxide used in bleaching is so small that no measurable increase in acidity of flour results from ordinary treatment. Balland (1904b) found only a slight increase in acidity in consequence of bleaching, and it is probable that the dosages used by him were fairly heavy. Brahm (1904) reported an acidity of .171 per cent in unbleached and .393 per cent in flour treated with nitrogen peroxide; but it is probable that the treatment was excessive, since the bread made from the treated flour was yellow in color. There was no difference in the titratable acidity of water extracts of natural or unbleached flours, and flours bleached with nitrogen peroxide in the series examined by Alway (1907b). Large quantities of the bleaching reagent increased the acidity, according to Alway and Pinckney (1908), but such treatments were excessive compared with commercial practice. The difference in the acidity of lightly bleached and heavily bleached flours examined by Winton and Hansen (1912) was slight. Peterson (1921) found the hydrogen ion concentration to be unaltered after bleaching flour with a normal dosage of nitrogen peroxide, while Weaver (1925) observed a small increase in hydrogen ion concentration.

Numerous investigations have resulted in the accumulation of data which indicate that, aside from the change in visual appearance or color, bleaching with nitrogen peroxide has little measurable effect upon the general baking properties of flour. This conclusion was reached by Arpin (1905), Ladd and Stallings (1906), Alway (1907b), Alway and Pinckney (1908), Hoffman (1908), Buchwald and Neumann (1909), and Weaver (1922). Harcourt (1910, 1915) reported comparative baking tests in which the loaves baked from the bleached flours were larger, bolder, and the crumb texture was superior to those baked from unbleached flours. A greater improvement was noted on bleaching flour milled from new or freshly harvested wheat than on similar treatment of flour milled from old wheat several months after harvest.

While it is beyond the scope of this treatment of the subject of bleaching to present any extended discussion of the effect of nitrogen peroxide bleaching on the digestibility of flour, a brief reference to certain biochemical studies will be included. Nitrous acid was found by Shepard (1908) to inhibit the hydrolysis of starch by diastase, concentrations of 1 part of nitrogen peroxide to 25,000 of solution (40 p.p.m.), increasing materially the time required to thus hydrolyse a unit quantity of starch. A concentration of 1 part of nitrogen peroxide to 100,000 (10 p.p.m.) slowed down peptic digestion of egg albumen, and the digestion of starch with pancreatin.

Digestion experiments with pepsin and pancreatin resulted in a slower rate of digestion of flour in consequence of bleaching, according to Ladd (1908a), and Ladd and Bassett (1909). In Mann's testimony (Notice of Judgment No. 722), experiments were described in which "gastric digestion" tests apparently showed a reduced rate of digestion of bleached flour gluten as compared with gluten from untreated flour. Drosera required about two-thirds the time to digest unbleached flour that was required when the flour had been treated with nitrogen peroxide.

Gluten from unbleached flour was found by Hale (1910) to be more easily digested by artificial gastric juice than gluten from flour bleached with varying quantities of nitrogen peroxide. It was further suggested that various medicinal substances may be modified and rendered toxic or otherwise altered in contact with nitrites.

Diastatic enzymes of flour were not affected by bleaching in the experiments reported by Arpin (1905). The digestibility of gluten of bread baked from bleached and unbleached flour was studied by Rockwood (1910), and no appreciable difference could be detected. Moist gluten from bleached flour digested as rapidly, and in some cases more rapidly with pepsin than that from unbleached flour. With pancreatin little difference was noted.

No inhibitory action of digestion of starch with saliva was noted by Monier-Williams on addition of sodium nitrite. The rate of digestion was retarded if the starch had been previously treated with nitrogen peroxide gas. Experiments conducted at the Lister Institute were reported by Monier-Williams in which bleaching failed to decrease the rate of tryptic digestion. A decided though small inhibition of peptic digestion by bleaching was noted.

The presence of nitrous acid hindered enzyme action, and previous treatment with nitrous acid reduced the rate of digestion of a protein, in the experiments of Halliburton (1909). The difference observed in the digestibility of bread prepared from bleached and unbleached flour was less marked than when the digestion experiments were made with the raw flour.

The weakness of the contention that bleaching constitutes a hazard in the nutrition of man because bleached flour is less digestible than unbleached, lies in part in the fact that flour is not eaten in the raw state, and experiments with raw flour are hardly valid as evidence. In such experiments as have been reported in which a baked product or bread was used, the resulting data justify the conclusion that the effect of bleaching on digestibility is too slight to merit consideration.

Numerous opinions have been announced to the effect that nitrite in ingested bread would be absorbed from the intestinal tract of man, and on appearing in the blood stream would give rise to an irreversible formation of methemoglobin from hemoglobin. Since the former has no oxygen-carrying capacity, the formation of a sufficient quantity would seriously impair that important capacity of the blood. Yet there has been little experimental evidence offered in support of such a contention, and Haley (1914) found the quantity of nitrite in bleached flour to be too small to cause formation of methemoglobin.

The general toxicity of bleached flour has also been discussed at length. An experiment in which alcoholic extracts of commercially bleached flour was held to have caused the death of rabbits to which it was fed, while similar extracts of bleached flour were without effect, was reported by Ladd and White (1909). This experiment was repeated by witnesses for the flour millers at the time of a hearing on an application for an injunction in Federal Court at Fargo, North Dakota, and rabbits were not harmed by extracts from bleached flour. Alcoholic extracts of bleached flour proved toxic to white mice and rats when administered subcutaneously by Hale (1910), while unbleached flour extracts were without effect. No toxic effect on rabbits was observed by Hale when alcoholic extracts of bleached flour were introduced into their stomachs. Extracts from commercially bleached and over-bleached flours were found by Monier-Williams (1911) to be apparently harmless when fed to animals. It was contended by Hamill (1911) that additions to and alterations in flour which result from a high degree of bleaching cannot be regarded as free from risk to the consumer, while Steensma (1916) feared the possibility that the vitamine of flour might be attacked. Yet these contentions lack support in the form of experimental evidence, particularly since, as pointed out by Wesener and Teller (1907) and others, the concentration of nitrites or nitrous acid in bread is so low that enormous quantities must be ingested to introduce a significant quantity of toxic material into the human system.

The study of nitrogen peroxide bleaching has been facilitated by the fact that analytical methods were available which made possible the detection and estimation of extremely small quantities of nitrous acid and nitrites. Griess (1879) described the reaction of nitrous acid with a solution of napthylamine and sulfanilic acid in dilute sulfuric acid solution, with the formation of a diazo compound rose-red in color. Warington (1881) called attention to the sensibility of the Griess reagent in detecting 1 part of nitrous acid in 100 million parts of water, using a reagent acidulated with hydrochloric acid. Ilosvay (1889) modified the reagent further, using a dilute solution of acetic acid in water to dissolve the organic reagents. The latter were dissolved separately, and the two solutions mixed at the time of making the test. This reagent has since been extensively used under the name of the Griess-Ilosvay reagent.

In applying the test to flour, several methods have been described for preparing the flour extracts, including those suggested by Alway (1907b), Winton and Shanley (1908), and Sheringa (1916). Mitchell's method is included in the Book of Methods of the Association of Official Agricultural Chemists (1919). Care must be taken to insure that the water used in extracting the flour is free from nitrites. Often tap water is to be preferred to ordinary distilled water, particularly when the latter is produced in a gas-heated water still. It was indicated by Buchwald and Treml (1909) that a few drops of the mixed Griess-Ilsovay reagent could be applied directly to the surface of dry flour in testing qualitatively for nitrites.

Unbleached flour was found by Weil (1909) to react positively with the Griess-Ilsovay reagent if the reaction period was extended to 30 minutes. Another procedure was accordingly suggested, which involved exposing the suspected flour to dry hydrogen sulfide for an hour, and then comparing its color with that of the sample before treatment. If bleached flour was thus treated, the appearance was altered, the flour acquiring the color which it possessed before it was bleached. With unbleached flour no change in color resulted from the treatment with hydrogen sulfide.

Another test for nitrogen peroxide treatment was suggested by Fleurent (1906). Fifty grams of flour was extracted with benzine, the extract evaporated at a low temperature, and the crude fat dissolved in 3 cc. of amyl alcohol. To this was added 1 cc. of a 1 per cent solution of potassium hydroxide in alcohol. With a natural or unbleached flour the only change was the development of a yellow coloration, while the extract of bleached flour acquired a red-orange hue. Ladd (1907) decided that Fleurent's method was superior to the Griess-Ilsovay test.

Still another procedure was suggested by Shaw (1906), who extracted 1 kilo of flour with 95 per cent alcohol, evaporated the extract nearly to dryness, and then extracted the residue with a mixture of equal parts of alcohol and ether. This re-extracted material was evaporated to a syrupy consistency and to it was added a drop of a sulfuric acid solution of diphenylamine. A blue coloration resulted when the extract was prepared from bleached flour, which did not appear when an unbleached flour was thus tested. This method was regarded as unsatisfactory by Alway and Gortner (1907).

E. H. Miller (1912) proposed the use of dimethylaniline hydrochloride as a reagent for the determination of nitrous acid. Paranitroso dimethylaniline is produced in the reaction and imparts a yellow color to the medium. It was stated that 1 part of nitrous acid per million of solution can be detected, and that nitrates do not interfere.

Nitrosyl chloride, NOCl, as a flour bleaching agent is referred to in U. S. Patent No. 863,684, granted John A. Wesener (1907). Two processes for producing this reagent are described in the patent. In the first process nitric oxide and chlorine in mixture are to be passed over heated animal charcoal to effect their chemical combination according to the equation: $Cl + NO = NOCl$. In the second process air is to be blown through a mixture of hydrochloric and nitric acids, with the formation of nitrosyl chloride and chlorine in equi-molecular proportions according to the reaction $3HCl + HNO_3 = NOCl, + 2 Cl + H_2O$. The presence of the chlorine in mixture with the nitrosyl chloride was not deemed objectionable by Wesener. This mixture was to be diluted with air and directed into a suspension of flour particles suspended in air by means of some suitable agitator.

Nitrosyl chloride is regarded by Dunlap (1922) as one of the most effective in bleaching flour, 1 pound being required to treat 2500 barrels of flour, which is equivalent to about 2 parts of the reagent to one million of flour. It was subsequently found impracticable to handle and ship nitrosyl chloride, so this reagent has not been used extensively in flour bleaching, except in mixture with chlorine. Patents covering its use in such a mixture were granted Wesener (U. S. Patent No. 1,096,480 1914).

Chlorine as a bleaching agent was not regarded favorably by those who conducted the earlier experiments with this reagent. It is probable, however, that the dosage was not properly controlled at the outset, with a resulting excessive increase in the acidity and chlorine content of the treated flour. For subsequent experience has shown that with adequate control the dosage of chlorine can be so adjusted as to chlorinate the requisite fraction of the carotin of flour without unfavorably affecting the baking quality of the flour.

In the Wesener patent mentioned above reference is made to the use of gaseous chlorine in mixtures with nitrosyl chloride in flour bleaching. The commercial mixture marketed under this patent consists of 0.5 per cent NOCl and 99.5 per cent of anhydrous chlorine. The patent granted Williams (No. 963,970, July 12, 1910) at about the same time apparently covers the use of chlorine in a flour bleaching process. For while no reference is made directly to chlorine as such, an electrolytic cell is described in this patent, and it is suggested that the electrolyte may consist of "A solution of any suitable salt, such as sodium chloride . . .," amongst other items named. Obviously if the electrolytic cell was charged with a NaCl solution, and an electric current passed through the electrolyte, chlorine would be liberated at the anode, appear in gaseous form in the atmosphere above the liquid in the cell, and could then be pumped off and directed into the flour stream to be treated. Such electrolytic chlorine generators have been used in certain American flour mills, in producing chlorine for the treatment of the flour. The quantity of chlorine liberated per unit of time varies with the current flowing between the cell electrodes, which current can be adjusted by suitable rheostats. A manufacturer of such cells states that 1.8 kilowatts are required to produce one pound of chlorine, which is sufficient to moderately bleach 30 barrels of flour.

Liquid chlorine is extensively purchased by flour millers for use in flour bleaching. In one such process one-half of one per cent of nitrosyl chloride is mixed with the chlorine as it is compressed and liquified in the containers. On relieving the pressure in the chlorine tanks the liquid boils, and the vapor is conducted from the tanks through lead tubes. These tubes are attached to the inlet side of a constant pressure device provided with a gauge and adjusting valve, so arranged that the gas may be discharged through the outlet at the desired pressure. The gas then passes through a standardized aperture in a silver disc. The volume which passes per unit of time is a function of the pressure of the gas as it leaves the constant pressure device. The gauge of the latter can be calibrated to read in terms of ounces of gas per hour which will pass through the aperture selected. By varying the bore of the aperture the same constant pressure device will serve through a wide range in the volume of chlorine discharged per unit of time. The gauge calibration must likewise be accommodated to the size of the aperture employed.

Another type of chlorine measuring device has recently made an appearance. This consists of a vertical glass tube dipping into sulfuric acid, the acid being sealed in an outer and larger tube provided with a single outlet. The tube which dips into the acid is provided with a number of holes of uniform size, spaced regularly along its length. These holes are submerged in the acid until pressure of the gas entering the tube forces the acid down inside the tube and out through the openings. On progressively increasing the gas pressure it will bubble through an increasing number of the openings in the vertical tube, and the presumption is that a unit quantity of gas will pass through each opening that is thus functioning. Thus if bubbles are observed to be discharged from only the upper hole of the tube, one unit of gas is being passed through the device. If bubbles discharge through two holes, twice as much gas is discharged, and so on. A manual control of the gas valve must be resorted to in adjusting the flow of chlorine through this device; but the entire mechanism of the gas measuring system is visible to the operator.

The chlorine gas leaving the controlling mechanism is mixed with air, and the mixture discharged into a suitable agitator.

Owing to the avidity with which chlorine combines with flour on coming in contact with it and ceases to exist as such, one company exploiting a chlorine process strongly urges the use of the vertical type of agitator, where the flour is showered through an atmosphere mixed with the chlorine gas, so that each particle gets its proportionate treatment. An agitator of this type is constructed of materials not affected by the gas. The treating chamber is also sealed at the inlet and outlet to prevent any escape of free gas. This type of agitator has been adopted after thorough tests, and has been perfected through experience in the use of the gas.

To accomplish the degree of bleaching commonly desired, a relatively large dosage of chlorine is required. In treating hard wheat patent flours which are destined to be used in yeast bread production, about one-half to three-quarters of an ounce of chlorine per 100 pounds of flour are now being recommended. This is equivalent to from 312 to 468 parts of chlorine per million parts of flour, as compared with 3 to 11 parts per million of nitrogen peroxide used in ordinary bleaching. With soft wheat flours intended for biscuit and pastry production, about the same dosage of chlorine is recommended.

Carotin presumably reacts with chlorine to form a dichlor addition compound $C_{40}H_{56}Cl_2$, which is white in color. Bleaching patent flour with chlorine was observed by Buck (1917) to reduce the gasoline color value (and accordingly the carotin content of the flour) to from 64 to 71 (average 66) per cent, of that of the natural or unbleached flour. The Association of Operative Millers (1923) published several answers to the query raised by a member as to whether chlorine would continue to bleach flour after the flour is packed. R. S. Herman expressed the belief that a gradual change in color would occur as the flour ages. Edgar Miller supported this view, as did an anonymous correspondent. J. C. Wood answered categorically *No*. His opinion was that chlorine immediately effects all the change for which it is responsible, and that further change in color is due to the processes of natural aging. This was also the opinion of M. J. Blish.

In addition to the formation of non-pigmented compounds of carotin, chlorine produces other changes in the composition of flour. The total chlorine content of the flour is increased in consequence of treatment with chlorine. The chlorine content of natural or untreated flour was observed by Utt (1914) to range from 442 to 576 parts per million of flour, while after chlorine treatment this was increased to from 648 to 972 parts per million. The difference between the maximum chlorine content of the unbleached and the minimum content of the bleached flour is too small in proportion to the total to render the total chlorine content of flour a suitable index of the quantity of reagent used in bleaching.

Unsaturated fats present in flour are apparently chlorinated in consequence of treatment of the flour with chlorine. A collaborative study of the chlorine content of fat extracted from natural flour and from the same flour after treatment with chlorine at the rate of 14 ounces per 26 barrels of flour, was reported by Rask (1922). The most satisfactory of the analytical methods employed indicated an average of 16.6 parts of chlorine in the crude fat per million of unbleached flour, and 93.7 p.p.m. in the fat of the chlorine treated flour.

Chlorine in solution in water yields hydrochloric and hypochlorous acids, and the latter may undergo reduction to hydrochloric acid. It accordingly follows that treatment of flour with chlorine increases the acidity and hydrogen ion concentration of water extracts of the flour. This increase in acidity was noted by Utt (1914), who found the average acidity of 12 unbleached flours to be .182 per cent, while that of the same flours after treatment with chlorine was .214 per cent. Dunlap (1922) reported that the acidity of a natural flour was .112 per cent, and after treatment with chlorine this was increased to .151 per cent. The acidity of the treated flour was observed to remain constant for 259 days, while that of the untreated increased to .135 per cent in the same period of time. It was suggested by Dunlap (1923) that this stability in the acidity of the chlorine treated flour might be attributed to the inactivation of the flour enzymes by the reagent, and that a similar stability in other properties of the treated flour might be anticipated.

Increased hydrogen ion concentration of flour as a result of treatment with chlorine was demonstrated by Bailey and Johnson (1922). The change in hydrogen ion concentration was found to be proportional to the dosage of chlorine. The pH of clear flour was changed less than that of patent flour by each unit of chlorine used, because of the higher buffer index of the clear flour. On continuing this study Bailey and Johnson (1924a) treated a straight grade flour with varying quantities of chlorine, with the results shown in Table 131.

TABLE 131

Hydrogen Ion Concentration (as pH) of Straight Flour Treated With Varying Dosages of Chlorine (Bailey and Johnson, 1924a)

Chlorine per Barrel of Flour, Ounces	Hydrogen Ion Concentration as pH
.00	6.04
.42	5.93
.50	5.90
.58	5.88
.67	5.87
.75	5.83
.83	5.80

Data substantiating these observations have been privately communicated by Dunlap, and are recorded in Table 133, together with the results of baking tests. In addition, the titratable acidity was determined, and it appears that each half ounce of chlorine per barrel increases the acidity by 0.02 per cent uniformly with increased dosages of chlorine within the limits studied.

The greater dosage of chlorine now being recommended in treating bread flour, as compared with the earlier recommendation of one-quarter ounce per 100 pounds of flour, is designed to effect an increase in hydrogen ion concentration, which brings the latter within the range believed by those promoting this process to be optimum for the production of bread of the highest quality.

The changes occurring in the hydrogen ion concentration of flours bleached by Bailey and Johnson were followed over a period of 29 months, and these observations were discussed in the preceding sections (Note Table 114).

The electrolytic resistance of water extracts of flour was found to be reduced after treatment with chlorine, Bailey and Johnson (1922) showing that the specific conductivity of such extracts varied with the dosage of chlorine. An examination of several samples of commercially bleached flour showed the following differences in the specific conductivity of natural and chlorine treated flour extracts ($K_{30} \times 10^4$).

| | Specific Conductivity of Water Extract | |
Sample	Untreated	Chlorine Treated
Patent A	5.29	5.53
" B	5.50	5.99
Straight	5.85	6.03

Since the conductivity of natural flour extracts apparently varies directly with the ash content (see Bailey and Collatz, 1921), the relative increase over the normal for a flour of known ash content may serve to indicate the approximate dosage of chlorine used in bleaching the flour. This assumes that the flour is free from other added electrolytes, such as flour improvers, and chemical leavening agents.

Since natural or untreated flour increases in acidity in the course of natural aging, while chlorine treatment effects an immediate increase in the hydrogen ion concentration of the treated flour, it has been claimed that such treatment parallels natural aging in effecting those increases in hydrogen ion concentration believed to be essential to the full development of the baking value of flour. It is doubtful, however, if all of the significant changes occurring in flour on aging can be attributed to the increase in hydrogen ion concentration alone. It does appear, however, that such acidulation as results from treatment of freshly milled flour with chlorine serves to improve its baking qualities. This is shown in the results of tests reported by Dunlap (1922), in which the loaf volume of the chlorine treated flour immediately after treatment was approximately that of the untreated flour after being held in storage for 116 days. These data are reported in Table 132.

TABLE 132

Loaf Volume of Bread Baked from Untreated and Chlorine Treated Flour, Reported by Dunlap (1922)

	Loaf Volume, in cc.	
Days After Treatment	Untreated Flour	Chlorine Treated Flour
0	2,060	2,340
116	2,300	2,350
259	2,370	2,350

The results of a series of experiments privately communicated by Dunlap indicate that the observed increase in bread quality as measured in terms of loaf volume and texture score might be due in part to the increased diastatic activity of the treated flour. The baking data show the flour receiving the heaviest treatment with chlorine to have yielded the best bread. This treatment increased the hydrogen ion concentration to pH = 5.22, as contrasted with pH = 5.77 in the original flour. The titratable acidity was likewise increased. The diastatic activity (in degrees Lintner) of the flour treated with ¾ ounces of chlorine per 100 pounds of flour was doubled by the treatment. These data are found in Table 133.

TABLE 133

Acidity, Diastatic Activity, and Baking Tests of Flour Treated With Graduated Dosages of Chlorine

Chlorine per 100 Lbs. Flour, Ounces	H-ion Concentration as pH	Titratable Acidity	Diastatic Activity (Lintner Value)	Loaf Volume, cc.	Baking Test		
					Texture	Color	"Value"
0.00	5.77	0.119	16.1	2060	100	100	100.0
0.25	5.50	0.139	20.6	2150	103	103	103.4
0.50	5.33	0.158	27.5	2180	104	104	104.6
0.75	5.22	0.178	32.1	2230	105	105	106.0

The Kota wheat flour treated with chlorine by Bailey and Sherwood (1924) was evidently improved in baking qualities by the treatment. Not only was the loaf volume increased in proportion to the quantity of chlorine used, but the deep yellow color characteristic of flour milled from Kota wheat was largely dissipated. This is shown by the data in Table 134 resulting from baking tests of untreated Kota wheat flour, and similar flour treated with two dosages of chlorine.

TABLE 134

Results of Baking Tests of Straight Grade Flour Milled from Kota Wheat, Bleached With Chlorine, and Unbleached

	Untreated	Treated With Chlorine at Rate per Barrel of	
		0.60 Oz.	0.87 Oz.
Loaf volume, cc.	2400	2500	2650
Color score	96	99	100
Texture score	98	99	99

Hydrogen ion concentration is believed to play an important rôle in the course of dough fermentation, and a control of this factor may be desirable in fermentation practice. It has accordingly been suggested that the production of flour with a standardized hydrogen ion concentration would prove of advantage, and that such standardization can be effected by graduating the dosage of chlorine. Clear flours, because of their higher buffer index, will require a heavier treatment with chlorine to bring their concentration of hydrogen ions within the range which is deemed desirable. Certain flour mills are accordingly treating collections of flour streams representing the various primary grades of flour rather than the finished flour, applying more chlorine to those groups of streams which are known to have the highest buffer index.

In addition to the process in which the gaseous chlorine is applied directly to the flour, another process has been recently developed in which heavily chlorinated flour is mixed with natural flour. The chlorinated fraction may have been treated in advance with more than one hundred times the quantity of chlorine ordinarily used in bleaching, and is presumably mixed with natural flour in a ratio of one part in 100 of the latter.

Arpin (1921) did not observe any beneficial effect from the treatment of flour with chlorine, basing his judgment upon the characteristics of the bread produced. It appears that the flours with which he was working were not highly refined, since they were reported to contain from 1.04 to 1.42 per cent of ash, and this may account for the small effect that was observed.

An improvement in the baking quality of flour treated with chlorine was noted by Weaver (1922), and the fermentation period was reduced in consequence of such treatment.

No extensive studies of the influence of chlorine treatment upon the digestibility, nutritive value, or possible toxicity of treated flour, have been published. In a private communication Dr. H. Gideon Wells detailed a study of the toxicity of flour treated with chlorine which led to the conclusion that water extracts of such flours exert practically the same local toxic effects as extracts of untreated flour. With alcoholic extracts the local toxicity and chronic intoxication were not distinctly different with the treated and untreated flours. Acute intoxication with large doses intraperitoneally showed a slight but definitely less toxicity with the extract of the treated flour.

The effect of chlorine treatment upon the enzymes of flour and of yeast has not been discussed in the literature at any length. Reference has been made to Dunlap's suggestion that the observed stability of treated flour with respect to acidity might be attributed to the inactivation of the flour enzymes, but no direct evidence is available to support this contention. In fact, the opposite conclusion was reached by Bailey and Johnson (1922) from the fact that the buffer index of treated flours increased regularly with the chlorine dosage. This was attributed to the increased activity of phytase in consequence of the higher hydrogen ion concentration of the treated flours. Here again it must be admitted that direct evidence is lacking to prove that the phytase is still active in the treated flours, and doubt as to the effect of the treatment upon the dough enzymes must continue to be expressed until a direct attack is made upon the problem.

That the dichlor derivative of carotin is not toxic has been shown by Wells and Hedenburg (1916), who administered dosages equivalent to the quantity of carotin in 200 kilograms of flour without harmful effects. Carotin was likewise found to be practically devoid of toxicity.

A combination treatment of flour with chlorine, followed by anhydrous ammonia, was patented by Fegan and Sasse (U. S. Patent 1,330,937, dated Feb. 17, 1920). It was claimed that the hydrochloric acid, formed by reaction of chlorine with the moisture of the flour, reacts with the ammonia to form ammonium chloride. The latter furnishes available nitrogenous yeast nutrients, and thus serves as a flour improver.

Nitrogen trichloride is used in flour bleaching under the trade name of the Agene process. The basic patents covering this application of the reagent were granted to Baker (U. S. Patent No. 1,367,530, issued Feb. 8, 1921). It is interesting to note that this is probably the first successful commercial process in which nitrogen trichloride has been used.

The reagent is produced by passing chlorine gas through a solution of an ammonium salt. The reaction proceeds in two stages, hypochlorous acid being formed in the first stage, which reacts with the ammonium salt to form nitrogen trichloride. The equations representing these reactions may be written as follows:

(1). $6Cl_2 + 6H_2O = 6HCl + 6HOCl$
(2). $6HOCl + (NH_4)_2SO_4 = 2NCl_3 + H_2SO_4 + 6H_2O$

From 1.5 to 5 grams are required to treat one barrel of flour, according to Baker (1922). A 4-gram treatment would, therefore, involve the use of 7 grams of chlorine. In bleaching with chlorine 28 grams or more are used per barrel of flour, or at least 4 times as much chlorine as is required to produce the requisite quantity of nitrogen trichloride. Since half the chlorine used in producing nitrogen trichloride is lost in the hydrochloric acid formed in the first stage of the reaction, it follows that the chlorine applied in the form of nitrogen trichloride is several times as effective as is chlorine in the elemental form.

When applied to flour, the reactions shown above may be reversed, and hydrochloric acid and ammonia will result. Baker found traces of nitrites in flour treated with nitrogen trichloride, and a trace of chlorine in the fats of the flour when the higher rates of treatment were applied. Acidity of the flour was not affected by ordinary treatments. Baker also asserted that gluten of the flour was improved and the viscosity of water suspensions of flour was increased. It has been indicated that the dosage of nitrogen trichloride may best be adjusted by progressively increasing the quantity applied until no further increase in the water-imbibing capacity of acidulated flour suspensions can be observed. This modification of the gluten, it is claimed, is reflected in an improved loaf texture, and increased loaf volume on baking. No deterioration was observed on storing treated flour for a period of 11 months.

Bread made from flour treated with nitrogen trichloride was observed by Weaver (1922) to show no increase in size or volume, but it had a better texture. The color of the crumb of bread from flour so treated was much whiter, and sufficient treatment would result in the complete bleaching of the yellow coloring matter of the flour to white derivatives. Weaver (1925) later reported that treatment of flour with nitrogen trichloride did not modify the hydrogen ion concentration of the flour.

Baker discussed flour bleaching before the 1923 convention of the American Association of Cereal Chemists, and pointed out that, other things being equal, a bleaching gas is more effective (1) the lower the temperature; (2) the finer the granulation of the flour (unless this produces stickiness); (3) the more intimate the contact of flour and gas; (4) the longer the contact; (5) the lower the gas concentration. Drier flour apparently bleached more effectively than moist flour, except with chlorine. Viscosity of acidulated flour suspensions was increased by treatment of the flour with nitrogen trichloride. Lawellin (1924) stated that the optimum dosage of nitrogen trichloride was best determined by the viscosity method, and that the maximum viscosity was not obtained by treating all flours at the same rate.

Goebel (1923) stated that in commercial treatment of flour with nitrogen trichloride the humidified air from the generator contains 0.1 to 0.4 per cent of the reagent.

Nitrogen trichloride is commonly regarded as a hazardous substance with which to deal because of its tendency to explode, and the student of chemistry is always cautioned to avoid compounding substances which may react with the formation of this substance. The promotors of the Agene process state, however, that under the conditions which prevail in the proper use of their generators, the dilution with air of the nitrogen trichloride formed will be so great as to obviate any possibility of explosions.

Benzoyl peroxide, $(C_6H_5CO)_2O_2$, is the organic peroxide which of this class of reagents has been most extensively used in the commercial bleaching of flour. This reagent has previously been employed in the bleaching of fats and oils. In America it has been used in the milling industry under the trade name of Novadelox B, which is a mixture of benzoyl peroxide (25 parts), and calcium phosphate (75 parts). The last named substance is used in the mixture to reduce its inflammability, and the liability of spontaneous ignition. When first introduced to the milling industry it was recommended that the mixed reagent be fed into the second break stock, but later the suggestion was made that it be mixed with the finished flour. Since the reagent is a solid rather than a gas, the finely divided powder must be thoroughly mixed with the flour to secure a uniform treatment, and a special agitator has been devised for use in this connection. One pound of the mixed reagent is intended to bleach about 40 barrels or 7,840 pounds of flour, and since the mixture is only one-fourth benzoyl peroxide, it follows that one pound of the latter is used with each 31,360 pounds of flour, or the equivalent of 32 parts of benzoyl peroxide to each million parts of flour.

The reaction of benzoyl peroxide with the carotin of the flour is not immediately apparent, and it requires about 48 hours for the maximum bleaching effect to occur. Presumably the carotin is oxidized to one of its colorless derivatives by oxygen released from the peroxide. The derivatives of benzoyl peroxide which may appear in the mixture with flour have not been reported in the literature, but it might be anticipated that among these benzaldehyde and benzoic acid would predominate.

Since benzoyl peroxide is not such a familiar reagent as the gaseous bleaching agents that have been discussed, a brief summary of its properties may be of interest. This substance is produced as glistening white crystals, which melt at 103.5°. On heating somewhat above this temperature the substance detonates with a slight explosion. The crystals are insoluble in water, and slightly soluble in hot alcohol. They are freely soluble in chloroform, somewhat less soluble in ether and benzine, and still less soluble in carbon disulfid. They are also soluble in liquid fats.

Benzoyl peroxide may be synthesized by treating benzoyl chloride with a peroxide. Barium peroxide has apparently been most extensively used; reactions involving the use of hydrogen peroxide in a potassium hydroxide solution, and of sodium peroxide have also been described. The researches of Brodie and of Sonnenschein have been of chief interest in this connection. Erlenmeyer observed that when benzaldehyde and acetic anhydride were mixed with fine sand to extend the surface, a formation of benzoyl peroxide occurred, the yield being somewhat less than the theoretical.

Benzoyl peroxide reacts with phenyl hydrazine to form monobenzoylphenylhydrazine, benzoic acid, and nitrogen. Crystals dropped into warm concentrated sulfuric acid explode, with a puff of white vapor. With a solution of ferric chloride a brown precipitate of ferric benzoate will be produced. Iodine is liberated from a solution of potassium iodide by benzoyl peroxide; iodine and bromine are liberated from the corresponding halogen acids. Benzoyl peroxide blues guiacic acid; powdered gum guaiac mixed with an alcoholic solution of benzoyl peroxide and warmed results in the appearance of a blue color when the latter is diluted to 10 parts in each million of solution. Flour treated with benzoyl peroxide at the rate of 25 parts of the reagent per million of flour, and at once extracted with 95 per cent alcohol yielded an extract which reacted with gum guaiac. The intensity of the reaction diminished with the lapse of time, however, due probably to the spontaneous reduction of the peroxide.

Scant reference to the effect of benzoyl peroxide upon the baking properties of flour has appeared in the chemical literature. It was stated by Stork (1922) that this reagent will increase the water absorbing capacity of flour by 1 per cent, and that the colloidal condition of gluten is favorably affected, resulting in larger loaf volume and texture, but no data in support of this contention have been published. The author's observations have tended to indicate that the effect of the reagent upon these properties of flour is imperceptible. Weaver (1925) found that benzoyl peroxide did not alter the hydrogen ion concentration of flour.

The use of peraldehydes, ozonides, perozonides, peroxozonides, ozonideperoxides or polymers of the ozonides as flour bleaching reagents are discussed in U. S. Patent No. 1,483,546, granted H. C. J. H. Gelissen (Feb. 12, 1924). As examples of these materials he mentions a peraldehyde of peroxidic character such as $C_3H_4CHO_2$; perozonides such as

$$CH_3(CH_2)_7 - CH - CH - (CH_2)_7 - CO_2 - OH$$
$$\diagdown \diagup$$
$$O_4$$

peroxozonides such as:

$$CH_3(CH_2)_7 - CH - CH - (CH_2)_7 - \underset{\underset{O}{\overset{\overset{O}{\|}}{\underset{\|}{C}}}}{C} - OH$$
$$\diagdown \diagup$$
$$O_4$$

and as an example of an ozonide peroxide, that of oleic acid is mentioned:

$$CH_3(CH_2)_7 - CH - CH - (CH_2)_7 - \underset{\underset{O}{\overset{\overset{O}{\|}}{\underset{\|}{C}}}}{C} - OH$$
$$\diagdown \diagup$$
$$O_3$$

A dosage of 0.01 per cent (100 parts per million) is mentioned as adequate in bleaching most substances. Peroxidase of the flour is believed to effect the liberation of active oxygen from these substances, which in turn oxidizes the carotin. Catalase of the flour, on the other hand, is detrimental to the extent that it liberates molecular oxygen from the reagents, which form of oxygen is without effect on the carotin. It is suggested in the patent that the catalase may be first inactivated by treating the flour with .002 per cent of chlorine, which quantity of the latter is insufficient to bleach the flour. The organic bleaching reagent may then be added, and the maximum effect thus secured.

Chapter 10.

Flour Strength and Enzyme Phenomena.

Flour is used more extensively as the principal ingredient of yeast leavened bread than in any other form of food in common use in Europe and America. In consequence, numerous studies have been conducted in an effort to establish a correlation between the composition of flour and its bread baking qualities. These researches have involved not alone the percentage of the numerous constituents of flour, but the physical condition and physico-chemical properties of these constituents as well. Over the past thirty years voluminous data have been accumulated, and it is difficult to fit these into a coherent consideration of the chemistry of baking qualities. As previously pointed out by Bailey (1913c), this is due in part to the fact that many investigators have endeavored to find a single factor which could be correlated with the findings of the baker. It should be obvious, however, that in as complex a material as flour, and with a living organism, yeast, involved in the fermentation of the dough, several factors must be considered in their single and joint effect upon baking properties.

Before proceeding with a discussion of the baking qualities of flour it is necessary to define the term "flour strength." This expression has been employed in describing several properties of flour, including (a) the quantity of water absorbed per unit of flour in preparing a dough of standard consistency; (b) the quantity of bread produced per unit of flour; (c) the physical extensibility of dough as indicated by the manner in which it handles in the bake-shop; and (d) the capacity of the flour to make large, well-piled loaves. The last definition was suggested by Humphries (1905), and is most commonly used. The term "pile" is borrowed from the textile industries, and is used to describe the relative silkiness and resiliency of the crumb. An exception to this definition has been noted by Kent-Jones (1924), who maintains that flours yielding doughs in which gas production is at a low level should not be regarded as "weak." The acceptance of this view would involve an addition to Humphries' definition, which would accordingly read: "Strength is the ability of flour to be converted into large, well-piled loaves, provided any deficiency in the rate of gas production in the dough stage is supplemented in a suitable manner." A general application of this definition would simplify the discussion of strength, but would necessitate the addition of another distinct consideration in the description of flour properties, namely, potential gas production when converted into dough.

Kent-Jones' limitation of the term has a sound basis in view of the possibilities of manipulating the properties of flour, and of dough made therefrom, since, in the present state of our knowledge, it is doubtless true that a deficiency in the rate of gas production can be supplemented more readily than certain other deficiencies that may be manifested by the flour. Unfortunately many bakers either are not capable of adjusting deficiencies of this kind in using flour, or do not desire to be obliged to superimpose diastatic preparations or yeast foods upon their basic dough formula. Regardless of the definition suggested at this time, these bakers will continue to regard as "weak" those flours which do not respond properly in fermenting and baking, even though the lack of response might be corrected through the use of diastase.

It will accordingly be difficult to depart substantially from the conception of strength implied in the definition of Humphries, although for purposes of scientific discussion this may perhaps be stated in more definite terms. Thus Bailey (1916c) suggested that the strength of flour is determined by the ratio between (a) the rate of production in, and (b) the rate of loss of carbon dioxide from the fermenting mass of dough. This involves a consideration of the factors which contribute to the rate of gas production in dough, and also of the dough properties which function in retaining the gases of fermentation. These will be discussed in that order, and, as nearly as possible, under the two general headings included in the definition.

It must be recognized, however, that in many practical baking trials no sharp distinction has been made between these two properties of flour and of dough made therefrom. Facilities have not been available in most instances to determine the nature of the deficiency in doughs which yielded subnormal bread. Conclusions have doubtless been drawn in many cases which were not supported by adequate evidence. Such conclusions have appeared in reports and published papers which were quoted until they were accepted as established facts. Hypotheses based on seemingly analogous cases have been advanced without experimental evidence to establish the soundness of the reasoning. While the evolution of theories is stimulating to the processes of thought, and sometimes serves to inspire research, it is hazardous to be content with the theory without determining its application in practice. And finally there has always existed the difficulty of recording the findings of the final tests, namely, the characteristics of the baked loaf of bread. Certain characteristics have lent themselves to measurement and mathematical expression, including the cubical dimensions or "volume" of the loaf as indicated by displacement. Briggs (1913) defended the acceptability of loaf volume as a convenient and accurate expression of flour strength. Other properties of the loaf are deemed by many technologists to be of importance, however, including the shape or contour of the vertical section of the crust, the texture and resiliency of the crumb as registered by the sense of touch (the "pile" of Humphries), and the grain of the crumb as indicated by the size and shape of the vescicles, and the thickness of their walls.

These properties of texture and grain have generally been "scored" by the baking expert on the basis of observations involving the feeling and appearance of the crumb, a procedure which does not appeal to the scientist, who prefers physical measurement with a suitable instrument. The system of recording crumb texture suggested by Mohs (1924), that involves coating the vesicle walls with a black paste, and impressing these upon unsized paper, which thus retains a picture of their outlines, may be of service in arriving at a generally accepted basis of scoring texture and reporting such findings. Other physical tests of bread properties are sadly needed in further developing such studies of flour qualities.

In the majority of baking tests reported by American chemists, the volume or cubical displacement of a loaf of bread baked under standard conditions and with a unit weight of ingredients, and in many instances a texture score, are the only data recorded from which the flour strength can be estimated. In certain instances, as in the experiments conducted by Swanson, Willard and Fitz (1915), the maximum expansion of the dough (before baking) was determined and recorded. Bailey (1916c) described a mechanical device, or "expansimeter," which registers automatically the maximum volume attained by a fermenting dough as one index of strength.

The baking strength data reported by Saunders (1907) are computed from observations of the characteristics of the baked loaf, and the water absorption of the flour. The method of computing baking strength as employed by Saunders was as follows:

		Specimen Calculation
Water in cc. used in making dough, per 100 grams of flour, $\times 2$	....	127
Water in cc. retained in bread, per 100 grams of flour, $\times 5$	....	200
Volume of loaf in cc. per 100 grams of flour...........	....	526
Shape of loaf (height/diameter) $\times 500$	....	320
Form of crust (100 = first class; 70 = extremely poor) $\times 2$	....	170
Texture score $\times 3$	....	264
Total	....	1607
Subtract from total	700	700
and divide difference by 10........................	..../10	907/10 =
= "strength"		90.7

In thus determining baking strength, consideration is given to water absorption and retention, although the latter is weighted heavier than the former, since the quantity retained per unit of flour is multiplied by 5, while the quantity of water absorbed in making the dough is multiplied by 2. It is evident from the specimen calculation recorded above that the loaf volume and shape are weighted the heaviest in this computation, with texture occupying third place.

The "baker's marks" used by Humphries (1905) evidently represent a score assigned by an expert baker after appraising the several important characteristics of the baked loaf. The exact manner in which these marks are derived is not indicated by Humphries, but from the general comments in the text of his paper it appears safe to conclude that size, shape and texture of the baked loaves were given principal consideration. A mark of 80 represented the average of loaves baked from London "households" flours, while 100 was the maximum strength obtained from the strongest wheats in ordinary use.

Two characteristics of the baked loaf, (1) size, and (2) shape, were considered by Wood (1907) in his discussion of the chemistry of flour strength. The correlations which he presented are discussed in a later section. It seems probable that certain properties of flour which were reflected in his index of "shape" might be responsible for the texture of loaves baked in high pans such as are used in America.

Various and sundry modifications of these general methods of computing and recording strength have been employed by technologists, but in the majority of cases the criteria enumerated in the preceding paragraphs have been used in measuring this property of flour. In the several researches summarized in the paragraphs which follow, such criteria are generally involved in the effort to correlate strength with the composition, and the physical properties of flour, dough and flour constituents.

Yeast Nutrition and Gas Production.

Certain strains of "distillery" yeasts have been selected and propagated for use in the production of yeast leavened bread. Such yeasts are characterized by a relatively high optimum temperature when contrasted with beer yeasts, and by their vigorous production of carbon dioxide and alcohol in the fermenting bread dough. To facilitate the rapid fermentation that is desired, a variety of substances must be present in the dough. In addition to other nutrients it appears that diffusible nitrogenous compounds are essential to the propagation of the yeast cells, while carbohydrates and phosphates must be present in the nutrient medium if fermentation is to proceed normally. An extended consideration of fermentation in its relation to baking practice cannot be included in this volume, but it must be recognized that if provision is not made for adequate nutrition of the yeast cells in dough, satisfactory leavened bread cannot be produced.

Certain "flour improvers" are in reality merely yeast nutrients which serve to stimulate the rate of gas production by this organism. Prominent among the materials present in these improver combinations are ammonium salts, phosphates, calcium compounds, bromates, salts of peracids, and other substances, many of which have long been known to accelerate yeast propagation and fermentation. The usefulness of these combinations is contingent upon the deficiency of the ordinary flour dough in the substances which are superimposed on the dough when the improvers are added. It is not the practice at present in America to mix these improvers with flour intended for bread production before the flour is offered for sale to the baker. In fact it appears doubtful if a mixture of an improver with wheat flour could be labeled merely "flour" and still conform to the definition of flour included in the present Federal food definitions and standards. A discussion of improvers fits more properly into a consideration of the chemistry of baking, and hence cannot be included in this volume.

Fermentable sugars must be present in dough if yeast fermentation is to proceed at the vigorous rate ordinarily demanded in yeast bread production. Sound flour ordinarily contains a small quantity of sugars, chiefly sucrose. Girard and Fleurent (1903) and Liebig (1909) have shown that the sugar content is in the neighborhood of $\pm$ 1.5 per cent. This quantity is not sufficient to yield the carbon dioxide necessary to leaven the dough, and leave an excess to impart the slightly sweet flavor demanded by the American consumer, and also to carmelize in and delicately brown the crust. It has been shown that in doughs made without the use of added sugar, and fermented in the usual manner with yeast, the sugar content at the conclusion of the normal fermentation period may be nearly as great as at the outset. There may have been a loss of nearly 2 per cent of dry matter during the fermentation of such a dough. It is thus evident that in such instances sugar must have made its appearance in the dough during fermentation in consequence of hydrolytic cleavage of the starch. Even in those baking processes in which sugar as such is one of the dough ingredients, it is doubtless advantageous to maintain a fairly uniform concentration of sugar in the dough throughout the fermentation period. This should facilitate the control of the carbohydrate metabolism of the yeast, maintain a uniformly high rate of gas production, leave the dough at the end of the fermentation period with a sufficiently high concentration of sugars so the crust will brown properly during the oven baking, and impart a sweet flavor to the crumb. A pale crust and a flat flavor are rough indications of low diastatic activity in the dough from which the bread was baked.

Two advantages accrue from a fairly high diastatic activity of bread doughs. The first involves the maintenance of a fairly constant and reasonably high sugar level. The second involves the economy resulting from using the starch of the flour as a source of fermentable sugars, since the necessary sugar can be secured more cheaply in this than in any other form. Experience and the results of researches appear to indicate that diastatic activity may play an important rôle in determining the behavior of flour in its conversion into yeast leavened bread.

The size of the baked loaf was found by Wood (1907) to be positively correlated with the gas evolved from a fermented flour suspension, and the latter in turn with the increase in sugar resulting from incubating aqueous flour suspensions (without yeast) for 3 hours at 40° C. This increase of sugar content was manifestly due to diastatic activity of the flour. Flour deficient in diastatic activity could not be baked into a large loaf of bread unless sugar was added to the dough. Experiments conducted by Maurizio (1902) were cited by Wood, from which it is evident that the mere addition of sugar to flour will not result in the production of loaves of uniform size from all flours thus treated. Wood points out that in such instances there is no assurance that the concentration of sugar in the various doughs would be constant throughout the fermentation period. The added sugar may be used up early in the period, and unless diastatic activity was uniform in the flours, there might be varying concentrations of sugars toward the close of the period.

These observations of Woods' evidently inspired considerable thought, and further researches on the part of other students of baking chemistry. A study of flour diastases was reported by Baker and Hulton (1908) in which several considerations involved in the properties of these enzymes were developed. Maltosazone could be synthesized from the extract of autolysed flour, demonstrating maltose to be one of the principal sugars produced by the action of diastases on starch. In studying the action of a flour extract on potato starch it was observed that the rate of action did not exactly follow Kjeldahl's law of proportionality. Diastase was differentiated into liquefying and saccharogenic enzymes, and it was observed that the starch-liquefying enzyme either was not present in certain flours, or was unable to act. Tissues (flour suspensions) exhibited a greater activity than flour extracts. While gas production increased with "bakers' marks," weak flours were encountered which possessed a higher diastatic power than normal strong flours. Gas production was observed to parallel the maltose content of 3-hour doughs. The general conclusion of Baker and Hulton was that gas retention appeared to be a more important factor than gas production.

Much the same conclusion was reached by Ford and Guthrie (1908). They contributed an interesting observation on the behavior of diastase in the presence of normal flour proteins, showing that when the proteins were partially hydrolysed with papain a pronounced increase in diastatic activity resulted. Unpublished researches at the Minnesota Agricultural

Experiment Station tend to confirm this observation. It appears that either the diastase, or an activator of diastase, may be adsorbed on gluten, from which it can be released by hydrolysis or dispersion of the gluten. Ford and Guthrie opened up the question of the possible significance of flour proteases in their relation to gas retention, a phase of this consideration of flour strength that will be discussed in the section dealing with gluten properties.

The total gas evolved during fermentation was not regarded by Humphries and Simpson (1909) as being of as much significance as that produced during the last stages of fermentation. Diastatic activity was regarded as responsible for the maintenance of the necessary sugar level throughout the fermentation period. More sugar was found in a fermented dough (made without added sugar) than was originally present. Liebig (1909) observed that the diastase of flour produced maltose in a dough to the extent of 4.6 per cent in 14 hours, and rated the diastatic activity (saccharogenic) of fine flour as about one-seventh that of malt, although the starch liquefying power of the flour appeared to be negligible. Neumann and Mohs (1909) maintained that the diastase present in flour is sufficient to produce the sugar necessary for fermentation, and is capable after a few hours of producing as much sugar as was originally present in the dough.

The evolution of carbon dioxide in fermenting flour suspensions was determined by Alway and Hartzell (1909), but the resulting data could not be correlated with the size of loaves baked from the same flours when a number of flour samples were thus tested. A survey of the data resulting from these studies of gas evolution indicated that the total volume of gas evolved in the first 7 hours was greater, in case of the clear flours, than when the patent flours were tested in a similar manner. The gas evolution from the clear grade flour suspensions averaged about 50 per cent greater than from the patent, although not all of the pairs from each mill stood in this relation.

A microscopic study of the changes occurring during the germination of wheat was reported by Whymper (1909), who observed that the large starch granules (30 to 35 μ in diameter) were more readily attacked by the diastase of the germinating grain than were the smaller granules (3 to 5 μ). Armstrong (1910) has commented on the large variation in the proportion of the two sizes of granules in flours examined. Thus in certain samples of flour as much as 40 per cent of the starch by weight was in the form of the large granules, while in other flours only 7 to 10 per cent was in this condition. Yet the observations of Buchanan and Naudain (1923) led them to conclude that the stronger flours contained the largest proportion of small grains. If the opposite had been the case, we might have reasoned that strong flours possibly owed their strength in part to their larger content of large starch granules, which in turn would be more readily hydrolysed by the flour diastases during fermentation. As matters now stand, we are left without any adequate basis for correlating the physical characteristics of the starch granules with flour strength in so far as the latter is concerned with diastatic activity of the flour.

In determining the relative activity of amylolytic enzymes in wheat flour, Swanson and Calvin (1913) found that the proportion of flour to water in the digestion mixture influenced the rate at which sugar was produced from starch. A ratio of 1 part of flour to 1 of water exhibited a lower rate of diastatic activity than a ratio of 1:3. Between the ratio of 1:4 and 1:10 there appeared to be little difference in rate, which reached its highest level in these proportions. A further increase in the proportion of water resulted in a decrease in rate of sugar production per unit of flour. The optimum temperature for the maximum production of reducing sugar over the period of one hour was 65° C. Sulfuric acid of normality 0.005 had a marked inhibiting effect on the diastatic activity of straight grade flour, but practically none on the low grade. The author ascribes this difference to the lower buffer index of the straight grade, the hydrogen ion concentration of which was increased by the sulfuric acid to a level which inactivated the enzyme. The same quantity of sulfuric acid probably changed the hydrogen ion concentration of the low grade flour through a narrower range, and did not increase it sufficiently to seriously interfere with the activity of the enzyme.

A simple procedure was followed by Stockham (1920) in determining the relative "fermentation possibilities" of various flour samples. The flour was shaken with water for 30 minutes, filtered, and the refractive index of the filtrate determined with an immersion refractometer. From this reading the index of refraction of water (15) was subtracted, and the difference recorded as an approximation of the comparative content of soluble extract. These values were observed to be correlated with the quantity of gas generated in the fermentation of dough made from the same flour samples. In these studies instances were observed in which the quantity and quality of the protein in the flour was adequate, but deficient gas production resulted in an inferior loaf of bread.

A convenient method for the estimation of the relative diastatic activity of flour was devised by Rumsey (1922). A 10 gram charge of flour was digested in 100 cc. of water at 27° C. for 60 minutes. This was then clarified by the addition of 3 cc. of a 15 per cent solution of sodium tungstate. Five drops of thymol blue indicator solution (0.04 per cent) was added, followed by sufficient sulfuric acid to produce a distinct pink coloration of the indicator. At this hydrogen ion concentration not only was the action of the diastase arrested, but the tungstate was most effective as a clarifying agent. The mixture was diluted to 200 cc. in a volumetric flask, clarified by centrifuging, and 50 cc. aliquots withdrawn for the determination of reducing sugars in the conventional manner. The reagents used in this method did not interfere with the sugar determination when the Soxhlet or similar reagents were employed.

The diastatic activity of a number of flours was determined by Rumsey, and a range of 34.8 units (in terms of milligrams of maltose produced per 10 grams of flour) to 308.1 units was encountered. The

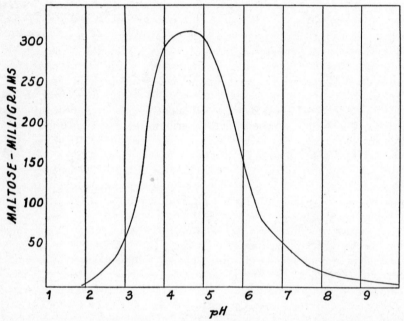

FIG. 18.—Effect of varying pH on the diastatic activity of flour. (Rumsey, 1922.)

highest diastatic values were shown by flours milled from Red River Valley spring wheat, Saskatchewan (Canada) spring wheat, and Kansas turkey red winter wheat; the lowest values in flours milled from soft white wheats from Utah and Washington, and soft red winter wheat from Ohio. In general, those flours which gave superior loaves when baked were possessed of relatively high diastatic activity, while the inferior loaves, as judged by loaf volume and texture, were baked from the flours low in diastatic activity.

These diastatic activity determinations were made in the flour suspensions, the hydrogen ion concentration of which ranged between $pH = 5.73$ to $pH = 6.13$. Rumsey determined the effect of varying the hydrogen ion concentration of the digestion mixture on the activity of flour diastase in the instance of a patent flour milled from a Red

River Valley spring wheat. It appears from his data, which are recorded in Table 135, and presented graphically in Figure 18, that the maximum activity was reached at a concentration of hydrogen ions slightly on the acid side of pH = 4.8. From the data in the graph it might be concluded that the relative diastatic activity of a dough should approximately double during the fermentation period, if the hydrogen ion concentration increased from pH = 5.9 to pH = 5.2. Such a change in pH of commercial doughs may occur, as shown by Bailey and Sherwood (1923).

It is conceivable that this effect of increasing hydrogen ion concentration on the activity of diastase in dough constitutes the chief benefit derived from the increased acidity. It is also possible that one of the principal advantages attached to the sponge method of fermentation, which is enjoying such a vogue in America at the present time, is due to the relatively rapid rate of increase in the degree of acidity of the sponge when compared with the straight dough. This increased acidity in turn probably accelerates diastatic activity in the sponge, which is reflected in a more vigorous fermentation and improved size and texture of the baked loaf.

TABLE 135

The Influence of Hydrogen Ion Concentration on the Activity of the Diastase of Wheat Flour Milled from Red River Valley Spring Wheat, and Digested at 27° C. for One Hour (Rumsey, 1922)

Hydrogen Ion Concentration After 1 Hour Digestion, pH	Maltose Produced per 10 Grams of Flour, Milligrams
1.95	+
2.77	34.8
3.52	196.1
4.01	295.1
4.03	302.2
4.40	313.5
4.81	316.8
5.11	300.6
5.19	304.7
5.71	213.4
6.35	90.7
7.02	57.4
7.53	27.4
9.07	9.9
9.13	+
9.97	+

Diastase of barley malt flour was observed by Collatz (1922) to reach its maximum activity in a somewhat more acid medium than the flour diastase of Rumsey, the optimum for the malt flour being about pH = ± 4.3. Gore (1925) found a wider range of hydrogen ion concentration through which malt diastase reached a maximum activity than did Collatz, the optima observed by Gore, in terms of pH, ranging from 4.5 to 5.5. The decrease in activity between pH 5.5 and pH 6.5 was not very great.

A graph of Alder's data was presented by Sorenson (1924), which showed that malt diastase had a marked optimum of activity at pH = 5 ±. Sorenson criticized Rumsey's method for the determination of the activity of flour diastase to the extent that it failed to make provision for comparing all flours at the same hydrogen ion concentration. It was suggested that, after determining the buffer value of a flour, it would be well to determine diastatic power at 3 different concentrations of hydrogen ions, and then draw a curve, which would indicate the relative activity through the range of pH covered by the determinations. Such a procedure has much to recommend it, not only in this, but in other enzyme studies.

The influence of temperature on the apparent optimum hydrogen ion concentration of diastases was discussed by Olsen and Fine (1924), who showed that the optimum pH appears to be a linear function of the temperature. Thus an increase in the temperature of digestion results in a lower optimum hydrogen ion concentration. An equation was evolved for determining the optimum pH at any temperature when the optimum pH at only one temperature is known. This takes the form: $p = \dfrac{82.5 + t}{25}$ where p is the optimum pH at any temperature, t. It is accordingly evident that in the case of this (and possibly any) enzyme it is essential that the temperature of observation be reported together with the observed optimum hydrogen ion concentration.

With temperature as the variable, Rumsey (1922) found that the maximum weight of maltose was produced by flour diastase at 63.5° C. when the digestion period was one hour. In another experiment, in which both time and temperature were included as variables, more maltose was formed at a temperature of 64-65° C., during a digestion period of 180 minutes, than at a temperature of 55°; but the difference in the quantity of maltose produced at the two temperatures during this period is less great than was the case during the first 120 minutes. This suggests that an even more protracted period of digestion might have resulted in a lower "optimum" temperature than appears from observations over a relatively short period of time. It is probable, therefore, that in reporting the apparent optimum temperature for an enzyme, all the limits involved in the observation should be recorded, including time and hydrogen ion concentration.

No correlation between the total gas produced in a dough, or the quantity of gas liberated during the proving period (the period while the dough rises in the pan) and the size of the baked loaf could be detected by Martin (1920b). It was observed, however, that when an active diastatic preparation was added to a dough deficient in diastatic activity, a much larger loaf could be produced. A sample of north Russian flour which baked into a loaf, the volume of which was 2100 cc., yielded loaves displacing 2550 to 2725 cc. when certain commercial diastases were incorporated in the dough.

The stimulation of fermentation and gas production in doughs resulting from the addition of diastatic malt preparations was also demonstrated by Collatz (1922). In many instances such additions increased the size or volume of the resulting bread. Using the viscometric procedure of Sharp and Gortner (1923) as a criterion of the hydrolytic cleavage of gluten, Collatz indicated that the proteases of malt were responsible for an appreciable modification of the gluten. Such changes may be regarded as undesirable in the aggregate. It is probable that the recent tendency in American bakery practice to avoid the use of highly diastatic malt extracts has not been due to any undesirable effect of the true amylolytic and saccharogenic enzymes of such preparations, but rather because active proteases accompany the diastases, which proteases may be responsible for a reduction in the gas retaining power of the dough in which they are present.

How to supplement diastatic activity in flours milled from certain types of wheat accordingly becomes a real problem. Wheats which exhibit this deficiency are not uncommon. In the writer's experience they are produced chiefly in wheat growing regions where the rainfall is comparatively light during the latter part of the growing period. In certain seasons a large part of the wheat crop grown in the northern and western portions of the Great Plains Area appears to be low in diastatic activity. The difficulty seems to be most acute in wheats from western Dakota, north central Montana, and portions of southern Canada. Instances have been known where millers have deliberately purchased parcels of wheat containing kernels which were sprouted, but otherwise sound, and added small percentages of such wheat to the usual wheat blend. Flour mill chemists agree that a hazard attends such practices, because of the increased tendency of flours milled from such a mixture to spoil when stored. Again, it is not always an easy matter to find a quantity of commercial grain containing sprouted kernels which is otherwise sound. Sprouted grain from different sources may not be uniform in properties and composition, which further complicates the production of a uniform flour.

This has turned the attention of certain mill chemists to the possibility of actually malting or sprouting wheat under controlled conditions, to be used in the wheat mixture. In collaboration with the author, Sherwood (1925) determined the effect of using varying quantities of sprouted wheat kernels in the mill mixture upon the diastatic activity, baking strength and keeping qualities of flour milled from such mixtures. The wheat used for this purpose was typical hard, vitreous winter wheat grown in the Judith Basin, in the vicinity of Lewistown, Montana. Such wheat has been long regarded by the author as deficient in diastatic activity, and the straight grade flour milled from the sample that was studied proved to have a diastatic activity, in terms of the units proposed by Rumsey, of 126. This is substantially lower in this particular than the strong flours examined by Rumsey (1922), which ranged between 200 and 300 + units in diastatic activity. The protein content of the flour was fairly high, 12.23 per cent, which is higher than the average protein content of the spring wheat flours milled in the Minnesota State Experimental Flour Mill during the crop seasons of 1922 and 1923, and about equal to that of 1921. When baked the apparent baking strength was good, but it appeared probable that improvement might be effected. After preliminary trials a portion was sprouted for 3 days at 18° C. ±, a longer germinating period having proven undesirable, probably because of the greater proteolytic activity of flour milled from such wheat. The sprouted kernels were then mixed with the normal or ungerminated wheat, and the mixture was milled into straight grade flour in the usual manner. A control sample without germinated kernels was also milled.

The effect of these additions of sprouted wheat on the baking quality of the flour is shown by the data in Table 136. An addition of 2 per cent of sprouted kernels increased the loaf volume 5 per cent, improved the texture and visual appearance and shortened the fermentation period. The diastatic activity of the flour was increased at the same time to 204 units. The addition of 3 per cent of sprouted kernels further accentuated this effect.

TABLE 136

RESULTS OF BAKING TESTS OF FLOURS MILLED FROM THE MONTANA HARD WINTER WHEAT, WITH AND WITHOUT THE ADDITION OF SPROUTED KERNELS

Wheat Mixture	Loaf Volume, cc.	Texture Score	Color Score	Fermentation Period, Mins.	Diastatic Activity
Control without sprouted wheat	2,000	98	97	340	126
With 2% sprouted kernels....	2,100	99	98	332	204
With 3% " "	2,150	100	99	323	242

The comparative gas producing and gas retaining power of doughs made from these flours was determined, using the method described by Bailey and Johnson (1924b). The flour milled from wheat containing 2 per cent of sprouted kernels yielded a dough in which the rate of gas production was about 2 per cent greater than the control. There was no appreciable difference in the loss of gas from this and the control doughs, consequently the dough expanded more when the sprouted wheat flour was used. With the flour milled from wheat containing 3 per cent of sprouted kernels, a still greater increase in gas production was noted, amounting to about 18 per cent more than the control. About two-thirds of this additional gas was lost from the dough, however, the volume of which was about 5 per cent greater than the control.

An analysis of the three flours failed to reveal any substantial increase in non-protein nitrogen in consequence of the admixture of these small percentage of sprouted kernels. When larger proportions were used, as in certain preliminary studies, the non-protein nitrogen was increased by one-half, but such large proportions (40 per cent) of sprouted kernels are neither necessary nor desirable. It appears from these studies that the rational use of properly germinated wheat kernels may serve to advantageously increase the diastatic activity of the flour milled from low-diastase wheat.

Mechanical treatment of the flour particles and starch granules may increase the apparent diastatic activity of flour. This has been demonstrated by the work of Alsberg and Perry (1925). When flour was reground in a ball mill, the quantity of sugar produced autolytically in a flour suspension was substantially increased as compared with the control on normally ground flour. Cold water extract of the overground flour was likewise increased.

The evident effect of several variables upon the rate of sugar production in a dough suggests that the expression "diastatic activity" may not be entirely appropriate. *Saccharogenesis* may be a more inclusive term to describe the totality of combined effects of the actual content of enzyme, and of the environment of the enzyme including hydrogen ion concentration, salt concentration, and the content and condition of the substratum or starch. These, and possibly other variables, determine the rate of sugar formation in which we are primarily interested. Such a term as saccharogenesis would likewise exclude those properties attributed to amylases of partially hydrolyzing starch to non-saccharine derivatives which are soluble in water.

Chapter 11.

Flour Strength and the Colloidal Behavior of Dough.

A consideration of the colloidal behavior of dough should be prefaced by a brief consideration of certain chemical and physico-chemical characteristics of its chief constituents. Prominent among these are starch, fat, and the flour proteins.

The observation of Beccari that wheat flour can be separated into two portions, one of them "vegetable," and the other "animal" in character was orally communicated to the Academy of Bologna in 1728. Later Beccari (1745) published a description of the method of separating gluten from flour by washing out the non-glutinous material of a dough with water. This is the first recorded observations of the existence of gluten in wheat flour.

During the two centuries that have since elapsed gluten has probably been subjected to more extensive and detailed study than any other food protein with the possible exception of casein. This has been due, not alone to the importance of flour as a food, but also to the comparative ease with which gluten can be prepared in crude form by washing out with water most of the other constituents of a flour dough.

That dilute ethyl alcohol will extract protein from flour was observed by Einhof (1805), who considered that the alcoholic extract contained all of the gluten. Taddei (1820) found that part of the gluten was insoluble in the alcohol; the substance which was extracted by alcohol was named "gliadin." The latter fraction was called "plant gelatin" by Liebig (1841), while the alcohol-insoluble fraction of gluten was called "plant fibrin" because of certain supposed resemblances to blood fibrin. This name for the glutenin was used by several chemists, although Ritthausen (1872) termed it "gluten casein." Other protein fractions were separated and named by several chemists, the names being borrowed in many instances from those previously assigned certain animal proteins. Their studies over a period of a century, have been reviewed by Osborne (1907), and need not be detailed here. It was not until Osborne and his co-workers painstakingly separated these proteins from flour, and determined certain of their chemical characteristics, that our knowledge of them was at all adequate. These studies, originally reported in several journal articles, were presented in organized form in Osborne's monograph. The principal proteins identified in flour were gliadin, glutenin, leucosin, a globulin, and one or more proteoses present in small proportions. The gliadin and glutenin are the chief protein constituents of the "gluten" to which reference has already been made. A small quantity of a nucleo-protein is also present in flour.

The Chemical Characteristics of Gliadin.

Gliadin is the principal prolamin of wheat. It is insoluble in absolute ethyl alcohol, but is soluble in dilute ethyl alcohol. Osborne employed 70 per cent alcohol by volume in extracting it from flour. A critical study of the quantitative determination of gliadin was reported by Greaves (1911) who recommended that 74 per cent alcohol by volume be used in extracting this protein from flour. Greaves concluded that the gliadin thus extracted was more nearly pure than when other concentrations of alcohol were used as the solvent. The criterion of purity of the gliadin in these alcoholic extracts was the constancy of the ratio of nitrogen content to optical rotation. A more dilute aqueous alcohol solution extracted an appreciably larger quantity of protein than the 74 per cent alcohol, however. This had also been observed by Shutt (1907), who found 60 per cent alcohol by weight (67.7 per cent by volume) extracted 5.36 per cent of gliadin from a certain flour, while 70 per cent alcohol by weight (76.9 per cent by volume) extracted 4.71 per cent. Similar relations of the concentration of alcohol to the quantity of protein extracted were noted by Kjeldahl (1896), Teller (1896), Snyder (1906), and Hoagland (1911). Thus Hoagland found that alcohol of 45 to 55 per cent by weight (52.3 to 62.8 per cent by volume) extracted more protein from hard and soft wheat flours than any other concentration of ethyl alcohol, and recommended that it be used as a solvent in the determination of gliadin.

Alcoholic solutions (70 per cent by volume in water) extracted nitrogenous substances from flour which Olson (1913) believed to be a mixture of gliadin and other proteins. The fraction which separated out when the alcohol of the extract was evaporated off and replaced by water was assumed to constitute all of the gliadin of the extract. It is probable that not all of the gliadin was coagulated by this treatment, however. Bailey and Blish (1915) failed to detect any substantial difference in the relative purity of the gliadin extracted from flour when alcoholic solutions of 50 and 70 per cent by volume were used as the solvents. Using the ammonia fraction of the hydrolysate of the proteins thus extracted as an indication of the purity of the gliadin in the extract, it was estimated that about 93 per cent of the protein in these alcoholic extracts was gliadin. When the flour was digested in a pressure flask at 83° C. with 50 per cent alcohol, 1.58 per cent of nitrogen was present in the proteins of the extract. The same concentration of cold alcohol extracted only 1.27 per cent of protein nitrogen. Greaves (1911) also observed that a larger quantity of protein was extracted by hot alcohol than by cold alcohol of the same concentration.

Ratio of flour to alcohol appears to influence the percentage of protein extracted from flour with aqueous alcohol solutions. Thus Greaves (1911) recorded data which indicated that in general a ratio of 2 grams of flour to each 100 cc. of alcohol yielded a larger percentage of protein in the extract than when more flour was used. In working with purified gliadin preparations, Tague (1925b) found that increasing the ratio of gliadin to alcohol resulted in an increased concentration of gliadin in solution in the alcohol after several hours.

Gliadin is soluble not only in aqueous ethyl alcohol solutions, but in methyl, propyl, and benzyl alcohol, phenol, paracresol, glacial acetic acid, and possibly in other organic solvents. Schleimer (1911) compared aqueous methyl alcohol solutions of varying concentration as solvents for gliadin and found the maximum solubility in 60 per cent solutions by weight. A similar comparison of aqueous propyl alcohol solutions indicated that the maximum quantity of protein was extracted from flour with a 40 per cent solution by weight. The quantity of protein in the propyl alcohol extract was somewhat greater than when ethyl alcohol solutions were used as solvents. Thus with a certain flour the maximum quantity of protein extracted with ethyl alcohol was 59.1 per cent of the total protein of the flour, while with propyl alcohol solutions 64.7 per cent of the total protein was extracted. Amyl alcohol, because of its immiscibility with water, was used in the pure state as a solvent for gliadin, and only 1.9 to 2.5 per cent of the total protein of the flour appeared in the extract.

Tague (1925b) observed a greater solubility of gliadin in 60 per cent methyl alcohol by volume than in more dilute aqueous solutions of the same alcohol. These observations were made with a purified gliadin preparation. A progressive increase in quantity of gliadin dissolved with the lapse of time up to 240 hours was noted.

Gliadin is dispersed as a colloidal sol in dilute aqueous solutions of acids and alkalis. In studying the solubility of a purified gliadin preparation in aqueous solutions of acids and alkalis, Tague (1925b) noted a greater solubility in acetic acid solutions than in solutions of hydrochloric and sulfuric acids. With all three acids the maximum solubility was in a solution with a concentration of hydrogen ions equivalent to $pH = 2 \pm$. The minimum solubility in a buffered phosphate solution was at $pH = 6.5 \pm$. On the alkaline side of neutrality the quantity of gliadin dissolved increased with the concentration of alkali up to a degree of alkalinity equivalent to $pH = 13.1$, which was the most alkaline solution used. A greater solubility of gliadin was encountered in 0.009

molar sodium carbonate solution than in more dilute or more concentrated solutions of the same alkali.

Gliadin can be precipitated from the sols produced on dispersing the protein in dilute acid and alcohol solutions, by bringing the solution to the isoelectric point of the protein (pH = 6.5 — 6.6). Fleurent (1896b) determined the gliadin-glutenin ratio of crude gluten by dispersing the gluten in an alcoholic solution of potassium hydroxide (3 grams per liter), and precipitating the glutenin by converting the hydroxide to carbonate with a current of carbon dioxide. Under these conditions the gliadin was presumed to remain dissolved in the alcohol, and an aliquot of this solution was evaporated to dryness, and the residue minus the weight of carbonate was considered to be gliadin.

The rapidity of precipitation of gliadin from hydrochloric acid and sodium hydroxide solutions when 10 per cent solutions of gliadin in alcohol were added thereto was used by Eto (1924) as an index of the region of the isoelectric point. This method indicated the isoelectric point to be about pH = 6.6. In mixtures of monopotassium phosphate and sodium hydroxide of varying hydrogen ion concentration the isoelectric point appeared to be in the region of pH = 7.2.

The isoelectric point of the gliadin prepared by Hoffman and Gortner (1925) was found to be pH = 5.76, when determined by suspending 1 per cent of the protein in pure water. The logarithms of the equivalents of acid or alkali bound were extrapolated to the point of intersection of the two curves for acid binding and alkali binding respectively. This point of intersection was considered as the isoelectric point. Such data were subjected to mathematical treatment in three different manners, and yielded values ranging from pH = 6.97 to pH = 7.60. These values are considered to represent the isoelectric points for the binding of acid above a hydrogen ion concentration of pH = 2.5, and the binding of alkali above a hydroxyl ion concentration of pH = 10.5. The true isoelectric point was believed to be somewhere near the measured value as determined potentiometrically (pH = 5.76), and is determined by the predominance of acidic groups in the molecule.

While Günsberg (1862) observed that gliadin was slightly soluble in water, its relative solubility in that solvent is ordinarily rated as low. Sharp and Gortner (1923) found, however, that repeated extractions of flour with water removed most of the alcohol-soluble protein from flour. Thus after 8, 12, and 16 extractions with water, 1.85, 1.25, and 1.12 per cent of alcohol-soluble protein remained in the extracted residue. Direct extraction of the same flour yielded 9.04 per cent of alcohol-soluble protein.

The presence of most salts in the aqueous solution used as solvent diminishes the solubility of gliadin. Bailey and Blish (1915) estimated that approximately half of the protein extracted from a certain flour sample by 1 per cent sodium chloride solution (in water) was gliadin, while less than 20 per cent of the protein in the extract was gliadin when 10 per cent sodium chloride solution was used as the solvent. Since the quantity of non-gliadin proteins in the two extracts was approximately the same, the decrease in gliadin solubility on increasing the concentration of salt in the solution used in extracting the proteins is evident. This diminishing solubility with increasing concentration of most salts is of particular importance in estimating the non-gluten protein content of flour, and reference is made to this fact in the discussion of the albumin and globulin content of flour.

In Tague's (1925b) study of the solubility of gliadin in various aqueous salt solutions, one notable exception to the general rule that increasing salt concentration diminishes the solubility of gliadin was encountered. This was in the solutions of magnesium chloride; the 1.0 molar solution of this salt dissolved 59.2 milligrams of gliadin per 100 cc. of solution, while the 0.01 and 0.001 molar solutions dissolved 5.4 and 4.0 milligrams respectively.

The products of hydroylsis of gliadin, as well as of glutenin, and leucosin, as determined by Osborne (1907) are shown in Table 137. All of the nitrogen of the original proteins cannot be accounted for by the Fischer esterification method as used by Osborne. It was established, however, that gliadin on hydrolysis yields a larger amount of ammonia, glutamic acid, and proline than other proteins. Lysine and glycocoll were not identified in the hydrolysate, and the relative proportion of histidine and arginine was small. Gliadin gave a strong Molisch reaction, suggesting the possible presence of a carbohydrate group in the purified substance. The idea advanced by Kosutany (1907) that gliadin is a hydrate of glutenin was definitely set aside as the result of these researches of Osborne et al., which showed that gliadin and glutenin differed substantially in their chemical constitution as indicated by the differences in the composition of the acid hydrolysate of the two proteins.

More recently several investigators have employed the Van Slyke (1911) method in the study of flour proteins. The results of such studies of gliadin and glutenin by Van Slyke (1911), Osborne, Van Slyke, Leavenworth and Vinograd (1915), Blish (1916), Cross and Swain (1924), and Hoffman and Gortner (1925) are recorded in Table 138. The ammonia fraction was found to be higher in gliadin than in glutenin, the arginine and lysine fraction was highest in the glutenin. Cross and Swain determined the tyrosine and tryptophan content of their gliadin and glutenin preparations by the Folin and Looney method with the following results:

	Gliadin	Glutenin
Tyrosine, per cent	5.04–5.10	5.34–5.92
Tryptophan, per cent	1.03–1.19	1.55–1.61

TABLE 137

PRODUCTS OF HYDROLYSIS AND ULTIMATE COMPOSITION OF WHEAT PROTEINS
(OSBORNE, 1907)

	Gliadin, Per Cent	Glutenin, Per Cent	Leucosin, Per Cent	Globulin, Per Cent
Products of Hydrolysis				
Glycocoll	0.00	0.89	0.94	....
Alanine	2.00	4.65	4.45	....
Amino-valerianic acid	0.21	0.24	0.18	....
Leucine	5.61	5.95	11.34	....
Proline	7.06	4.23	3.18	....
Phenyl-alanine	2.35	1.97	3.83	....
Aspartic acid	0.58	0.91	3.35	....
Glutamic acid	37.33	23.42	6.73	....
Serine	0.13	0.74	....	....
Tyrosine	1.20	4.25	3.34	....
Cystine	0.45	0.02	....	....
Lysine	0.00	1.92	2.75	....
Histidine	0.61	1.76	2.83	....
Arginine	3.16	4.72	5.94	....
Ammonia	5.11	4.01	1.41	....
Tryptophan	present	present	present	....
Total	65.81	59.66	50.32	....
Ultimate Composition				
Carbon	52.72	52.34	53.02	51.03
Hydrogen	6.86	6.83	6.84	6.85
Nitrogen	17.66	17.49	16.80	18.39
Sulfur	1.03	1.08	1.28	0.69
Oxygen	21.73	22.26	22.06	23.04
Ratio of N to total protein	1:5.66	5.72	5.95	5.44

Varying percentages of nitrogen are contained in the flour proteins, as shown by the data reported by Osborne (1907) that will be found in Table 137. The ratio of nitrogen to the dry weight of the protein ranges from 1:5.44 in case of the globulin, to 1:5.95 in the leucosin. The mean ratio in these two proteins is so nearly that of the gliadin-glutenin combination as to render the factor $N \times 5.7$ the suitable and most accurate factor for the calculation of the total protein content of flour from the percentage of nitrogen as determined by the Kjeldahl method or one of its modifications.

The rate of hydrolysis of gliadin was studied by Vickery (1922), who found that the liberation of ammonia, presumably amide hydrolysis, was readily effected at boiling temperatures by very dilute acid or alkali. The ammonia was set free with great rapidity by the stronger acid reagents. Peptide hydrolysis did not proceed so rapidly as amide hydrolysis, and 20 hours were required to carry it to completion in 20 per cent hydrochloric acid. Much longer was required when less concentrated acid was used. Vickery (1923) could not detect a differential hydrolysis of gliadin involving the breaking of the amide

TABLE 138

Distribution of Nitrogen in the Acid Hydrolysate of Gliadin and Glutenin, as Determined by the Van Slyke Method by Several Chemists

Percentage of Nitrogen in	Van Slyke (1911)	Gliadin				Glutenin	
		Osborne, et al. (1915)	Blish (1916)	Cross and Swain (1924)	Hoffman and Gortner (1925)	Blish (1916)	Cross and Swain (1924)
Ammonia	25.52	24.61	25.90–26.13	26.20–26.79	24.61	16.17–16.50	13.11–15.98
Humin	0.86	0.58	0.50– 0.57	....	0.87	1.66– 1.84	....
Cystine	0.68	0.80	0.29– 0.37	0.71– 0.84	1.68	0.18	0.65– 0.72
Arginine	5.01	5.45	4.47– 4.55	4.78– 5.21	6.38	9.27– 9.69	8.17–12.94
Histidine	4.36	3.39	5.62– 6.77	5.54– 6.41	5.41	5.47– 7.59	6.20–11.42
Lysine	0.64	1.33	0.65– 0.97	0.56– 0.66	0.57	1.90– 2.61	4.72– 6.42
In filtrate from bases:							
Amino—N	53.13	51.95	53.46–54.10	51.44–53.26	53.49	53.38–53.59	54.02–56.55
Non-amino—N	9.58	10.70	7.44– 7.55	4.44– 5.65	6.14	9.35– 9.52	2.64– 3.98

linkages only. Hydrolysis of the amide and peptide bonds proceeded simultaneously although the latter linkages were broken at a slower rate.

An extensive study of the specific rotation of gliadin in various organic solvents was reported by Mathewson (1906), while Osborne and Harris (1903b), Kjeldahl (1896), Lindet and Ammann (1907), Osborne (1907), Woodman (1922), Blish and Pinckney (1924) and Dingwall (1924) have added observations. Certain of their data are included in Table 139.

TABLE 138

Specific Rotation of Gliadin in Various Organic Solvents

Observer	Solvent	$(\alpha)_D$
Kjeldahl (1896)	55% ethyl alcohol	— 92.0°
	Phenol	— 130.0°
	Glacial acetic acid	— 81.0°
	.1–5% acetic acid	— 111.0°
Osborne and Harris (1903b)	80% ethyl alcohol	— 92.28°
Mathewson (1906)	50% ethyl alcohol	— 98.45°
	60% " "	— 96.66°
	70% " "	— 91.95°
	70% methyl "	— 95.65°
	60% propyl "	— 101.10°
	70% phenol	— 123.15°
	Anhydrous phenol	— 131.77°
	Glacial acetic acid	— 78.60°
	Paracresol	— 121.00°
	Benzyl alcohol	— 53.10°
Osborne (1907)	80% ethyl alcohol by volume	— 92.3°
Lindet and Ammann (1907)	70% ethyl alcohol (α fraction)	— 81.6°
	70% " " (β ")	— 95.0°
Woodman (1922)	70% " " by volume..	— 93.7°
Blish and Pinckney (1924)	70% " " " "	93.6° [1]–100.0°
	Glacial acetic acid	79.5° [1]– 85.8°
Dingwall (1924)	70% alcohol—wave length of light	6500 Ång. — 72.4
		5890 " — 92.3
		5300 " — 119.9
		4800 " — 151.4

[1] Gliadin from Polish wheat, T. polonicum.

A method for the quantitative determination of the gliadin content of flour, based on the optical rotation of polarised light by its alcoholic solutions was proposed by Snyder (1904b).

Gliadin, like other proteins, may be racemized by warming it in dilute alkali solutions. Dakin (1912) maintains that racemization probably involves a keto-enol transformation within the protein molecule, as

$$R - \overset{|}{C}H - CO \rightleftharpoons R - \overset{|}{C} = C(OH) -,$$ the central carbon atom losing its assymetry. This effects a progressive decrease in the specific rotation of the protein. Such a change in specific rotation in gliadin solu-

tions incubated in 0.5 N sodium hydroxide solutions at 37° C. was observed by Woodman (1922).

Fleurent (1901, 1905b) proposed that the gliadin content of flour be estimated from the density of an alcoholic extract, but objections to this procedure were raised, since the differences in density encountered were rather narrow and experimental error might give rise to substantial variations in the apparent gliadin content of the same material. The refractive index of gliadin in various solvents was investigated by Robertson and Greaves (1911), but no extensive use has been made of this property in the quantitative estimation of the gliadin content of flour.

On the assumption that the gliadin molecule contains 5 atoms of sulfur, Osborne (1907) assigned to it a minimum molecular weight of 15,568. Woodman (1922) found that gliadin has a combining weight of approximately 5,000, and its molecule probably contains 3, or a multiple of 3 carboxyl groups. Cohn (1925) concluded that the minimum molecular weight of gliadin is 20,700 and the probable molecular weight 125,000. The combining weight of gliadin for HCl was calculated by Hoffman and Gortner (1925) to be 5,014 at $pH = 2.8$, and 2,507 at $pH = 2.5$; for NaOH at $pH = 10.2$ it was calculated to be 5,014.

The Chemical Characteristics of Glutenin

The properties of glutenin differ markedly from those of gliadin. It is practically insoluble in water, alcohol, and aqueous saline solutions, and is only slightly soluble in hot dilute alcohol, from which it separates on cooling. Glutenin can be dispersed by dilute acids and alkalis. The studies of Sharp and Gortner (1923) indicate that the maximum dispersion with acids may be effected at $pH = 3$, and with alkalis at $pH = 11$. The same observations indicate that glutenin may have an isoelectric range between pH 6 and pH 8, since there was a minimum of swelling and imbibition of water by flour through this range of hydrogen ion concentration. This swelling, as indicated by viscosity of the flour suspension, was shown to be almost wholly attributable to the imbibition of water by the glutenin present. Tague (1925a) determined the isoelectric point of glutenin by observing the change in hydrogen ion concentration resulting from the addition of a unit quantity of glutenin to phosphate buffer solutions of varying pH. The isoelectric point was regarded as the concentration of hydrogen ions in which the glutenin did not alter that concentration, and this was observed to be $pH = 6.8$ to 7.0.

A fractional precipitation of glutenin by addition of acid to its alkaline solution indicated to Halton (1924) that at least two proteins were present in the "glutenin" preparation. The separation was effected by adding normal hydrochloric acid to the alkaline solution of glutenin until protein separated out in a flocculent state. This protein was allowed to settle out, and further additions of acid were then made to the clear supernatant liquid freed from the first precipitate. The liquid became progressively more and more milky in appearance with additions of acid until the equivalent of 3.8 cc. of normal hydrochloric acid had been added to each liter of solution. A second flocculation of protein then occurred. About five times as much protein was obtained in the first separation when the least acid was present as in the second fraction. The two glutenin fractions were found to be unlike when their racemisation curves were compared.

Unfortunately Halton apparently did not determine the hydrogen ion concentration of the solution when the precipitation of protein occurred and it thus becomes difficult to exactly repeat his work. Blish (1925) attempted to fractionate glutenins separated from five flour samples, but without success. It was found that if glutenin was partially racemized, it precipitated at a different hydrogen ion concentration than the original glutenin. This suggests that a portion of Halton's glutenin may have been partially racemized which would account for the two fractions he obtained.

The rotation dispersion of glutenin was determined by Dingwall (1924) using light of three different wave-lengths. The specific rotation in 0.5 N sodium hydroxide solution was as follows:

Wave Length	Specific Rotation
6500 Ång. units	$-60.1°$
5890 " "	$-74.5°$
5300 " "	$-94.9°$

The specific rotation in 0.04 N sodium hydroxide of the glutenin which Woodman (1922) prepared from two flours was as follows:

Glutenin from Manitoba flour	$= -99.5°$
" " English "	$= -78.8°$

Blish and Pinckney (1924) examined several preparations from hard wheats and found their specific optical rotation in 0.1 N sodium hydroxide after 2 hours to be 76.2°.

Glutenin is the most distinctive of the proteins of flour. It apparently does not occur in any species other than those of the genus Triticum. Within this genus its properties may vary widely among the several species, and even in individual specimens of the same species. This is particularly true of its colloidal properties, to which reference will be made in another connection.

The products of acid hydrolysis were referred to in the discussion of the data in Tables 136, and 137, where they were compared with those of gliadin.

The quantitative estimation of the glutenin content of flour has heretofore been an indirect procedure, and it is probable that many of the statements in the literature regarding the percentage of this constituent in flour are of doubtful accuracy. In the majority of instances the glutenin content of flour has been estimated by deducting the sum of the alcohol-soluble protein and the protein extracted by dilute saline solutions from the total protein content. The alcoholic extract may contain proteins other than gliadin. The saline solutions may extract appreciable quantities of gliadin, depending on the kind and concentration of the salt in the aqueous solution used as the solvent. It accordingly follows that an accumulation of errors appears in the final calculation, and it is almost inevitable that such a calculation will yield too low a value for the glutenin content. Sharp and Gortner (1923) modified this procedure, extracting the flour with a 5 per cent aqueous solution of potassium sulfate to remove the albumin and globulin, and then extracting the residue with 70 per cent alcohol. The sum of the protein fractions extracted by both solvents was deducted from the total protein, and the difference recorded as glutenin. Another modification of the older methods has been suggested by Blish and Sandstedt (1925). Flour was shaken with a mixture consisting of 10 parts of water and 1 part of normal sodium hydroxide solution for 1 hour; 30 parts of acetone-free methyl alcohol was then added. This solution contained approximately 70 per cent methyl alcohol by volume and was somewhat less than 0.025 normal with respect to sodium hydroxide. The suspension was then decanted through a cotton plug. The glutenin was precipitated from an aliquot by the addition of dilute hydrochloric acid until the hydrogen ion concentration corresponded to a pH of about 6.4, as shown by the color change of brom thymol blue indicator. The glutenin settled out, leaving a clear solution containing the other proteins that had been extracted from the flour. A compact disk of glutenin could be secured by whirling the suspension, after precipitation with acid, in a centrifuge tube. This glutenin was removed, and the quantity determined by the Kjeldahl nitrogen method. Results obtained by this method agreed well with those obtained when the procedure outlined by Sharp and Gortner was employed.

The minimum molecular weight of glutenin is 36,300, in the opinion of Cohn (1925), and the probable molecular weight is 108,900.

The Chemical Characteristics of Leucosin

Leucosin, the albumin of wheat, was found by Osborne (1907) to constitute about 0.3 to 0.4 per cent of the wheat kernel. The proportion in the embryo of the kernel is much larger than in the endosperm, since about 10 per cent of leucosin was found in commercial "germ meal."

An albumin was found by Jones and Gersdorff (1923) to constitute about one-sixth of the total protein of washed bran.

Like albumins generally, leucosin is soluble in pure water, and is coagulated when its solutions are boiled. It is soluble in dilute saline solutions, and is relatively insoluble in aqueous alcohol solutions. It may be precipitated from aqueous solutions by saturating them with sodium chloride or magnesium sulfate, or by half saturating with ammonium sulfate. In this particular it is more readily precipitated than albumins of animal origin; in its other characteristics it resembles animal albumins more closely than most vegetable proteins.

The products of hydrolysis of leucosin as determined by Osborne (1907) are shown in Table 136. The ammonia fraction in the acid hydrolysate is much smaller than was encountered in hydrolysed gliadin and glutenin. Osborne and Harris (1903a) reported that 6.8 per cent of the nitrogen in the hydrolysate of leucosin was in the form of ammonia.

The isoelectric point of leucosin was found by Lüers and Landauer (1922) to be in the range pH = 4.6.

The Chemical Characteristics of Wheat Globulin

It has proven difficult to separate and purify a large quantity of the globulin of wheat flour, and less is known of its chemical constitution and properties than of the gluten proteins and leucosin. Osborne (1907) found about 0.6 per cent of globulin in the wheat which he examined. The embryo evidently contains more than the endosperm, 5 per cent being extracted from the former. The ultimate composition of the globulin separated by Osborne showed it to contain a larger percentage of nitrogen than the other flour proteins. This is shown by the data in Table 136. Like leucosin, the ammonia fraction in the acid hydrolysate was low, Osborne and Harris (1903a) finding 7.7 per cent of the total nitrogen of the hydrolysate in the form of ammonia.

Globulin is regarded as relatively insoluble in pure water, but freely soluble in dilute saline solutions. In fractionating the proteins of flour, the usual practice is to extract both the globulin and leucosin with 5 per cent aqueous potassium sulfate solution, or 10 per cent aqueous sodium chloride solution. Gliadin is appreciably soluble in less concentrated aqueous saline solutions which accordingly must be avoided.

Globulin is precipitated by saturating its solutions with magnesium sulfate, but not with sodium chloride. It is not readily coagulated by heating its solution in 10 per cent aqueous sodium chloride at temperatures below 100°.

The Chemical Characteristics of Tritico-Nucleic Acid

It is probable that a small proportion of a nucleo protein is present in flour. The percentage of this protein is low because of the relatively small proportion of nuclear material in the starchy parenchyma cells of the endosperm of wheat. Consequently little attention is given to this constituent of flour. Osborne and Harris (1902) examined the triticonucleic acid of wheat germ. The molecule was of large size and its sols exhibit colloidal properties. In cold water the free acid is practically insoluble; in hot water it forms a pasty mass. Salts of triticonucleic acid with the alkali metals are soluble. The acid appears to be polybasic, and its salts are fairly stable, difficulty being experienced in again recovering the free acid. The sodium salts of nucleic acids from wheat germ were prepared and studied by Thomas and Dox (1925).

The Characteristics of Crude Gluten

The fact that the percentage of crude protein in flour usually varied directly with the percentage of crude gluten that could be washed from it, was first indicated by Millon (1854). A high positive correlation coefficient of dry gluten and protein content was found in most of the groups of samples in which the coefficient was computed by Zinn (1923). With the exception of the Colorado spring wheats, the correlation coefficient was greater than $+0.8$. This serves to establish the fact that the protein content of flour serves to indicate the gluten content, and since the percentage of protein can be determined with greater precision than the percentage of dry gluten, the former determination is to be preferred from the quantitative standpoint.

TABLE 139

Coefficient of Correlation of Dry Gluten and Protein Content as Computed by Zinn (1923)

Kind of Wheat	Number of Samples	Coefficient of Correlation
Colorado commercial spring wheats	48	0.6951 ± 0.0503
North Dakota " " "	37	0.8300 ± 0.0345
Idaho " winter wheats	60	0.9824 ± 0.0031
Ohio " " "	58	0.8695 ± 0.0217
" pure strains of winter wheats	13	0.9483 ± 0.0190
Maine " " " spring wheats	31	0.9603 ± 0.0095

In the hands of a skilled technician, the crude gluten can be separated approximately quantitatively and with a fairly uniform degree of purity. The physical properties of this glutinous mass can then be observed. The quantity of the wet gluten obtained from a unit of flour serves as an index of the percentage of gluten in the flour. The dried crude gluten, after desiccation of the wet material in a suitable drying oven, is a more precise measure of the actual gluten content of the flour.

As the percentage of crude gluten came to be extensively used as a measure of flour properties, other investigators came to appreciate that numerous possible sources of error in this determination must be recognized, and measures of control were introduced to insure accurate and acceptable results. Benard and Girardin (1881) indicated that the quantity of gluten that can be washed from flour dough increases to a maximum and then decreases with the length of the period that the dough is allowed to stand before washing. This was confirmed by Balland (1884), who found, moreover, that this maximum varied with different flours. The size of the flour sample, and the proportion of water used in making the dough likewise influenced the yield of gluten. Later Balland (1893) discussed the significance of the temperature at which the dough was maintained in its relation to yield of gluten. At $2°$, $15°$, and $60°$ C. the wet gluten separated from a certain flour sample represented 27.0, 27.6 and 30.0 per cent respectively of the quantity of flour. Arpin (1902) likewise obtained an increased quantity of dry crude gluten by increasing the temperature of the dough.

Hard water, when used in preparing and washing the dough resulted in a greater yield of dry crude gluten than when distilled water was used. Increasing the time that the gluten was submerged in water was observed by Arpin (1902) to result in the wet gluten absorbing additional water. When the time of washing was increased beyond the normal, the yield of wet and dry gluten diminished.

Increasing the proportion of water in the dough, the length of time the dough was allowed to stand, or the temperature of the wash water was found by Kepner (1914) to affect the weight of wet gluten recovered more than the weight of dry gluten. Over-washing decreased the percentage of both wet and dry gluten.

The influence of the salts in hard waters upon the yield and physical properties of wet gluten having been made apparent by the studies of Balland (1884, 1893), Stein (1904), Fleurent (1904) and others, several suggestions were made relative to the use of synthetic hard waters in washing gluten. Fleurent (1905a) recommended that a solution of 0.01 per cent of calcium carbonate in distilled water be used in washing the gluten from flour, and that the wet gluten thus obtained be then washed for 2 minutes in distilled water. Jacobs (1915) suggested a solution of 0.3 grams of calcium chloride and 0.3 grams of magnesium chloride in each liter of distilled water. LeClerc (1920) reported that washing gluten with distilled water containing 0.1 per cent of sodium chloride gave results comparable with those obtained through the use of tap water.

A critical study of the gluten washing problem was made by Dill and Alsberg (1924). A progressive decrease in the non-nitrogenous constituents of the gluten with prolongation of the washing period was observed. Loss of gluten proteins resulted from the use of distilled water instead of tap water. The use of buffer solutions with varying hydrogen ion concentration in washing resulted in the selection of 0.1 per cent sodium phosphate solution (pH = 6.8) as the most suitable for washing gluten. This was prepared by mixing 4 per cent disodium phosphate solution with monosodium phosphate solution of the same concentration until the desired hydrogen ion concentration was reached as shown by testing with brom-cresol-purple, and phenol red. The mixture was then diluted with water to a concentration of 0.1 per cent of solute.

When flour suspensions were decanted to remove the aqueous extract, and the extracted residue was dried, Olson (1912a) found it impossible to wash gluten from a dough made from the dried residue. Dialysing a flour suspension had much less effect upon the quantity of dry gluten that could be washed from the dried residue although an appreciable reduction in the yield of gluten resulted from such treatment.

That dry crude gluten contains considerable non-protein material was shown by Macfarlane (1905). The protein content of dry crude gluten specimens analysed by Norton (1906) ranged from 68 to 83 per cent, the average being 80.91 per cent. Of this, the gliadin represented 39.09 per cent, glutenin 35.07 per cent, and protein soluble in 10 per cent sodium chloride solution, 6.75 per cent. Ash, ether extract, fiber, and carbohydrates other than fiber in the dry crude gluten averaged 2.48, 4.20, 2.02, and 9.44 per cent respectively. The average total protein content of dry crude glutens analysed by Olson (1912b) was 72.79 per cent. Gerum and Metzer (1923) observed that prolonged washing of the wet crude gluten tended to render it starch-free. A dry crude gluten produced in this manner contained 89.4 [2] per cent of total protein (N $\times$ 5.7). Dill and Alsberg (1924) also produced a dry crude gluten which contained less non-protein material than the control in consequence of prolonged washing. Thus with tap water the protein content of dry gluten which had been washed 25 minutes was 78.8 per cent, and after 60 minutes washing it was increased to 83.5 per cent. This increase in relative purity of the crude gluten resulted from the mechanical removal of starch and other impurities. It is evident from these studies, however, that it must be practically impossible to free the wet crude gluten from all impurities without excessive loss of gluten proteins, and that starch, ash, fiber, and fat will almost invariably be present, and in varying proportions depending upon the treatment accorded the crude gluten in washing.

Dill (1925) analysed crude gluten separated from soft winter wheat

[2] Correcting their data by the use of the factor N $\times$ 5.7, instead of N $\times$ 6.25, which latter factor was employed by them.

flour, and a sample of commercial gluten. The principal carbohydrate found was starch. The proportion in different crude gluten preparations was evidently quite variable. Appreciable quantities of lipoids were found, amounting to about half the lipoid of the equivalent weight of flour from which the gluten was derived. About one-sixth of the ash, one-fifth of the total P_2O_5, one-fourth of the lipoid P_2O_5, and one-seventh of the ash present in the flour appeared in the gluten. This is shown by the data in Table 140.

TABLE 140

Composition of Dry Crude Gluten (Dry Basis) Reported by Dill (1925)

	Gluten from Soft Winter Wheat	Percentage of Each Constituent Appearing in Gluten	Commercial Gluten
Protein (N × 5.7)	72.67	84.5	81.00
Ether extract	0.75 [a]	....	....
Lipoids (neutral extraction method)	7.05	4.75	11.56
Ash	0.63	15.6	0.91
Carbohydrates (acid hydrolysis)	18.82	....	4.93
	99.17		98.40
Carbohydrates (Pflüger glycogen method)	17.52	....	3.50
Total P_2O_5	0.38	20.4	0.46
Phytin P_2O_5	0.05	13.0	0.22
Lipoid P_2O_5	0.10	26.0	0.10
Calculated phosphatide	1.13	....	1.13
Lipoids (acid digestion method)	5.01	....	10.76

[a] Not included in total.

When patent or high grade flours containing an unusually low percentage of gluten are mixed into a paste or dough and washed with water, it proves impossible at times to recover an appreciable quantity of wet crude gluten. The dilution of the gluten by starch is so great in such instances that the gluten particles are not afforded an opportunity to agglutinate into a coherent mass in the washing process. When this difficulty is encountered, a high gluten flour of known composition and properties may be mixed in definite proportions with the low gluten flour, as suggested by Olson (1912b). This mixture may then be subjected to the usual treatment, and the yield of gluten from the mixture corrected by deducting the gluten contributed by the strong flour.

Gluten Content and Flour Strength

The discovery of the presence in flour of the peculiar and characteristic protein mixture known as gluten served to focus attention upon the correlation between the concentration of this substance and baking strength of the flour. Numerous investigators including Balland (1896), Guess (1900), Hall (1904), Humphries (1905), Bremer (1907), Shutt (1907), Schneidewind (1909) and others, recognized that gluten content must play an important rôle in determining flour strength. The views expressed varied somewhat, but those of Hall are fairly representative. Hall stated that while the concentration of protein fails to measure flour strength in any absolute sense, the order of protein content will be the order of the strength of flours or nearly so. Shutt (1907) apparently concurred in this opinion, since he stated that the best breadmaking flours are, as a rule, those richest in gluten. All of these scientists recognized that variations in the "quality" or physical properties of the glutens from different flours were encountered, and were searching for a measure of these properties that could be accurately determined and expressed.

Several studies have been reported recently which serve to indicate the extent to which gluten content can be depended upon as an index of flour strength. Bailey (1913c) subjected several hundred flour samples to baking tests, and their ash and protein content was determined. These flours were grouped on the basis of their content of ash and protein, and the volume or cubical displacement of test loaves baked from the flours in each group was then averaged. The resulting data are presented in Table 141, and these indicate that when the averages of a sufficient number of samples are involved, the order of gluten content is the order of loaf volume. Comparisons must be restricted to the same grade of flour however, since the same data show that the loaf volume diminished with increasing ash content. Individual flours in each group varied somewhat from the mean of the group.

TABLE 141

Relation Between the Percentage of Ash and Crude Protein in Flours, and the Volume of Loaves Baked Therefrom, Reported by Bailey (1913c)

Range of Ash Content in Each Group	Range of Crude Protein Content in Each Group						
	9.0 to 10.0	10.1 to 11.0	11.1 to 12.0	12.1 to 13.0	13.1 to 14.0	14.1 to 15.0	15.1 to 16.0
0.35 to 0.40	....	....	....	2560	2610	....	....
0.41 to 0.50	2130	2270	2410	2425	2500	2520	....
0.51 to 0.60	1980	2090	2220	2360	2390	2410	....
0.61 to 0.75	1895	1925	1960	2110	2195	2250	....
0.75 to 0.90	1600	1855	1925	2010	2050	2060	2120
0.91 to 1.10	....	....	1890	1900	1910	1875	2010
1.11 to 1.25	....	....	....	....	....	1670	1770

Continuing the same study, Thomas (1917b) presented a graph in which the loaf volume of test bakes of flours, grouped on the basis of protein content, was plotted against the average percentage of protein in the wheats from which they were milled. This curve established a positive correlation between average baking strength, as indicated by loaf volume, and protein content. One exception was noted in the instance of hard spring wheats having a crude protein content of over 15 per cent. Much the same correlation was shown by Stockham's (1920) graphs, although the protein of soft winter wheat flours seemed to be less effective in producing large loaves than did the same concentration of protein in hard wheat flours.

A similar analysis of the data resulting from the tests of hard spring

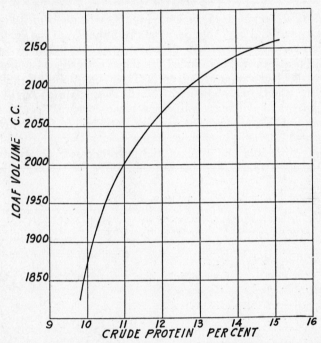

Fig. 19.—Graph showing the correlation between protein content and baking strength (in terms of loaf volume) of flours grouped on the basis of protein content. (Bailey, Minn. State Dept. Agr. Bul. 34.)

wheat straight grade flours milled in the Minnesota State Experimental Flour Mill was made by Bailey (1924). The relation between average loaf volume, and average protein content of the groups of flours classified on the basis of the percentage of crude protein, is shown by the data in Table 142, presented graphically in Figure 19. From the slope of the curve it is evident that the loaf volume did not increase regularly with protein content; each increment of increase in percentage of protein was progressively less effective in increasing the size of the loaf. The curve accordingly approaches a logarithmic rather than a linear form, and would necessitate the use of an exponential value in an equation representing the function of one variable in terms of the other.

TABLE 142

Relation of Average Loaf Volume and Average Crude Protein Content of Flours Grouped on the Basis of the Percentage of Crude Protein in the Flour (Bailey, 1924)

Crude protein, per cent	9.81	10.58	11.35	12.46	13.27	15.15
Loaf volume, cc.	1,823	1,963	2,024	2,093	2,041	2,160

An application of statistical methods to the analysis of the flour strength and protein content data in the literature was made by Zinn (1923), who computed the coefficient of correlation of the protein content of flour and loaf volume. Data resulting from the examination of commercial and pure strains of wheat grown in five states and one Canadian province were included in these correlation studies. The coefficients of correlation are given in Table 143, omitting the correlation for the commercial varieties of North Dakota spring wheats which was negative and without significance because it was less than three times the probable error. The positive correlations found in the instance of the Colorado commercial spring wheats, the commercial winter wheats of Minnesota and Kansas, the pure lines of spring wheats of Minnesota and Ontario, and the pure lines of Ohio winter wheats are

TABLE 143

Correlation Coefficients of Protein Content of Flour and Loaf Volume of Test Bakes

Kind of Wheat	Number of Samples	Coefficient of Correlation
Computed by Zinn (1923)		
Commercial varieties of spring wheat		
Minnesota	203	$.2586 \pm .0442$
Montana	34	$.3448 \pm .1019$
Colorado	48	$.6130 \pm .0608$
Commercial varieties of winter wheat		
Minnesota	43	$.6396 \pm .0594$
Montana	91	$.3620 \pm .0614$
Kansas	43	$.7956 \pm .0377$
Ohio	99	$.4709 \pm .0528$
Pure lines of spring wheat		
North Dakota	28	$.3018 \pm .1158$
Minnesota	48	$.5469 \pm .0689$
Ontario	16	$.5752 \pm .1129$
Pure lines of winter wheat		
Wisconsin	25	$.3990 \pm .1134$
Ohio	25	$.5560 \pm .0932$
Reported by Mangels and Sanderson (1925)		
North Dakota crop of 1921 spring wheat	128	$.307 \pm .053$
" " " " 1922 " "	136	$.427 \pm .047$
" " " " 1923 " "	194	$.345 \pm .043$

notably high, and establish the significance of protein content in estimating baking strength. Similar correlation coefficients computed by Mangels and Sanderson (1925) from the data of three crop seasons, included in Table 143, show a significant positive correlation.

Durum wheat flour commonly contains a higher percentage of protein than common or vulgare wheat flour. The baking strength of the durum flour is inferior to that of spring wheat, or hard winter wheat flour, however, due to some inherent and characteristic difference in their properties as yet undetermined. This is shown by the data of Ladd and Bailey (1911), Thomas (1917b), and Shollenberger and Clark (1924).

The significance of protein concentration in determining baking strength was recognized by Sharp and Gortner (1923), who include the percentage of glutenin as one of the factors, together with the gluten quality factor, (b), in an equation, the other member of which is loaf volume multiplied by a constant (k). Since their data shows a relatively uniform ratio of glutenin to total protein, it follows, that the percentage of protein might be inserted in the same equation in place of glutenin content by increasing the numerical value of the constant (k). Their equation would then take the form:

$$\text{(Percentage of protein)} \times \text{(gluten quality factor, b)} = \text{loaf volume} \times (k).$$

In addition to these studies involving large numbers of samples, a direct attack has been made upon the problem of determining the relation of gluten content to flour strength by actually modifying the percentage of gluten in flour. Two procedures have been followed, in the first of which gluten has been added to the flour, or to the dough. Stein (1904), and Stockham (1920) reported an increase in loaf volume in consequence of such additions of gluten. Fenyvessy (1911) found that the loaf volume was increased by the addition of gliadin to flour, and also indicated that rye gliadin would produce the same effect. Snyder (1901) increased the percentage of gluten in doughs by washing out a portion of the starch, but observed no effect of such manipulation upon the characteristics of the bread. It should be remarked that such treatments as must have been accorded the gluten in these several experiments probably modified its properties, and it is doubtful if gluten can be separated from flour by any means now known without impairing those properties which are of importance, and thus diminishing its effectiveness in a dough.

A more satisfactory, and equally convenient procedure involved the dilution of the gluten of flour with starch. Snyder attempted this, and remarked (1901) that the addition of 10 per cent of starch to flour had little effect upon its baking qualities. When 20 per cent of starch was added the quality of the bread was impaired. The author repeated this experiment a few years later, and noted a steady decrease in loaf volume and texture as the proportion of starch was increased. Jago (1915) reported the results of baking experiments to a committee of the 63rd Congress which indicated that each addition of starch of 5 per cent or more tended in the direction of reducing the size of the baked loaf. The data recorded in Table 144 are fairly typical of the three series of such tests reported by him.

TABLE 144

Baking Tests of Canadian Patent Flour Containing 16.1 Per Cent of Dry Crude Gluten, to Which Varying Percentages of Starch Were Added Before Baking (Jago, 1915)

Mixture as Baked		Loaf Volume, cc.
Flour, Per Cent	Starch, Per Cent	
100	0	2,316
95	5	2,275
90	10	2,190
85	15	2,057
80	20	1,920
75	25	1,914

The extensibility of dough, as measured by the Chopin extensimeter, was substantially reduced by mixing starch with flour, as shown by Bailey and LeVesconte (1924). Such a reduction in extensibility as is here evident implies an impairment of the elasticity and gas retaining power of the dough. The measurements as made with the Chopin instrument are recorded in Table 145.

TABLE 145

Extensibility of Doughs Made from Flour, and from Flour and Starch Mixtures, as Determined by Bailey and Le Vesconte (1924)

Starch in Flour Mixture, Per Cent	Extensibility as Measured with the Chopin Device
0	18.03
10	14.92
20	13.89
30	11.82
40	10.56

The gas-retaining capacity of strong flour doughs was found by Bailey and Weigley (1922) to be substantially greater than that of weak flours, and in the flours compared the strong flour contained the highest percentage of crude protein. Johnson and Bailey (1925) proceeded to reduce the gluten content of flour by adding starch, and found a regular decrease in gas-retaining capacity of the dough with each increment of starch. This is shown by the curves C in Figure 20, which represent the loss of carbon dioxide on fermenting doughs made with a strong

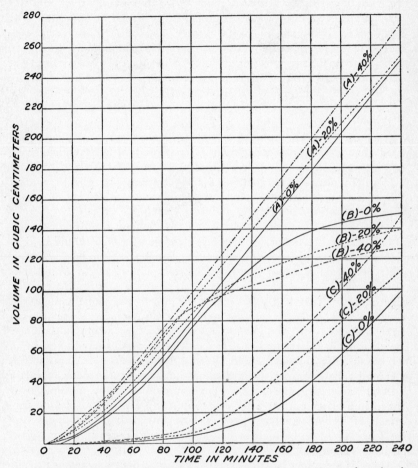

Fig. 20.—Changes in gas production and gas retention of fermenting doughs effected by diluting flour with 20 per cent and 40 per cent of starch. Curves *A* represent the sum of the increase in volume of the dough, plus the volume of gas lost from the dough; curves *B*, the increase in volume of the dough; and curves *C*, the volume of carbon dioxide lost from the dough. (Johnson and Bailey, 1925.)

flour, and the same flour mixed with 20 per cent and 40 per cent of starch.

The Gliadin-Glutenin Ratio and Flour Strength

Gliadin and glutenin separated from flour have been observed to possess widely different physical properties, so far as these could be determined by casual observation. Their solubilities and chemical constitution were known to be unlike. It was not difficult to believe that variations in the ratio of gliadin to glutenin might account for corresponding variations in the properties of flour. This was suggested as a possibility by Fleurent (1896a, 1896b), and referred to frequently in subsequent publications by himself, and Girard. Fleurent's method for separating gliadin was quite different from that which was subsequently used for a like purpose, and it is not surprising to note a lack of agreement among different investigators as to the optimum ratio. Thus Fleurent concluded that a ratio of 75 parts of gliadin to 25 of glutenin resulted in gluten of the best physical properties. Snyder (1899a) used a different method for the determination of gliadin and stated that a well-balanced gluten is composed of approximately 65 per cent of gliadin and 35 per cent of glutenin. He (1904b) later developed a polarimetric method for determining the gliadin content of flour and indicated the limits of gliadin in good baking flours as being from 55 to 65 per cent of the total protein.

Guess (1900) suggested that the gliadin-glutenin ratio might be regarded as a quality factor, while the percentage of gluten should be considered as a quantity factor in evaluating flours from the baking standpoint. The quality factor multiplied by the quantity factor would thus afford a more satisfactory expression of flour properties than either value alone. Norton (1906) came to much the same conclusion. Kosutany (1903, 1909) lent the weight of his opinion to the support of the contention that the gliadin-glutenin ratio is of prime importance. The gliadin which Fenyvessy (1911) added to flour appeared to effect an increase in loaf volume, while added glutenin either had no effect, or caused a deterioration in the baking qualities of the flour. Martin (1920b) "amended" the gliadin figure by deducting the percentage of protein in a water extract after a long extraction period, from the percentage of alcohol soluble protein. This amended gliadin figure was found by Martin to be high in flours with high gas-retaining capacities and bakers' marks.

There is a distinct lack of agreement among flour chemists as to the value of the gliadin-glutenin ratio, or the gliadin content of flour as criteria of flour strength. König and Rintelen (1904), Wood (1907), Shutt (1910), Rammstedt (1909) and others concluded that the gliadin-glutenin ratio was not of particular significance, or at least was not the determining factor in the strength of flour. Part of this lack of agreement may have been occasioned by the difficulties which existed in accurately determining the percentages of these proteins in flour. Mention of certain of these difficulties has already been made in the discussion of the properties of the flour proteins. Blish (1916) concluded that the ratio of gliadin to glutenin is much more nearly constant than had previously been supposed. Zinn (1923) calculated the coefficient of correlation of gliadin and total protein, which was found to approach unity so closely as to indicate that one varies with the other. Thus in two of the four groups of samples involved in these calculations the correlation coefficient was $0.9244 \pm .0160$, and $0.9295 \pm .0181$ respectively, while the minimum was $0.7863 \pm .0243$. Sharp and Gortner's (1923) data indicate a surprisingly constant ratio of glutenin to total protein in flours of widely different characteristics. Blish and Sandstedt (1925b) reported the glutenin and total protein content of a number of vulgare wheat flours. The author calculated the glutenin/protein ratio and found it to vary within the comparatively narrow limits of 0.35 to 0.40 in all instances but one. Similar ratios were observed in a series of commercial flours recently examined in the author's laboratory. To dispose of this problem, a critical study of the proportions of the several proteins in flour must be conducted to ascertain whether the variations in the proportions of these constituents are of sufficient magnitude to account for the observed differences in flour properties.

Water-Soluble Proteins and Baking Strength

No correlation between baking strength and the water-soluble protein of flours could be detected by Bremer (1907), Shutt (1910), Rammstedt (1909), and Olson (1917b). Rousseaux and Sirot (1913, 1918) believed that the ratio of water-soluble to total protein is not without significance. It seems that some of the variations from the normal in this ratio which they encountered may have been due to unsoundness of the flour, since the largest proportion of water-soluble protein observed was in a flour which had been stored for an extended period.

In the preceding discussions of various forms of unsoundness of wheat, such as the sprouted, immature, and frosted condition it was shown that the proportion of soluble protein and other forms of soluble nitrogenous constituents increased with the degree of unsoundness. Unsoundness of grain from which flour is milled may thus account for abnormally high percentage of the water-soluble forms of nitrogenous compounds.

Lower grades of flour also contain a larger proportion of the total protein soluble in water and dilute saline solutions than is found in the higher or patent grades. The percentage of these "soluble proteins" accordingly is an indication of the relative grade of the flour, providing the flour is milled from normal or sound wheat.

Proteases and Flour Strength

Little is known at present concerning the rôle of the proteases in fermentation of bread doughs, or the effect of proteases on the gas-retaining properties of dough. The lack of precise knowledge in this connection is due in large part to the complex character of the substratum or flour proteins. It is conceivable that substantial modification of the gluten proteins in contact with active proteases may result without any material increase in the simpler degradation products of proteolysis which can be estimated by analytical methods now available. The methods must be improved before much progress can be made in the study of the flour proteases.

It is probable that the lower grades or clear flours will manifest greater proteolytic activity than the highly refined or patent grades. This is suggested by the work of Stockham (1920), who reported that the time required for a unit quantity of patent and second clear flour to liquify gelatin was 354 hours, and 144 hours respectively. Flour milled from the germ end of the wheat kernel exhibited a higher gelatin-liquifying power than flour milled from the opposite or blossom end. The latter baked into bread much superior to the bread baked from straight grade flour milled from the germ end of the kernel.

No correlation between baking strength and gelatin-liquifying power of flour is evident from the data reported by Stockham, which data resulted from the examination of a considerable number of normal flour samples. It is possible, however, that the time required to liquify gelatin is not an adequate measure of the rate of autolytic cleavage of the gluten proteins in doughs made from the same flours, and it is the latter in which we are chiefly interested.

Reference has already been made to the suggestions made by Ford and Guthrie (1908) that the weakness or lack of strength manifested by certain doughs might be attributed to the activity of proteolytic enzymes.

In the fermentation of sponges used in the production of soda crackers, the hydrolytic cleavage of the gluten is apparently essential to the manufacture of satisfactory biscuit. This has been discussed by Johnson and Bailey (1924).

Traces of pepsin and trypsin impaired the properties of dough made from normally aged flour, in the experiments of Weaver and Wood (1920). Stockham (1920) also remarked that no improvement was effected by the addition of active proteases to dough.

The formol titration procedure as developed by Sörenson was used by Swanson and Tague (1916) in a study of certain conditions which affect the activity of proteolytic enzymes in wheat flour. Ammonium chloride and calcium chloride appeared to accelerate the autolytic cleavage of the flour proteins. No conclusions were presented relating to the significance of flour proteases in dough fermentation or the variations in proteolytic activity of various types of flour.

Chemical Constitution of the Gluten Proteins From Strong and Weak Flours

It might be assumed that inherent differences in the physical properties of the gluten proteins could be traced to differences in the chemical constitution of these proteins. Yet it has not been possible to detect any substantial differences in the chemical constitution of gliadins separated from strong and weak flours, nor in glutenins from the same sources, so far as is revealed by the distribution of nitrogen in the products of acid hydrolysis. This was the experience of Wood (1907b), who followed the Osborne and Harris modification of Hausmann's method in studying gliadins, as well as crude gluten preparations from flours of varying baking strength. Blish (1916) used the Van Slyke method, and observed no significant differences among the gliadin preparations, and among the glutenin preparations from strong and weak flours. Cross and Swain (1924) reported a like experience.

It must be recognized, however, that the relatively gross and inaccurate procedures involved in these methods may fail to reveal either the linkages and configuration of the protein molecule, or the size of the aggregates. That the latter may vary is indicated by the researches of Gortner and his colleagues, to which reference is made in the discussion of the colloidal properties of gluten.

Immunological Methods as a Means of Differentiating Proteins

Gliadin of wheat, and of rye were either identical, or at least closely related so far as could be determined by the anaphylactic reaction in the experiments of Wells and Osborne (1911). Gliadin and zein did not react, nor did hordein with zein. Wells and Osborne (1913) later observed that gliadin and glutenin reacted anaphylactically with one another, although less strongly than with themselves, and that these chemically distinct proteins contain common reacting groups. Complement fixative tests applied by Lewis, Wells, Hoffman, and Gortner (1924) indicated that the prolamines from emmer, einkorn, spelt, and durum are closely related to gliadin and glutenin from T. vulgare or common wheat. No reactions were obtained between antisera of the corn group with those of the wheat group. These conclusions were confirmed through the use of uterus strip, anaphylaxis, and bronchospasm methods. It is evident from these observations that these immunological reactions did not permit of as sharp differentiation between wheat proteins as is afforded by chemical or physico-chemical methods.

The Physical and Physico-Chemical Properties of Flour

Gluten. Crude gluten triturated with a solution of sodium chloride by Balland (1883d, 1884) was observed to diminish in weight, and to approach the consistency of rubber. Other salts exerted similar effects which varied in degree. Dilute acids dispersed the gluten. These were the earliest recorded observations on the colloidal properties of crude gluten, although it was not until a score of years had passed that the significance of this work was appreciated.

The important pioneer work in this field was done by Wood (1907), who applied the methods that were then being used in the study of the emulsoid colloids. In an effort to determine the effect of acids, bases, and salts on the crude gluten, portions of the latter were suspended by V-shaped rods in beakers containing the several solutions. Gluten which retained its coherence in distilled water began to disperse when acids were added to the medium. After a certain concentration of hydrochloric acid was exceeded the gluten again became coherent, until a coherence greater than that of the original gluten was reached in 0.083 normal hydrochloric acid. In aqueous solutions of sulfuric, phosphoric, and oxalic acids dispersion of the gluten occurred in concentrated solutions. The addition of salts to these aqueous solutions of acid tended to counteract the dispersion effect. Variations in coherence and elasticity of gluten were attributed to the chemical environment of the gluten rather than to differences in the composition of the gluten. The salts which are responsible for the properties of gluten are laid down in the wheat kernel while it is maturing, in the opinion of Wood. Salts added to a dough are believed to exert less effect than the salts of the natural flour, due to the relatively long time which is apparently required to induce the maximum effect.

An interesting discussion of the colloidal behavior of gluten was presented by Wood and Hardy (1909). They postulated that the formation of a hydrosol in the presence of dilute acids and alkalis is due to the development of electric charges around the particles of gluten. These charges result from the chemical interaction of the protein, the acid (or alkali), and water. A physical hypothesis which accounts for this potential difference between the colloid particles, and the fluid maintains "that the colloid particles at any moment of time contain within themselves an excess of the most penetrating and rapidly moving ions present, and they therefore have the charge of that ion. In the presence of acid they have the charge of the hydrogen ion, in the presence of alkali that of the hydroxyl ion." Coagulation or precipitation of such a solution is approximately coincident with the reduction of the potential difference to zero, the most coherent coagulum being formed at the isoelectric point. The tenacity, ductility, and water content of a solid mass of moist gluten depends upon the partial or total disappearance of the electric double layers, and the reappearance of the adhesion of the colloid particles, that attraction which makes them cohere when they come together. Hardy (1910) later stressed the fact that gluten of itself has neither ductility nor tenacity. These are conferred by electrolytes. Any salt confers the property of cohering, the quantity required depending upon the concentration of acid present.

Upson and Calvin (1915, 1916) employed a more refined procedure than Wood and reached similar conclusions. In their researches, Upson and Calvin determined the imbibition of water by gluten discs immersed in various solutions by weighing the discs after a definite period of immersion. It was noted, for example, that discs which had increased greatly in weight after immersion in dilute acid regained their original weight and properties of cohesion and elasticity when placed in 0.1 normal dipotassium phosphate solution for an hour.

Lüers and Ostwald (1920) indicated that their observations supported the conclusions reached by Wood, and by Upson and Calvin.

The relations of water to gluten in flour were discussed by Mohs (1922). Three states or conditions of water were suggested: (a) the lowest, which is found in flour that has been heated: (b) that which is found in normal flour: (c) that which is encountered when water is mixed with the flour. In the instances of excessively strong flours, Mohs contended that added water did not penetrate the gluten particles, but was absorbed on their surface. In the case of normal flours part of the water penetrated the particles of gluten causing them to swell. When all the water entered the particles and none remained on their surface the gluten would be characterized as weak.

Gortner and Doherty (1918) refuted certain of the conclusions of Upson and Calvin. Using essentially the same method, they found that glutens from strong and weak flours did not respond in the same manner when immersed in dilute acid solutions. The strong flour glutens imbibed more water in aqueous acid solutions of definite concentration, and dispersed less than glutens from weak flours. Gortner and Doherty accordingly concluded that "the difference between a strong and weak gluten is apparently that between a nearly perfect colloidal gel with highly pronounced physico-chemical properties, such as pertain to emulsoids, and that of a colloidal gel in which these properties are less marked. It is suggested that such differences may be due to the size of the gluten particles and that at least a part of the particles comprising the weak gluten may be nearer the boundary between the colloidal and crystalloidal states of matter than is the case with the stronger glutens."

The gluten preparations used by Upson and Calvin, and by Gortner and Doherty were not freed from ash. Their findings must accordingly be regarded as qualitative only. For the presence of electrolytes in solution in the imbibed water, or adsorbed on the gluten aggregates may have exerted a profound effect upon the imbibing capacity of the crude gluten. Henderson, Cohn, Cathcart, Wachmann and Fenn (1919) have indicated that the actual swelling of gluten in dilute acids varies with the amount of electrolyte in the system, as well as with the hydrogen ion concentration. The degree of swelling also tended to increase with the actual quantity of gluten (dry basis) under observation. They contended that in systems containing gluten and acids or bases, the formation of salts, in accordance with the requirements of the mass law, is the fundamental phenomenon. An approximate constancy of the ratio of corrected protein salt conductivity to combined acid was observed. Thus the effect upon conductivity of a given amount of protein chloride was constant.

Several efforts have been made to develop a device for testing the ductility and expansibility of gluten. One of the earliest of these was devised by Boland (1848). It consisted of a tubular vessel into which was fitted a piston with a graduated stem. Gluten washed from 50 grams of flour was placed in the tube, the piston placed on the gluten, and the tube and contents heated to 150° C. in an oil bath. The height to which the piston raised in consequence of the expansion of the gluten could be determined by the graduations on the stem of the piston. Several modifications of the Boland aleurometer were devised, of which the Liebermann (1901), and the Foster gluten testers are perhaps the best known. The latter was described by Wiley et al (1898).

Hankóczy's (1920) device was designed to test the ductility and extensibility of wet crude gluten in a direct manner. The gluten was clamped between two plates, each of which was perforated, and on raising the pressure of the air on one face of the sheet of gluten, it stretched and expanded into a bubble thru the perforation in the opposite plate. Provision was made for measuring the volume of air displaced by the expanding bubble of gluten. Hankóczy contended that baking strength of the flour was related to the volume of the bubble, and to the pressure which the bubble would maintain.

Kress (1924) described a gluten testing apparatus devised by James. In using this device, the washed gluten was clamped between two plates having an 18 mm. opening or circular perforation in the center of each. The sheet of gluten was compressed to a thickness of 3 mm. A round headed plunger 12 mm in diameter was then placed on the sheet of gluten through the perforation in the upper plate and a force of regularly increasing magnitude was applied to the plunger. This forced the gluten downward through the perforation in the lower plate. The distance that the gluten could be stretched before breaking was found to be correlated with the results of baking tests.

Chopin (1921) developed an extensimeter somewhat similar to the Hankóczy device, but designed to test the properties of dough, rather than of wet crude gluten. Its application to this purpose will be discussed in another section. An instrument designed to measure the ductility of dough was devised by Rejtö, and described by Kosutany (1907).

None of these methods are in common use in America. The great difficulty in the way of general use of the aleurometer has been the varying results obtained on replicating the tests of a single flour. The proportion of water in the gluten, rate of heating the gluten, and other variables difficult to control, affect the results of the test. In the direct tests of gluten extensibility with the Hankóczy, or the Kress-James instruments it is necessary to insure that the conditions under which the gluten is washed out of the dough are always uniform. Variations in the quantity and character of electrolytes in the water, temperature of the wash water, time of washing, and other conditions are known to affect the physical properties of wet crude gluten. It is not a simple matter to control all of these variables, nor to insure that the crude gluten preparations are always of the same degree of purity.

Certain French chemists, notably Arpin, have stressed the significance of the proportion of water in the wet gluten as an index of the "quality." This criterion of gluten quality was also mentioned by d'Andre (1922) who presented data to show that wet crude glutens in which the percentage of water exceeded 65 per cent were obtained from excellent and good flours, while the flours which were rated as fair and poor in baking value yielded wet crude glutens in which the water content was less than 65 per cent. American chemists and flour testers do not stress this property of gluten, and the ratio of dry gluten to moisture in the wet crude gluten rarely appears in the data reported in the routine testing of flour in the laboratories of the United States.

A turbidity method was evolved by Harvey and Wood (1911) for determining the strength of small samples of flour. The flour was shaken with cold water for an hour, and then filtered. To 10 cc. of the more or less opalescent extract was added 1 cc. of 0.1 per cent solution of iodine in potassium iodide. After standing one hour the solution was poured into a tube with a plane glass bottom. A small electric lamp was mounted directly below this tube. A plunger of tubing with a plane glass bottom was forced downwards into the turbid liquid until the filament of the lamp just became visible. The depth of the liquid was recorded as an index of strength of the flour under examination. As shown by the following data, the depth through which the lamp filaments could be seen was much less in the instance of strong Fife flour, than with the weaker English wheats. Karachi (East Indian) wheat was intermediate in these particulars. The method was deemed particularly suited to the examination of small wheat samples in wheat breeding practice.

Type of Wheat from Which Flour Was Milled	Depth of Liquid Through Which Lamp Filaments Could Be Seen
Fife	3– 4 cm.
Karachi	6– 8 cm.
Square Heads Master	10–12 cm.
Rivett	15–18 cm.

Gliadin. The effect of acids, bases, and salts on the viscosity of wheat gliadin sols was studied by Lüers (1919), using an Ostwald capillary viscosimeter. Viscosity of gliadin sols decreased slightly on additions of small quantities of hydrochloric acid, then increased to a maximum at 0.0004 normal, and on further acidulation fell to about the original value. With sulfuric acid the increase in viscosity was not so great. The maximum viscosity with this acid was reached at 0.0022 normal, and the gliadin precipitated at higher concentrations of the acid. Lactic acid effected a greater increase in viscosity than the sulfuric acid, the maximum being reached at 0.018 normal. At higher concentrations of lactic acid (0.084 normal), only a slight turbidity was produced. On additions of sodium hydroxide the viscosity increased to a maximum at 0.0022 normal, and remained at a high level when the concentration of alkali was increased to 0.07 normal. Salts caused a decrease in viscosity of the gliadin dispersed in 0.5 normal lactic acid. Chlorides were least effective in reducing viscosity, while phosphates, sulfates, and tartrates were increasingly effective in the order named. Rye gliadin behaved in a similar manner when treated with acids, alkalis, and salts.

In alcoholic solutions (71.5 per cent of ethyl alcohol) the viscosity of gliadin solutions was increased to a higher level on acidulation than on similar treatment of aqueous gliadin solutions. Alcoholic solutions of gliadin were not as sensitive to the additions of neutral salts as the water sols. From these differences in the properties of alcohol sols and water sols, Lüers concluded that the former may be regarded as emulsoids, while the latter have more of the characteristics of suspensoids.

The isoelectric point of gliadin was observed to be slightly on the acid side of neutrality. This observation has been confirmed by the findings of Eto, (1924), and Tague (1925a), who reported pH = 6.6, and pH = 6.5 respectively as the isoelectric point of gliadin.

Kataphoresis experiments conducted by Lüers (1919) showed that at pH = 9.05 the gliadin migrated to the anode, at pH = 6.7 there was no apparent migration, while at pH = 5.97 the migration was to the cathode.

In general, Lüers seemingly regarded gliadin as the most important colloid of the gluten, and emphasized its value as a protective colloid.

Reference has already been made to the racemisation of gliadin in contact with alkalis. Woodman (1922) could not detect any differences in the racemisation of gliadin prepared from strong Manitoba flour, and from a weak English flour. Blish and Pinckney (1924) prepared gliadin from Polish wheat (T. polonicum), hard winter, and hard red spring wheat flours, and found that the first named differed from the other two in specific rotation before and after racemisation.

The rotation dispersion, and refractive index dispersion of gliadin preparations from flours of varying baking strength were determined by Dingwall (1924). In the study of rotation dispersion a special lamp bulb was used, in which a flat strip of metallic tungsten was rendered luminous by passing a current of 25 amperes at a potential of 10 volts through it. The beam of light from the lamp bulb impinged upon the prism of a monochromatic illuminator. The light emerging from the illuminator was focussed upon the slit of the analyser of a polarimeter. This facilitated observations of optical rotation of light of various wave lengths. The observations of the three gliadin preparations failed to reveal any substantial differences in the gliadins. These results, using light of four different wave lengths, are shown in Table 146.

TABLE 146

Rotation Dispersion of Gliadin in 70 Per Cent Ethyl Alcohol at $21.1°$ C. (Dingwall, 1924)

Wave Length of Light, Ang. Units	Specific Rotation of Gliadin		
	Hard Spring Wheat Flour	Missouri Soft Winter Wheat Flour	Washington Soft Wheat Flour
6,500	72.7	72.5	72.1
5,890	92.9	91.6	92.3
5,300	120.7	118.9	120.0
4,800	151.7	151.3	151.3

These observations do not necessarily establish absolute uniformity in the molecular configuration of the several gliadins. They do indicate that those portions of the molecule which are responsible for optical activity are identical.

In the measurement of refractive index dispersion Dingwall used a Pulfrich refractometer, which measures the refraction of light in the surface film. Measurements were also made by the spectrometer method, which affords an expression of the optical properties of the body of the solution. The refractive index was determined with three wave lengths of light, namely 5,680, 5,461, and 4,358 Ång. units. Identical values were obtained from the examination of the three gliadin preparations.

Absorption spectra in the ultra-violet of the three gliadin preparations were photographed by Dingwall (1924), and no marked differences have thus far been disclosed by examination of the photographic plates.

Surface tension, viscosity, gold number, specific rotation, and refractive index of gliadins from different flours, and the time required to decompose a unit quantity of hydrogen peroxide in the presence of a platinum sol and gliadin, were determined by Groh and Friedl (1914). No differences in these properties could be detected in the gliadin preparations from different flours, or in several gliadin fractions from the same flour when these were free from impurities.

Glutenin. Early in their studies of the viscosity of flour suspensions, Sharp and Gortner (1923) observed that the change in viscosity induced in their manipulations of the material could be attributed largely, if not wholly, to the properties of the glutenin. The viscosity, as determined with the MacMichael viscosimeter, constituted a convenient measure of water-imbibing capacity of the material, since the internal friction of the suspensions, or of the hydrosols increased with the proportion of imbibed water. It accordingly appears appropriate to discuss their conclusions under the heading of this section, though most of their observations were made directly on flour suspensions.

Electrolytes in the flour served to depress the imbibitional capacity when flour suspensions were acidulated. In the comparisons of glutenins of different flours it proved desirable to remove the electrolytes by leaching the flour with several portions of distilled water. Incidentally, this leaching removed most of the gliadin, and without substantial modification of the imbibitional capacity, indicating the latter to be a function of glutenin concentration and properties. In a study of the effect of acids on viscosity, they found that metaphosphoric acid effected little change in viscosity of the leached flour suspensions. Sulfuric acid was somewhat more effective. The other acids that were used were effective in increasing viscosity in the following order: tartaric, oxalic, citric, lactic, nitric, hydrochloric, and ortho-phosphoric, the latter giving the greatest increase. Viscosity on acidulation approached or passed through a maximum at $pH = 3 \pm$ in the instance of all the acids. Further increases in hydrogen ion concentration resulted in a decrease in viscosity which approached that of the original suspension at about $pH = 1$ when strong acids were used. Lactic acid constituted a convenient reagent, since its solutions can be held with ease within the range of maximum viscosity.

When the flour was digested by Sharp and Gortner with 0.01 normal solutions of various salts, and subsequently acidulated with lactic acid, the imbibitional capacity was reduced below that of acidulated flour suspensions that had previously been digested with water alone. Of the salts that were tried $MgSO_4$, K_2SO_4, and $CaCl_2$ were most effective in reducing imbibitional capacity, and were similar in their effect, while $NaCl$, KCl, K_2HPO_4, and $MgCl_2$ constituted a second and less effective group. The addition of 0.25 cc. of normal $MgSO_4$ solution to an acidulated flour suspension reduced the viscosity to nearly that of the original or unacidulated suspension.

When flour suspensions were rendered alkaline by the addition of sodium hydroxide, the viscosity increased, passing through a maximum at $pH = 11 \pm$. When the hydroxyl ion concentration exceeded $pH = 11$, the viscosity tended to decrease, reaching a second minimum at pH = 11.8. Further additions of alkali caused another slight increase in viscosity, which may have been occasioned by swelling of the starch. Barium hydroxide was much less effective in inducing increases in viscosity than was sodium hydroxide. The salts present in natural flour repressed imbibition in alkalis inappreciably, in contrast to their pronounced depressing effect on imbibition in acids.

A typical series of data, resulting from observations of the viscosity of leached suspensions of a strong and a weak flour at various hydrogen ion concentrations are recorded graphically in Figure 21.

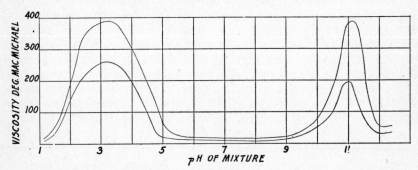

FIG. 21.—Relation of viscosity of extracted flour and hydrogen ion concentration. Flour suspensions acidulated with HCl, and rendered alkaline with NaOH. (Sharp and Gortner, 1923.)

Viscosity of acidulated flour suspensions was found to increase with concentration of flour, and hence of glutenin. The relation between viscosity and concentration of glutenin conformed to the expression:

$$\text{Log. of viscosity} = a + b \,(\text{log. of concentration})$$

in which "log. viscosity" is the logarithm of the maximum viscosity obtained with lactic acid. The imbibitional capacity of the glutenin is accordingly represented by the constant (b) of the equation. When this constant was computed from the results of the examination of a number of flours, its insertion into the equation: $\dfrac{(\text{Glutenin}) \times (b)}{\text{loaf volume}}$ = K established its relation to baking strength. The fact that K was of reasonably constant value when hard spring and hard winter flours were compared indicates that in the instance of flours of uniform glutenin content, the glutenin quality factor (b) was a linear function of loaf volume. In other words, the constant (b), when properly determined, was deemed to constitute an index of the quality of the glutenin. And since gliadins of different flours apparently are more or less similar, a measure of the quality of the glutenin of a flour should represent an index to the quality of the gluten.

The calculation of the quality factor (b) was detailed by Gortner (1924). It is necessary to make several viscosity determinations in examining each flour. Four or five determinations are advised, representing as many ratios of flour to water in the suspension. When the logarithms of concentration or ratio of flour to water are plotted as abscissas against logarithms of the maximum viscosity of the acidulated suspension of leached flour as ordinates, an approximately straight line should result when the coordinate points are connected. The slope of this line in respect to the axis of abscissa is a measure of the increment of increase of the logarithm of the viscosity per increment of increase of the logarithm of flour concentration. The tangent of the angle which this line forms with the axis of abscissa is the index of glutenin quality, or the quality constant (b). The angle can be measured with a protractor, or can be easily calculated if the points lie in a sufficiently straight line so that the curve can be drawn correctly. Since experimental error may result in some of the coordinate points lying slightly above or below the theoretical curve, it is convenient to simultaneously calculate the tangent or quality factor, and smooth the curve by the method of least squares. This can be done through the use of the following formula:

$$(b) = \frac{\Sigma(x) \cdot \Sigma(y) - n \Sigma(xy)}{(\Sigma x)^2 - n \Sigma(x^2)}$$

in which Σ = sum of values; x = logarithm of flour concentration; y = logarithm of viscosity; and n = number of observations.

More recent investigations at the Minnesota Agricultural Experiment Station suggest that the positive correlation between the constant "b" and baking strength is not of the magnitude that was at first believed to be the case. It may be that the constant (a) of the equation representing the relation between viscosity and concentration of glutenin may appropriately be inserted into the equation in which colloidal properties of flour are contrasted with baking strength. This constant is the calculated point at which the viscosity concentration curve intersects the axis of abscissa (or the axis on which the logarithm of viscosity is recorded) at zero concentration. It accordingly is related to the actual viscosity values of the acidulated flour-water suspension. This constant can be calculated by the method of least squares through the use of the equation:

$$a = \frac{\Sigma(x) \cdot \Sigma(xy) - \Sigma(x^2) \cdot \Sigma(y)}{(\Sigma x)^2 - n \Sigma(x^2)}$$

Blish and Sandstedt (1925) likewise observed that the quality constant "b" did not afford an adequate measure of gluten properties, so far as the latter are reflected in the results of baking tests.

Doughs which Sharp and Gortner (1923) brought to a pH = 3,

or pH = 11 by the addition of acid or alkali, and then restored to their original hydrogen ion concentration by neutralizing the acid or alkali were reduced in baking strength. Doughing flour with 70 per cent, or 95 per cent alcohol, followed by evaporation of the alcohol at room temperatures likewise were reduced in baking strength. It is thus evident that the properties of glutenin are easily modified, and the baking strength of flour is impaired by such manipulations.

Glutenins prepared by Woodman (1922) from a strong Manitoba wheat flour and from a weak English wheat flour were dispersed in 0.04 normal sodium hydroxide solution, and their specific rotation determined. The specific rotation of the first was $-99.5°$, and of the second $-78.8°$. The same glutenin preparations were racemized at $37°$ C. with .05 normal sodium hydroxide. The consistent differences in optical rotation during the racemisation led Woodman to believe that the glutenin of the strong flour was a different protein from that contained in the weak flour.

No appreciable change could be detected in specific rotation of glutenin dispersed in 0.036 normal (0.2 per cent) potassium hydroxide held at room temperature for a month. From this observation Woodman contended that the specific rotation of glutenin is not altered in the process of extracting it with 0.2 per cent potassium hydroxide in purifying the glutenin preparations.

Dingwall (1924) studied the rotation dispersion of glutenin in 0.5 normal sodium hydroxide. No differences could be detected in this property of three glutenin preparations isolated from flours of as many types. This is shown by the data in Table 147.

TABLE 147

Rotation Dispersion of Glutenin in 0.5 Normal Sodium Hydroxide

Wave Length of Light, Ång. Units	Specific Rotation of Glutenin Prepared from		
	Hard Spring Wheat Flour	Missouri Soft Winter Wheat Flour	Washington Soft Wheat Flour
6,500	60.0	60.4	59.7
5,890	75.3	75.3	74.2
5,300	93.4	93.5	93.3

The difference in refractive index dispersion of these three glutenin preparations, as determined by Dingwall, was too small to justify the assumption that the glutenins differed materially in this property. The dispersion value (α) as calculated after the procedure of Robertson, representing the alteration of refractive index of the solvent on dissolving 1 gram of protein, was somewhat greater in the instance of the glutenin preparations, than was observed when the corresponding gliadin preparations were compared. Absorption spectra in the ultra violet, as photographed by Dingwall (1924) reveal no marked differences in the three glutenin preparations.

Blish and Pinckney (1924) determined the specific rotation of glutenin preparations from a hard winter wheat flour of poor baking quality, and a hard spring wheat flour of excellent baking quality. No difference in this property was observed on comparing the two glutenin preparations, either in the initial specific rotation (after 2 hours in 0.1 normal sodium hydroxide) or after racemizing the preparations in 0.5 normal sodium hydroxide at a temperature of 37° C. Several additional preparations supplied by Cross and Swain were also examined by Blish and Pinckney, and the specific rotation of all were within the limits of experimental error. The specific rotation observed was approximately that of the glutenins from weak English wheat flour examined by Woodman (1922). A glutenin preparation from Polish wheat (T. polonicum) exhibited a lower optical rotation than the preparations from vulgare or common, and club wheat.

The Colloidal Behavior of Dough and Gas Retention

Flour was described by Ostwald (1919) as a coarse dispersion of several poor-in-water hydrogels. These include a mechanically divided and dried protein gel without definite structure; carbohydrate gels existing as starch granules; and the cellulose gel of the cell walls or fiber. The individual gel particles contain molecularly dispersed substances such as salts, sugar, acids, and so forth, and water in particular. Gases are present, not only dissolved, but also adsorbed on the surface. Dough was described as a polydispersoid, water being the dispersion medium in which molecularly dispersed, colloidal, and suspended particles are found. The molecularly dispersed substances include dextrins, sugar, alcohol, salts, acids, and gases. Particles of colloidal dimensions include the plant albumen, swollen starch grains, and the finer gas bubbles. Yeast cells and lactic acid bacteria constitute the coarsely dispersed portion. Dough exhibits properties of both liquids and solids such as are characteristic of strongly hydrated emulsoids. It is a viscuous liquid in that it will flow, and gradually fill a confining vessel. It may be cut like a solid, and when torn has a fibrous structure similar to torn woody material.

To yield satisfactory bread, in the opinion of Ostwald, the dough must behave primarily as a plastic substance. If too stiff it will comport itself like a solid substance in the baking. If not sufficiently viscous, the pores will tend to run together, forming large holes. A comparison was drawn between the gluten of bread, and rubber. The harmful effects of over-mixing may be due to colloidal behavior. In analogous processes as in paper-making, butter-making, and rubber-processing, excessive kneading results in mechanical depolymerization and an increase in the degree of dispersion.

Lüers and Ostwald (1919) studied the colloidal behavior of flour-in-water pastes. The viscosity of heated pastes containing varying proportions of flour was determined. A four-fold increase in concentration of flour effected a six-fold increase in viscosity (after deducting the viscosity of the water) through the lower limits of concentration. Upon standing the paste decreased in viscosity as a rule, which was explained as due to an increase in the degree of dispersion, as well as to a spontaneous dehydration of the starch particles. A short extraction (40 per cent) flour gave a more viscous paste than the flours of longer extraction (94 per cent).

The tendency of starch paste to decrease in viscosity with the lapse of time was studied in a detailed manner by Farrow and Lowe (1923).

When dough solutions studied by Lüers and Ostwald (1919) were compared (not heated) the long extraction flours gave the more viscous solutions. Lactic acid increased the viscosity of suspensions of high grade flours. The proportional increase was small in the instance of the lower grades. This is shown by the data previously presented in Table 104 in connection with the discussion of flour grades.

The viscosity of dough solutions was studied by Lüers and Schwarz (1925). Varying proportions of flour and water were employed, representing 6, 8, and 10 per cent of the dry flour substance in suspension. These suspensions were heated over a gas flame, and the maximum viscosity attained was measured by means of a torsion viscosimeter devised by A. van Stolkcz. The viscosity — concentration function of the dough solutions was computed using the formula:

$$\log \eta = \log a + n \log c$$

in which η = viscosity, c = concentration of flour, while n and a are constants. Rendering this equation free from logarithms gives:

$$\eta = a \cdot c^n$$

The numerical value of the exponent n was computed in the instance of 18 flour samples, which were also subjected to baking tests. The correlation of n, and loaf volume was made evident by dividing n by loaf volume in cc. In 14 of the 18 instances the quotient varied within comparatively narrow limits, (0.30 — 0.41) indicating a positive correlation between the exponential value n and baking strength as indicated by loaf volume.

It should be emphasized that these observations of Lüers and Schwarz were made under entirely different conditions than were those of Sharp and Gortner (1924). Thus Sharp and Gortner computed their quality constant b from viscosity measurements made with a *leached*, and *acidulated* flour suspension. The acidulation (in the absence of salts) swelled the glutenin of the flour and substantially increased its viscosity. The resulting magnitude of the viscosity measurement was believed to be conditioned largely if not entirely by the concentration and colloidal properties of the glutenin. Lüers and Schwarz computed the exponential value n from the measurements of the maximum viscosity attained on heating a flour suspension. The magnitude of such a measurement must be attributed in large measure to the characteristics of the starch paste rather than to the properties of the heated (and presumably denatured) protein.

After stressing the significance of the properties of wheat gliadin and glutenin which makes it possible to separate them from other dough constituents, of their sensitiveness to acids, bases, and salts in terms of elasticity, tenacity, and cohesiveness, Cohn and Henderson (1918) pointed out that these proteins (or gluten) impart to dough the physical properties that are peculiar to it. The swollen, coherent gluten in dough permits the stretching and distending of the mass in bread-making. Dough made from flour or meal produced from grains other than wheat lack these properties. Increasing the acidity of dough, and the content of certain salts such as calcium sulfate were believed to be responsible for increase in the elasticity of the dough. The best concentration of hydrogen ions for baking was deemed to be that indicated by the turning of methyl red from orange to red (pH = 5 ±) and it was stated that "the acidity of the dough at the time of baking seems to be the most important variable factor in bread-making." This agreed in general with the conclusion previously reached by Jessen-Hansen (1911). The author notes an objection to this sweeping contention at this point and reference will later be made to other data which do not support this contention of Cohn and Henderson.

Henderson, Fenn, and Cohn (1919) detailed the results of experiments which indicated the influence of electrolytes upon the viscosity of dough. "Viscosity" in this instance was in reality the resistance of dough to stirring, and involves plastic rather than viscous flow. The instrument used was a bowl with four equidistant pins on which the dough was impaled. Paddles attached to a vertical spindle were rotated through this dough, the spindle being turned by a cord wound around a drum on its upper end. This cord in turn passed over a pulley and was attached to a 150 gram weight which was allowed to fall through 2 meters. The time required for this fall was recorded and constituted the measure of "viscosity." When hydrogen ion concentration of the dough was varied from pH = 3.6, to pH = 7.6, the viscosity passed through a minimum a little on the acid side of pH = 5. Small quantities of NaCl, Na_2SO_4, $MgSO_4$, NH_4Cl, $MgCl_2$, $KBrO_3$, and sodium lactate diminished the viscosity. As the concentration of the salt was increased there was commonly, and probably with sufficiently high concentration always a later rise in viscosity. Sulfates of sodium and magnesium seemed to "slacken off" the dough appreciably. The simultaneous effect of salt and acid (or alkali) on the dough was studied and showed the influence of the salt to vary with the hydrogen ion concentration. Their investigations led them to conclude that in bread-making, the action of acids, bases, and salts (with the possible exception of potassium bromate) is favorable chiefly through the effect upon viscosity, and any influence upon the activity of yeast seemed to be a matter of secondary importance.

The author has failed to find any published data which involves a correlation between these observations of Henderson et al and the results of actual baking trials conducted by them. Efforts on the part of the author and his co-workers to establish such correlations have not met with success. Thus Bailey and LeVesconte (1924) found that increasing the hydrogen ion concentration of a dough (original $pH = 5.8$) decreased its extensibility as measured with the Chopin extensimeter. The addition of alkali to the dough increased extensibility, which passed through a maximum at $pH = 6.1$, and then decreased with further additions of alkali. This is shown by the data in Table 149.

It might be argued that the measurement of extensibility by this method is no more valid index of how the dough will comport itself in baking than is a measurement of the viscosity of the dough. It appears to the author that the extensimeter method has distinct advantages, however, since it measures certain of those peculiar and distinctive properties of a wheat flour dough which adapt such a dough to the production of yeast-leavened bread. These include its ability to stretch or extend its area many times without rupturing. This may be more evident if the method is described. The Chopin (1921) extensimeter provides facilities for producing a thin (3 mm.) sheet of dough, which is elevated from the plate on which it rests by means of air under slight compression. This air pressure is directed underneath the sheet of dough, which is forced upwards, and is caused to assume the appearance of a bubble. This bubble is expanded at a uniform rate until it finally bursts and collapses. At the instant that it ruptures the current of air used in inflating it is shut off. The manometer and gasometer connected with this air pressure device provide measurements of the actual volume of air used in inflating the dough. These in turn provide a basis for computing the increase in dough area or "extension" of surface effected by the inflation of the bubble, in which terms the results of observations are recorded. All of the measurements made by Bailey and LeVesconte were replicated at least 20 times, thus reducing the probable error of the mean of these tests.

That the optimum hydrogen ion concentration of the dough is not on the acid side of $pH = 5$, so far as gluten properties, and properties of the dough itself are concerned, is indicated by indirect evidence.

TABLE 149

Effect on Extensibility of Varying the Hydrogen Ion Concentration of the Dough (Bailey and Le Vesconte, 1924)

H-Ion Concentration of the Dough, pH	Percentage of Extensibility of Original Dough
7.9	66.7
7.1	78.5
6.7	87.3
6.5	94.6
6.2	95.8
6.1	103.1
5.9	100.5
5.8[a]	100.0
5.5	97.0
5.2	93.4
5.1	79.6
4.7	66.7
4.2	48.9
3.2	34.5

[a] Control or untreated dough.

The iso-electric point of gliadin has recently been shown to be pH = 6.5 — 6.6 by Eto (1924) and Tague (1925a, 1925b). Recent studies of Sharp and Gortner (1923) show that glutenin is least hydrated between pH = 6 and pH = 8, while Tague (1925a) places its iso-electric point at pH = 6.8 — 7.0. Since these two gluten proteins apparently have about the same iso-electric point,[5] ranging around pH = 6.5 — 7.0, it might be anticipated that maximum coherence and extensibility of gluten and of dough should be encountered in about the same range. While Jessen-Hansen's (1911) contention that the best bread resulted when dough acquired a hydrogen ion concentration equivalent to pH = 5.0 ± may be acceptable, it appears probable that if this is a fact, it is because the increased acidity accelerates the activity of the enzymes of the dough, and particularly of diastase and zymase. There is as yet no tangible evidence to prove that acidulation of the dough improves its physical properties in a direction that makes for better bread.

Mention should also be made of the oft-repeated contention that the addition of certain salts to dough improves the "quality" of the gluten, and thus improves the bread made from doughs to which these salts have been added. It is quite evident that crude gluten washed from dough containing certain salts, or washed from dough by means of a solution of such salts possesses physical properties differing from those of a gluten washed from a plain dough (without added salts) by means of distilled water. That the action of acids and alkalis on gluten

[5] The earlier observation of Bailey and Peterson (1921) that the iso-electric point of gluten was approximately pH = 5.1 was based on less refined methods than are now available and was probably incorrect.

is quite different in the presence and absence of salts is likewise very evident. That these salts, notably the sulfates and chlorides of the alkaline earth metals have a profound effect upon the physical properties of an ordinary dough is doubtful, however. In the first place, such a dough is normally loaded heavily with electrolytes. Not infrequently will the water of dough contain 3 per cent of sodium chloride, used as a normal ingredient of the dough batch in addition to electrolytes contributed by the flour and the water. There is, therefore, no dearth of neutral salts to counteract the effect of the increased hydrogen ion concentration (which results from fermentation) upon the water-imbibing capacity of the gluten. Whether small additions of other salts (excepting those which contribute tri-valent ions such as PO_4^{---}, and Al^{+++}) in the proportions of one or two tenths of one per cent or less will register an appreciable effect in such a mixture of electrolytes seems problematical. A direct attack upon this problem has thus far yielded negative results. Bailey and LeVesconte (1924) detected no increase in extensibility of dough on the addition of $CaSO_4$, or $MgSO_4$ in the proportions in which they are employed in commercial practice. Monocalcium phosphate effected a slight increase in extensibility. Johnson (1923) in collaboration with the author did not detect any substantial effect of these and certain other salts on the gas-retaining power of dough. Aluminum salts did increase gas-retention to an appreciable extent.

Several of these salts, notably those yielding ammonium, calcium and phosphate ions in solution, stimulated fermentation and gas production. Their principal effect as flour improvers is believed to be due to their usefulness as yeast nutrients.

Reference has been made in the preceding paragraphs to the gas-retaining power of dough. Before concluding this discussion, the manner in which this dough property has been measured should be described. Bailey and Weigley (1922) placed an aliquot of a dough batch, representing the equivalent of 50 grams of flour in a vessel that was closed except for tubulures at top and bottom. Air freed from carbon dioxide was passed into the vessel through the upper tubulure, over the dough surface, and out through the lower tubulure. The gas mixture was then led into an absorption tower where the carbon dioxide was absorbed in barium hydroxide solution. By frequently changing the absorption towers, the quantity of carbon dioxide lost from the dough per unit of time could be determined. This study revealed several differences in the properties of doughs made from weak soft wheat flour, and strong hard spring wheat flour. The rate of loss of carbon dioxide from the former was at a higher level throughout the fermentation period. The maximum volume of the weak flour dough was reached sooner in point of time, and was not as great as that of the strong wheat flour. Incidentally the study showed interesting differences in the rate of expansion, and of gas retention of freshly mixed doughs, and of doughs which had been normally fermented and were ready for the final raising in the bread pan. Bailey and Johnson (1924b) modified the apparatus, rendering it more convenient of operation, while retaining the essential features of the original method. This method of study yielded tangible evidence of the effect of various treatments on the properties of wheat flour doughs.

The colloidal behavior of dough can be modified by seemingly simple manipulations. When Sharp, Gortner and Johnson (1923) brought doughs to a pH of 3, or a pH of 11 by the addition of hydrochloric acid, or sodium hydroxide respectively and after 30 minutes restored the hydrogen ion concentration to that of the original dough, the baking properties of the dough were found to be greatly impaired. Flour was mixed with 70 per cent, and 90 per cent alcohol, the paste promptly dried, milled, and bolted, and the resulting flour made into dough and baked. Again the manipulation definitely impaired the baking quality of the dough. Johnson and Bailey (1925) treated flour in a similar manner with 96 per cent alcohol, and found that the gas retaining capacity of dough made from the dried and remilled paste was reduced. Flour kneaded with water, and dried at room temperature, was remilled, and made into a dough. This dough showed an even greater reduction in gas-retaining capacity than the dough made with the remilled alcoholic paste, the loss of gas in fermentation being nearly 60 per cent greater than from a normal dough. The extensibility of these two doughs was likewise measured with the Chopin device, and a decided reduction in this property was observed.

Prolonged mechanical kneading of an ordinary dough in a kneading machine was found by Bailey and LeVesconte (1924) to reduce the extensibility appreciably. Thus the extensibility after 4, 16 and 24 minutes was 17.5, 14.5 and 13.6 units respectively. This effect of overmixing has been attributed by Ostwald (1919) to a depolymerization in a mechanical manner, which results in an increase in the degree of dispersion. This in turn would effect a modification of the colloidal behavior of the dough. An increased dispersion of the gluten would confer the properties of a weaker gluten, according to the hypothesis of Gortner and Doherty (1918). These observations lead to the conclusion that vigorous mechanical kneading impairs rather than improves the baking strength of dough. The "gluten development" claimed by the manufacturers of high-speed dough mixers may in reality be merely a more uniform incorporation of the water. It is possible that if dough could be prepared without the strenuous kneading which it ordinarily receives, its colloidal properties would more nearly approach the ideal.

It has been generally recognized by skilled flour millers that deliberate or incidental attrition of middling stocks, resulting from heavy grinding, or handling in horizontal conveyors, tended to impair certain properties of the flour produced from the middlings. The diminished quality of the flour thus produced may be due in part to the pulverizing of branny structures present in the middling stocks, which normally escape disintegration in the initial grinding of the middlings. The pulverized bran thus produced finds its way into the flour and tends to unfavorably affect its appearance, and baking qualities. It is also possible that the physical treatment to which the endosperm particles are thus subjected may be responsible for the effects observed.

When flour was deliberately ground to varying degrees of fineness as evidenced by passing it through sieves with openings of progressively decreasing size, certain general differences in the properties of the resulting products were noted by Shollenberger (1921). Flours of fine granulation absorbed water more quickly, and doughs made from them fermented more rapidly than in the instance of the coarser flours. Size (volume) and color of bread were improved with increasing fineness of granulation, although flours of intermediate granulation baked into bread having the best texture. Doughs made from the finest flours were notably stickier than when the coarser particles predominated.

Five commercial flours were fractionated into particles of various sizes by LeClerc, Wessling, Bailey, and Gordon (1919) by bolting on Nos. 15, 18, 20 and 21 XX mill bolting silks. Five fractions were thus obtained representing the particles which remained on each of the flour sieves, and the finest fraction which passed through the No. 21 silk. Analyses and tests of each fraction were made, and certain of the averages of the resulting data are given in Table 150. The finer flour particles thus separated were higher in ash and acidity, and lower in protein content than the coarser particles. Bread made from the finer particles was inferior to that from the coarser particles in color, texture, and elasticity of crumb.

TABLE 150

RESULTS OF TESTS AND ANALYSES OF FLOUR PARTICLES OF VARYING DEGREES OF FINENESS AS SEPARATED BY SIFTING ON VARIOUS SIZES OF BOLTING CLOTH (LeClerc, Wessling, Bailey, and Gordon, 1919)

Fraction	Percentage of Original	Ash, Per Cent	Protein, Per Cent	Acidity, Per Cent	Maximum Expansion, cc.	Color Score	Texture Score
Original	100.0	.397	10.28	.114	856	97.9	97.7
On No. 15 sieve	24.8	.369	10.43	.105	826	98.5	97.5
" " 18 "	7.8	.398	11.00	.107	858	98.8	98.1
" " 20 "	12.0	.386	11.17	.112	850	98.6	98.1
" " 21 "	24.1	.414	10.72	.124	854	98.3	97.2
Thru No. 21 sieve	31.2	.425	8.95	.126	810	97.8	96.0

A mechanical injury of the starch granules of flour in consequence of over-grinding has been demonstrated by Alsberg and Perry (1925).

This manifests itself by the appearance of the starch granules under the microscope. The matrix in which they are normally held by the gluten is broken up, and a larger proportion of the granules can be stained with Congo Red, and swelled in cold water. Cold water extract of over-ground flour was substantially increased, the percentage of solids in such an extract being doubled in consequence of grinding flour for 53 hours in a ball mill. When flour was reground in a roller mill with smooth rolls the magnitude of the effect was less great, but in the same direction. Diastatic activity of the over-ground flour was likewise substantially greater than that of the normal flour. This may result from the mechanical effect of the grinding upon the starch aggregates, freeing the individual granules so that they are more accessible to the diastase, and are thus more rapidly attacked and converted into fermentable sugars. Or the effect may go so far as to partially disintegrate the individual starch granules, which disperse to a greater extent in the dough solution, as is evident by the increase in the percentage of solids in the cold water extract. This modification of the properties of the starch is reflected in an increased water-absorption in preparing a dough from flours subjected to over-grinding. Such doughs, when fermentd with yeast, expanded rapidly on the first rise, but lost a much larger fraction of the gases of fermentation than a normal dough.

The studies of Alsberg and Perry also indicate that the properties of the gluten of flour are likewise affected by over-grinding. The evidence is not as conclusive as in the instance of the starch, since the properties of the latter are reflected in the characteristics of the dough. Crude gluten washed from over-ground flours contained less non-protein impurities, a higher percentage of ash, and absorbed less water per gram of gluten than crude gluten from the same flour before it was reground. The physical properties of the wet crude gluten washed from the flours that were ground in a ball mill were quite different from the normal flour glutens. The former were characterised as "soft, soapy, sticky—vastly poorer in quality" than the normal. In certain instances over-grinding the flour modified it to such an extent that no gluten could be washed from it. Dough prepared from such flours was crumbly, and cheesy in texture; when fermented with yeast and baked, the resulting bread was smaller in volume than, and inferior in texture to bread baked from the normal flour. As in the instance of starch properties, the effect on baking qualities was of less magnitude when moderate regrinding between smooth steel rollers was resorted to, than when excessive and prolonged regrinding in a ball mill was involved.

The relation of starch to the colloidal behavior of dough is problematical. Ordinarily the normal raw starch granules are considered to be inert diluents of the gluten, and the other dough constituents in which we are chiefly interested. This is a rational concept insofar as the raw or unbaked dough is concerned. The concentration of starch would accordingly be of interest since it generally varies inversely with the gluten content. When the dough is baked into bread the starch undergoes significant changes during the heating or baking process, however. It "gelatinizes"; its suspension in the other dough ingredients is transformed into a viscous, strongly-hydrated gel. Any substantial differences in the viscosity of such gels resulting from the baking of different types of flour might be reflected in the properties of the baked loaf. That such differences exist is indicated by the work of Rask and Alsberg (1924). The starch which they prepared from common winter wheats yielded more viscous starch pastes (after heating) than did starch from the common spring wheats. In general, high starch viscosity appeared to be associated with two or more of the following conditions in the case of each flour examined: (1) low loaf volume; (2) low gluten content; (3) high temperature of locality of growth. It accordingly appears that starch should receive more consideration than has heretofore been accorded it in flour strength studies.

A discussion of the rôle of starch in the staling of bread belongs in a treatise on baking. References to this phenomenon may be found in the papers by Ostwald (1919), and Whymper (1920).

The effect of lipoids upon the colloidal behavior of dough has not been studied as extensively as the possibilities merit. Salamon (1908) stated that flour that had been extracted with ether, and the ether vaporized, baked into a loaf larger than that which could be baked from the original flour. This has recently been confirmed by Johnson, whose data have not yet been published. Stockham (1920) did not concur in this opinion, however, and reported a smaller loaf volume as a result of baking ether extracted flour in comparison with the original or untreated flour. He also observed an improved baking strength when 0.6 per cent of wheat fat was added to the original flour. The addition of lard and cotton-seed oil impaired baking strength somewhat. The experiments of Saunders, Nichols, and Cowan (1921) resulted in an increased loaf volume when lard was added to flour. Other fats, including butter, and several vegetable oils effected an improvement in bread properties.

Working (1924) studied the influence of lipoids on gluten quality. When wheat phosphatide was added in the proportion of 0.25 per cent to patent flour, the crude gluten washed from the flour was much softer than the normal or control gluten. With 0.50 per cent of phosphatide the gluten was as soft as that from a low grade flour. When 3 per cent was added the crude gluten could not be recovered at all. An addition of 0.5 per cent of wheat phosphatide reduced the viscosity of acidulated water suspensions of flour by 33 per cent, while a like quantity of egg lecithin reduced the viscosity 16 per cent. The same quantity of phosphatides, when added to flour used in a baking test reduced the baking strength. Its effect was particularly noticeable in the impaired crumb texture.

These data, resulting from the work of Salamon, Johnson, and Working have convinced the author that the natural glycerides, and the phosphatides of flour have a greater influence upon the colloidal behavior of flour than has been generally conceded. It appears possible that certain of the inferior properties of the lower grades of flour might be traced to their higher lipid content.

A study of the properties and chemical constants of fats extracted from flour, and from wheat germ has been recently reported by Ball (1924). Certain of the detailed data, together with similar data reported by other investigators, have been incorporated in Tables A and B in the appendix.

Two phytosterols were found in wheat by Anderson and Nabenhauer (1924), sitosterol, and dihydrosterol. The possible rôle of these sterols in determining the colloidal behavior of dough has not been subjected to careful study.

Recapitulation

By way of a summary of the present state of our knowledge concerning the colloidal behavior of dough, it should be noted that dough consists largely of water, starch, sugars, gluten, "soluble" proteins, lipoids, mineral salts, and yeast cells. The raw starch is more or less of an inert diluent of those constituents which contribute to the colloidal condition of raw or unbaked dough, but it gelatinizes and takes on the characteristics of a highly hydrated gel during the baking process. The gelatinizing properties of the starch in different wheat types apparently varies appreciably.

Gluten consists of two unlike proteins, gliadin, and glutenin. Gliadin of wheat appears to have many characteristics in common with the prolamins of certain other cereals. The gliadin preparations from different samples of flour do not vary sufficiently to indicate that such variations could account for the observed differences in the properties of gluten. Glutenin is the distinctive protein of wheat. The colloidal behavior of the glutenin in different wheats varies within rather wide limits. This variation appears to be due to the dimensions of the glutenin aggregates. These are determined largely by the inheritance of the wheat plant, and the conditions under which it is grown. Mechanical treatment of flour or dough, as in over-grinding flour, or overmixing dough may modify the colloidal behavior of the dough and tend to impair its baking qualities. The colloidal behavior of gluten and of dough is conditioned by the quantity and character of the electrolytes present. Since the iso-electric point of both gliadin and glutenin are within the range $pH = 6.5$ to $pH = 7.0$, it is probable that any substantial departure from this range of hydrogen ion concentration will be accompanied by an impairment of those colloidal properties which are of significance in bread-making. Dough is normally rich in neutral mineral salts, notably sodium chloride, in addition to salts contributed by the other ordinary dough ingredients, and it appears improbable that the addition of small quantities of other salts, such as calcium sulfate, has any profound effect upon the colloidal behavior of the dough. Salts contributing trivalent ions, such as PO_4^{---}, and Al^{+++} effect a measurable modification of the dough properties.

Dilution of the gluten with starch effects an impairment of baking strength. Gluten content is accordingly an important factor in determining the colloidal behavior of dough. This has been demonstrated in a variety of ways.

Gluten properties may be measurably and significantly modified by varying the lipoid content of dough. The quantity and characteristics of these lipoids may accordingly be one of the factors involved in the distinctive properties of high grade or patent, and the lower grade or clear flours milled from the same wheat.

APPENDIX

TABLE A

Physical and Chemical Constants of the Wheat Oils

Determination	Embryo Oil (Extracted)				Flour Oil (Extracted)			
	I	II	III	Average	I	II	III	Average
Sp. gr. sample 1	.....	.....	.....	0.92485 (25°/1°)	.....	.....	.....	0.006729 (28°/1°)
" " 2	.....	.....	.....	.....	.....	.....	.....	0.97145
Refract. Index (17.5°)	.....	.....	.....	1.4686	.....	.....	.....	1.4714
Saporification No.	183.91	183.93	184.54	184.13	159.49	160.77	162.49	160.86
Iodine No.	123.99	123.63	123.32	123.64	159.63	159.23	163.54	.....
Iodine No. of insol. acids	128.58	127.85	127.90	128.11	105.27	105.57	105.77	105.43
Iodine No. of liquid acids	145.93	146.01	.....	145.97	.....	.....	.....	.....
Unsapon., per cent								
Method (1)	3.51	3.51	.....	3.51	.....	.....	.....	.....
Method (2)	3.67	3.65	.....	3.66	2.35	2.67	2.50	2.51
Acid No., fresh oil								
After 3 years	21.41	21.55	.....	21.48				
Hehner No.	34.94	34.99	35.08	35.00				
Neut. No. of insol. acids	93.94	93.48	.....	93.71				
Neut. No. of solid acids	206.38	205.34	205.47	205.73				
Neut. No. of liquid acids	230.26	227.63	.....	228.94				
Mean mol. wgt. of insol. acids	202.97	202.34	.....	202.66				
Mean mol. wgt. of solid	.....	.....	.....	272.72				
Mean mol. wgt. of liquid	.....	.....	.....	245.08				
Ester No.	.....	.....	.....	276.85				
Reichert-Meissl No.	0.46	0.49	.....	162.65				
Polenske No.	0.26	0.24	.....	0.475				
Per cent Glycerol (approximate)	.....	.....	.....	0.25				
				8.90				

TABLE B

Chemical and Physical Constants of the Oils of Wheats as Given by Various Investigators

Investigator	Stellwag (1890)	Spaeth (1894)	de Negri (1898)	Frankforter and Harding (1899)	de Negri and Fabris and Plucker	Alpers (1918)
Part of wheat used	Bran	Flour	Embryo	Embryo	Flour	Embryo
Sp. gr.	.....	0.9068 (15°/15°)	0.9245 (15°/15°)	0.9202 (15°/15°)	.....	0.9320 (25°/25°)
Refract. Index	.....	1.4851 (25°)	1.4750	1.48325 (20°)	1.4851	1.4766
Saponification No.	183.1	166.5	182.81	188.83	182.8	180.0
Iodine No.	.....	101.5	115.17	115.64	96.1 to 112.5	122.6
Reichert-Meissl No.	.....	2.8	.....	.....	2.95 to 4.95	0.75
Hehner No.	.....	.....	.....	.....	.....	95.31
Unsapon., per cent	7.45	.....	.....	.....	.....	.....
Free acids	14.35	.....	5.65	.....	.....	.....
M. p. of solid acids	.....	34.0°	39.5°	.....	.....	.....
Iodine No. of insol. acids	.....	.....	123.27	.....	.....	.....
Neutral fats, per cent	78.73	.....	.....	.....	.....	.....
Total acids, per cent	89.71	.....	.....	.....	.....	.....
Mol. wgt. of fatty acids	285.	.....	.....	.....	.....	.....

BIBLIOGRAPHY

Adorjan, Joseph. 1902. "Die Nahrstoffaufnahme des Weizens." J. Landw., 50, 193-230.

Agr. Gaz. N. S. Wales. 1923. "The contamination of milling wheat. The effects of strong-scented weed seeds." Agr. Gaz. N. S. Wales, 34, 628.

Allen, R. M. 1910. "Bleached flour." Kentucky Agr. Exp. Sta. Bul. 149.

Alsberg, C. L., and Perry, E. E. 1925. "The effect of fine grinding upon flour." To be published in Cereal Chem.

Alway, F. J. 1907a. "The bleaching of wheat flour." Nebr. Agr. Exp. Sta., Press Bul. 24.

Alway, F. J. 1907b. "The effect of bleaching upon the quality of wheat flour." Nebr. Agr. Exp. Sta. Bul. 102.

Alway, F. J., and Clark, V. L. 1909. "The color and ash content of different grades of Nebraska flour." Nebr. Agr. Exp. Sta. 23rd Ann. Rept., 26-30.

Alway, F. J., and Gortner, Ross Aiken. 1907. "The detection of bleached flours." J. Am. Chem. Soc., 29, 1503-1513.

Alway, F. J., and Hartzell, Stella. 1909. "On the strength of wheat flour." Nebr. Agr. Exp. Sta. Rept., 1909, pp. 100-110.

Alway, F. J., and Pinckney, R. M. 1908. "The effect of nitrogen peroxide upon wheat flour." J. Am. Chem. Soc., 30, 81-85.

Ames, J. W. 1910. "The composition of wheat." Ohio Agr. Exp. Sta. Bul. 221.

Ames, J. W., Boltz, G. E., and Stenius, J. A. 1912. "Effect of fertilizers on the physical and chemical properties of wheat." Ohio Agr. Exp. Sta. Bul. 243.

Amos, P. A. 1912. "Flour manufacture." New York. 280 pp.

Anderson, R. J., and Nabenhauer, F. 1924. "The phytosterols of wheat endosperm." J. Am. Chem. Soc., 46, 1717-1721.

d'Andre, Henry. 1922. "Importance of the value of the 'absorptive capacity' of the true gluten and its practical use in flour analysis." J. Am. Assoc. Cereal Chemists, 7: 30-32.

Armstrong, E. F. 1910. "The chemical properties of wheaten flour." Sup. 4, J. Bd. Agr. (Eng.), 17, 45-52.

Arpin, M. 1902. "Dosage du gluten humide dans le farines." Ann. Chim. Anal. Chim. Appl. 7, 325-331, 376-381, 416-420.

Arpin, Marcel. 1905. "Blanchiment des farines." Rapport, Syndicat de la Boulangerie de Paris.

Arpin, Marcel. 1921. "Procede de blanchiment des farines par le chlore gaseux." Rapport. Conseil d'hygiene publique et d'salubrite du department la Seine.

Arpin, Marcel, and Pecaud, Mlle. M. T. 1923a. "Détermination de la valeur boulangère de différentes variétés pures de Blés cultivés en France." Compt. rend. trav. Semaine Nat. du Ble. (Paris), 104-137.

Arpin, M., and Pecaud, M. T. 1923b. "Variations dans le poids des farines." Ann. fals, 16, 586-597.

Ashton, John. 1904. "The history of bread." Religious Tract Soc'y., London.

Assoc. Off. Agr. Chemists. 1920. "Official and tentative methods of analysis." Pub. by Assoc. Off. Agr. Chem., Washington, D. C.

Association of Operative Millers. 1923. Bulletin of the A. O. M., October, 1923, 129, 130.

Association of Operative Millers. 1923-24. "Report of Operations." Weeks ending Aug. 11, 1923; Oct. 6, 1923; Aug. 16, 1924; Oct. 11, 1924.

(Australia) Advisory Council of Science and Industry. 1917. "Problems of wheat storage. 1. Damaged Grain. 2. Insect Pests." Advisory Council of Science and Industry (Aust.) Bul. 5. Melbourne.

Avery. 1907. "A contribution to the

chemistry of the bleaching of flour." J. Am. Chem. Soc., *29*, 571-574.

Azzi, G. 1921. "The critical period of wheat with respect to rain." Nuovi. Ann. (Italy) Min. Agr., *1*, 299-307. Abstract in Exp. Sta. Rec., *48*, 233. Original not seen.

Azzi, G. 1922. "Effect on wheat yields of variations in soil moisture during and after the critical period." Cultivatore (Italy), *68*, 308-312. Abstract in Exp. Sta. Rec., *49*, 738. Original not seen.

Bailey, C. H. 1913a. "Minnesota wheat investigations." Series I. Minn. Agr. Exp. Sta. Bul. 131.

Bailey, C. H. 1913b. "How wheat grows." Operative Miller, July, 1913, 454-455.

Bailey, C. H. 1913c. "Relation of the composition of flour to baking quality." Canadian Miller and Cerealist, *5*, 208-209.

Bailey, C. H. 1914a. "Marquis wheat. Part II. Milling quality." Minn. Agr. Exp. Sta. Bul. 137, 9-14.

Bailey, C. H. 1914b. "Composition and quality of spring and winter wheats, crops of 1912 and 1913." Minn. Agr. Exp. Sta. Bul. 143.

Bailey, C. H. 1914c. "The invisible loss in milling." Canadian Miller and Cerealist, *6*, 74-75.

Bailey, C. H. 1916a. "The relation of certain physical characteristics of the wheat kernel to milling quality." J. Agr. Sci., *7*, 432-442.

Bailey, C. H. 1916b. "Annual report of the chief Inspector of Grain to the Minnesota Ry. & Warehouse Comm." (pp. 28-36.)

Bailey, C. H. 1916c. "A method for the determination of the strength and baking qualities of wheat flour." J. Ind. Eng. Chem., *8*, 53-57.

Bailey, C. H. 1917a. "The moisture content of heating wheat." J. Am. Soc. Agron., *9*, 248-251.

Bailey, C. H. 1917b. "The handling and storage of spring wheat." J. Am. Soc. Agron., *9*, 275-281.

Bailey, C. H. 1917c. "The quality of western-grown spring wheat." J. Am. Soc. Agron., *9*, 155-161.

Bailey, C. H. 1917d. "The catalase activity of American wheat flours." J. Biol. Chem., *32*, 539-545.

Bailey, C. H. 1918. "Specific conductivity of water extracts of wheat flour." Science, *47*, 645-647.

Bailey, C. H. 1920. "The hygroscopic moisture of flour exposed to atmospheres of different relative humidity." J. Ind. Eng. Chem., *12*, 1102-1104.

Bailey, C. H. 1921. "Respiration of shelled corn." Minn. Agr. Exp. Sta., Tech. Bul. 3.

Bailey, C. H. 1922. "The character of 1919 crop spring wheat dockage." J. Am. Soc. Agron., *14*, 88-93.

Bailey, C. H. 1923. "Report of operation, State Testing Mill, Season 1921-1922." Minn. State Dept. Agr. Bul. 23.

Bailey, C. H. 1924. "Report of operation, State Testing Mill, crop season of 1922." Minn. State Dept. Agr. Bul. 34.

Bailey, C. H., and Blish, M. J. 1915. "Concerning the identity of the proteins extracted from wheat flour by the usual solvents." J. Biol. Chem., *23*, 345-357.

Bailey, C. H., and Collatz, F. A. 1921. "Studies of wheat flour grades. I. Electrical conductivity of water extracts." J. Ind. Eng. Chem., *13*, 319-321.

Bailey, C. H., and Gurjar, A. M. 1918. "Respiration of stored wheat." J. Agr. Research, *12*, 685-713.

Bailey, C. H., and Gurjar, A. M. 1920. "Respiration of cereal plants and grains. II. Respiration of sprouted wheat." J. Biol. Chem., *44*, 5-7. "III. Respiration of rice paddy and milled rice." *44*, 9-12. "IV. Respiration of frosted wheat plants." *44*, 13-15. "V. Note on the respiration of wheat plants infected with stem rust." *44*, 17-18.

Bailey, C. H., and Hendel, Julius. 1923. "Correlation of wheat kernel plumpness and protein content." J. Am. Soc. Agron., *15*, 345-350.

Bailey, C. H., and Johnson, A. H. 1922. "Studies of wheat flour grades. III. Effect of chlorine bleaching upon the electrolytic resistance and hydrogen ion concentration of water extracts." J. Assoc. Official Agr. Chemists, *6*, 63-68.

Bailey, C. H., and Johnson, A. H. 1924a. "Studies on wheat flour grades. IV. Changes in hydrogen ion concentration and electrolytic resistance of water extracts of natural and chlorine treated flour in storage." Cereal Chem., *1*, 133-137.

Bailey, C. H., and Johnson, A. H. 1924b. "Carbon dioxide diffusion ratio of wheat flour doughs as a measure of fermentation period." Cereal Chem., *1*, 293-304.

Bailey, C. H., and Le Vesconte, A. M. 1924. "Physical tests of flour quality with the Chopin extensimeter." Cereal Chem., *1*, 38-63.

Bailey, C. H., and Peterson, Anna C. 1921. "Studies of wheat flour grades. II. Buffer action of water extracts." J. Ind. Eng. Chem., *13*, 916-918.

Bailey, C. H., and Sherwood, R. C. 1923. "The march of hydrogen ion concentration in bread doughs." Ind. Eng. Chem., *15*, 624-627.

Bailey, C. H., and Sherwood, R. C. 1924. "Report of operation, State Testing Mill, crop season of 1923." Minn. State Dept. Agr. Bul. 37.

Bailey, C. H., and Weigley, Mildred. 1922. "Loss of carbon dioxide from doughs as an index of flour strength." J. Ind. Eng. Chem., *14*, 147-150.

Bailey, L. H., and Thom, C. 1920. "Some observations of corn meal in storage." Operative Miller, *25*, 368-371.

Baker, J. C. 1922. "Flour bleaching reagents." J. Am. Assoc. Cereal Chemists, *7*, 108-111.

Baker, J. L., and Hulton, H. F. E. 1908. "Considerations affecting the 'strength' of wheat flours." J. Soc. Chem. Ind., *27*, 368-376.

Ball, Chas. D. 1924. "A study of wheat oil." A thesis filed in the library of the University of Minnesota, Minneapolis.

Ball, C. R., and Leighty, C. E. 1916. "Alaska and Stoner or 'Miracle' wheats. Two varieties much misrepresented." U. S. Dept. Agr., Bul. 357.

Balland, A. 1883a. "Mémoire sur les farines." Compt. rend., *97*, 496-497.

Balland, A. 1883b. "Mémoire sur les farines. Des causes de l'altération des farines." Compt. rend., *97*, 651-652.

Balland, A. 1883c. "Altérations qu'éprouvent les farines en vieillissant." Compt. rend., *97*, 346-347.

Balland, A. 1883d. "Mémoire sur les farines. II. Experiences sur le gluten." J. pharm. chim. ser. 5, *8*, 433-440.

Balland, A. 1884. "Altérations qu'éprouvent les farines en vieillisant." Ann. chim. phys. ser. 6, *1*, 533-557.

Balland, A. 1893. "Sur la preexistence du gluten dans le blé." Compt. rend., *116*, 202-204.

Balland, A. 1896. "Sur le dosage du gluten dans les farines." Compt. rend., *123*, 136-137.

Balland, A. 1904a. "Sur la conservation des farines par le froid." Compt. rend., *139*, 473-475.

Balland, A. 1904b. "Sur le blanchiment des farines l'électricité." Compt. rend., *139*, 822-823.

Balland, A. 1914. "Sur la baisse du gluten des farines." J. pharm. chim., ser. 7, *9*, 510.

Baston, G. H. 1921. "Influence of weevils on temperature in wheat." Heating and spoiling of wheat, abstracted by C. Louise Phillips, U. S. Bur. Agr. Econ. (note pp. 12-13).

Bates, E. N., and Rush, G. L. 1922. "The bulk handling of grain." U. S. Dept. Agr. Farmers Bul. 1290.

Beccari. 1745. "De Fromento." De Bononiensi Scientiarium et Artium Instituto atque Academia Commentarii, 1745, II, part I, p. 122. Cited by T. B. Osborne (1912). Original not seen.

Bedford, S. A. 1893. "The cutting of wheat at different stages of ripeness." Cent. Exp. Farm. (Canada) Report 1893, 234.

Bedford, S. A. 1894. "Wheat cut at different stages of growth." Cent. Exp. Farm. (Canada) Report 1894, 284.

Bell, H. G. 1908. "Changes in stored flour." Operative Miller, *13*, 591-592.

Bell, H. G. 1909. "Fungus and bacterial growth on stored flour." American Miller, *37*, 280-281.

Bénard, and Girardin, J. 1881. "Sur le dosage du gluten dans les farines." J. pharm. chim. ser. 5, *4*, 127-128.

Bertrand, G., and Muttermilch, W. 1907a. "Sur l'existence d'une tyrosinase dans le son le froment." Compt. rend., *144*, 1285-1288.

Bertrand, G., and Muttermilch, W. 1907b. "Sur le phenomene de coloration du pain bis." Compt. rend., *144*, 1444-1446.

Beyer, C. 1913. German patent No. 270,909, Jan. 7th, 1913. Abstracted in Chem. Abst., *8*, 2204. Original not seen.

Biffin, R. H. 1905. "Mendel's laws of inheritance and wheat breeding." J. Agr. Sci. *1*, 4-48.

Biffin, R. H. 1908. "On the inheritance of strength in wheat." J. Agr. Sci. *3*, 86-101.

Biffin, R. H. 1909. "The inheritance

of 'strength' in wheat." J. Agr. Sci. *3*, 223-224.

Birchard, F. J. 1920. "Report of the Dominion Grain Research Laboratory." Ottawa, Canada.

Birchard, F. J., and Alcock, A. W. 1918. "Report of trial shipment of bulk wheat from Vancouver via the Panama Canal to the United Kingdom." Dominion Grain Res. Laby. (Canada), Bul. 1.

Blish, M. J. 1916. "On the chemical constitution of the proteins of wheat flour and its relation to baking strength." J. Ind. Eng. Chem., *8:* 138-144.

Blish, M. J. 1920. "Effect of premature freezing on composition of wheat." J. Agr. Res., *19*, 181-188.

Blish, M. J. 1925. "The individuality of glutenin." Cereal Chem., *2*, 127-131.

Blish, M. J., and Pinckney, A. J. 1924. "The identity of gluten proteins from various wheat flours." Cereal Chem., *1:* 309-316.

Blish, M. J., and Sandstedt, R. M. 1925a. "Glutenin—a simple method for its preparation and direct quantitative determination." Cereal Chem., *2*, 57-67.

Blish, M. J., and Sandstedt, R. M. 1925b. "Viscosity studies with Nebraska wheat flours." Cereal Chem. *2*, 191-201.

Bodenstein, M. 1922. "Bildung und Zersetzung der höheren Stickoxyde." Z. physik. Chem., *100*, 68-123.

Boerner, E. G. 1919. "Factors influencing the carrying qualities of American export corn." U. S. Dept. Agr. Bul. 764.

Bogdan. 1900. "Report of the Valniki Agr. Exp. Sta., Government of Samara, 1895-1896." Dept. Agr. (St. Petersburg) report 1900, 126 pp. Abstract in Exp. Sta. Rec. 13, 329. Original not seen.

Boland. 1848. "Mémoire sur les moyens de reconnaître et d'apprécier les propriétés panifiables de la farine de froment à l'aide de l'aleuromètre, instrument inventé par M. Boland, ancien boulangère à Paris." Bul. soc. encour. ind. nat. *47*, 704-709.

Bornand, M. 1921. "Catalases végétales, applications de la catalase en hygiène alimentaire." Mitt. Lebensm. Hyg., *12*, 125-133.

Bowen, J. C. 1914. "Wheat and flour prices from farmer to consumer." U. S. Bur. Labor Statistics Bul. 130.

Brahm, Carl. 1904. "Studien über das Alsop'sche Mehlbleich-Verfahren und über den Einfluss des Ozons in bezug auf Farbe und Backfähigheit von Weizenmehl." Mitteilung aus der Versuchs-Anstatt des Verbandes Deutscher Müller an die Königl. Landw. Hochschule zu Berlin. 18 pp.

Brandeis, E. 1908. "Die Anwendung der 'Kalte' in der Broterzeugung." 1st Cong. Internat. Froid (Paris), Rap. et Commun., *3*, 98-112.

Breazeale, J. F. 1916. "Effect of sodium salts on the absorption of plant food by wheat seedlings." J. Agr. Res., 7, 407-416.

Bremer, W. 1907. "Hat der Gehalt des Weizenmehles an wasserlöslichen Stickstoff einen Einfluss auf seinen Backwert." Z. Nahr. Genussm. *13;* 69-74.

Brenchley, W. E. 1909. "On the 'strength' and development of the grain of wheat." Ann. Bot. *23*, 117-139.

Brenchley, W. E., and Hall, A. D. 1909. "The development of the grain of wheat." J. Agr., Sci. *3*, 195-217.

Brewer, W. H. 1883. "Relations of grain to moisture." Tenth Census (U. S.) Reports, *3*, 28-31.

Briggs, C. H. 1913. "The significance of loaf volume." N. W. Miller, *95:* 79-80.

Briggs, C. H. 1923. "Has northwestern spring wheat deteriorated?" J. Am. Assoc. Cereal Chemists, 8, 117-119.

Briggs, L. J. 1895. "Harvesting wheat at successive stages of ripeness." Mich. Agr. Exp. Sta. Bul. 125.

Buchanan, J. H., and Naudain, G. G. 1923. "The influence of starch on strength of wheat flour." Ind. Eng. Chem., *15:* 1050-1051.

Buchwald, J. 1913. "Zur Beurteilung der Mehle durch die botanische Analyse." Z. ges. Getreidew., *5*, 50-56.

Buchwald, J. 1916. "Überfeuchtes Getreide." Z. ges. Getreidew., *8*, 57-70.

Buchwald, J., and Neumann, M. P. 1909a. "Über das Bleichen der Mehle." Z. ges. Getreidew., *1*, 137-144; 257-262.

Buchwald, J., and Neumann, M. P. 1913. "Untersuchungen über das Humphries-Thomas-Verfahren zur Feuchtbehandlung der Mahlprodukte." Z. ges. Getreidew., *5*, 24-41.

Buchwald, J., and Treml, Hans. 1909. "Über den Nachweis gebleichter Mehl." Z. ges. Getreidew., *1*, 96-97.

Buck, C. F. 1917. "The gasoline color of flour." J. Am. Assoc. Cereal Chemists, *2*, 17-19.

Burlakow, G. 1898. "Ueber Athmung des Keimes des Weizens, Triticum vulgare." (Abstract) Bot. Centbl., *74*, 323-324. (Original Article in Arb. Naturf. gesell. K. Univ. Charkow, 31, Beilage, 1897. Not seen.)

Calendoli, E. 1918. "Una reazione cromatica per l'esame della farina, specie per la determinazione del grado di abburattemento." Ann. Ig. (Rome), *28*, 76-77.

Cerkez, S. 1895. "Mehluntersuchungen." Z. angew. Chemie, 1895, 663-665.

Choate, Helen A. 1921. "Chemical changes in wheat during germination." Bot. Gaz., *71*, 409-425.

Chopin, Marcel. 1921. "Relations entre les propriétés méchaniques des pâtes de farine et la panification." Bul. soc. encour. ind. nat. *133*: 261-273.

Clark, J. A. 1924. "Improving quality of American grown durum wheat." The Macaroni J., *6*, 21-23.

Clark, J. Allen, Martin, John H., and Ball, C. R. 1922. "Classification of American wheat varieties." U. S. Dept. Agr. Bul. 1074.

Clark, J. Allen, Martin, J. H., and Smith, R. W. 1920. "Varietal experiments with spring wheat on the northern great plains." U. S. Dept. Agr. Bul. 878.

Clark, J. Allen, and Salmon, S. C. 1921. "Kanred wheat." U. S. Dept. Agr. Circular 194.

Clark, J. A., Stephens, D. E., and Florell, V. H. 1920. "Australian wheat varieties in the Pacific coast area." U. S. Dept. Agr. Bul. 877.

Clark, J. A., and Waldron, L. R. 1923. "Kota wheat." U. S. Dept. Agr. Circ. 280.

Clark, Rowland J. 1924. "Bread troubles in the light of hydrogen ion concentration." Cereal Chem., *1*: 161-167.

Cobb, N. A. 1905. "Universal nomenclature of wheat." Dept. Agr. N. S. Wales Misc. Pub. 539.

Cobb, N. L. 1908. "The inner structure of the grain as related to flour and bread." 13th Ann. Rept. Bur. Labor & Ind. Statistics (Wisconsin) Part V, 735-749.

Cohn, E. J. 1925. "Biochemical methods of characterizing proteins. VII. The molecular weights of the proteins." J. Biol. Chem., *63*, Supplement, xv-xvi.

Cohn, E. J., and Henderson, L. J. 1918. "The physical chemistry of bread making." Science, *48*: 501-505.

Coleman, D. A., and Fellows, H. C. 1925. "The hygroscopic moisture of cereal grains and flaxseed exposed to atmospheres of different relative humidity." Cereal Chemistry, *2*, (In Press).

Coleman, D. A., and Regan, S. A. 1918. "Nematode galls as a factor in the marketing and milling of wheat." U. S. Dept. Agr. Bul. 734.

Colin, H., and Belval, H. 1922. "La genese des hydrates de carbone le blé. Presence de lévulosanes dans la tige." Compt. rend, *175*, 1441-1443.

Colin, H., and Belval, H. 1923. "Les hydrocarbones solubles du grain de blé au cours du devélopement." Compt. rend., *177*, 343-346.

Collatz, F. A. 1922. "Flour strength as influenced by the addition of diastatic ferments." Amer. Inst. Baking Bull. 9.

Collatz, F. A., and Bailey, C. H. 1921. "The activity of phytase as determined by the conductivity of phytin-phytase solutions." J. Ind. Eng. Chem., *13*, 337-339.

Corbould, M. K. 1921. "Wheat, flour and bread." Ohio Agr. Exp. Sta. Bul. 350.

Coward, K. H. 1924. "The lipochromes of etiolated wheat seedlings." Biochem. J., *18*, 1123-1126.

Cox, J. H. 1916. "The drying for milling purposes of damp and garlicky wheat." U. S. Dept. Agr. Bul. 455.

Cross, R. J., and Swain, R. E. 1924. "The amino acid distribution in proteins of wheat flours with a note on an improved method for the preparation of aldehyde-free alcohol." Ind. Eng. Chem., *16*, 49-52.

Dakin, H. D. 1912. "The recemization of proteins and their derivatives resulting from tautomeric change." J. Biol. Chem., *13*, 357-362.

Davidson, J. 1922. "The effect of

nitrates applied at different stages of growth on the yields, composition and quality of wheat." J. Am. Soc. Agron., *14*, 118-122.

Davidson, J., and LeClerc, J. A. 1917. "The effect of sodium nitrate applied at different stages of growth on yield, composition and quality of wheat." J. Am. Soc. Agron., *9*, 145-154.

Davidson, J., and LeClerc, J. A. 1918. "The effect of sodium nitrate applied at different stages of growth on yield, composition, and quality of wheat. 2." J. Am. Soc. Agron., *10*, 193-198.

Davidson, J., and LeClerc, J. A. 1923. "The effect of various nitrogen compounds applied at different stages of growth on the yield, composition, and quality of wheat." J. Agr. Res., *23*, 55-68.

Dedrick, B. W. 1913. "Establish definite grades of flour, offals, and stock." Operative Miller, *18*, 441.

Dedrick, B. W. 1921. "Patent flour." Operative Miller, *26*, 61.

Dedrick, B. W. 1924. "Practical milling." Pub. by National Miller, Chicago, Ills.

Deherain, P. P., and Meyer. 1882. "Recherches sur le developpement du blé première anneé d'observation." Ann. Agron., *18*, 23-43.

Deherain, P. P., and Dupont, C. 1902. "Sur l'origine de l'amidon du grain de blé." Ann. Agron., *28*, 522-527.

Dempwolf, O. 1869. "Untersuchung der Ungarischen Weizens und Weizenmehls." Ann. Chem. Pharm., *149*, 343-350.

Dendy, A., and Elkington, H. D. 1920. "Report on the effect of airtight storage upon grain insects—Part III." Report of the Grain Pests (war) Comm., Royal Soc. (Eng.) No. 6, London, Eng.

Dienst, Karl. 1919. "Verfahren zum Lagern von Getreide." Z. ges. Getreidew., *11*, 10-11.

Dill, D. B. 1925. "The composition of crude gluten." Cereal Chem., *2*, 1-11.

Dill, D. B., and Alsberg, C. L. 1924. "Some critical considerations of the gluten washing problems." Cereal Chem., *1*, 223-246.

Dingwall, Andrew. 1924. "Studies of proteins of wheat flour in relation to flour strength." Thesis filed in the library of the University of Minnesota, Minneapolis.

Dobrescu, I. M. 1921. "Le climat et la blé roumain." Bul. Soc. Stiinte Cluj., *1*, 171-176.

Dunlap, F. L. 1922. "Flour facts—properties that affect baking efficiency." J. Am. Assoc. Cereal Chemists, *7*: 2-9.

Dunlap, F. L. 1923. "Bleaching and maturing of flour." J. Am. Assoc. Cereal Chemists, *8*, 9-19.

Duvel, J. W. T. 1907. "Garlicky wheat." U. S. Dept. Agr., Bur. Plant Ind. Bul. 100.

Duvel, J. W. T. 1909. "The deterioration of corn in storage." U. S. Dept. Agr., Bur. Pl. Ind. Circ. 43.

Duvel, J. W. T., and Duval, L. 1913. "The shrinkage of shelled corn while in cars in transit." U. S. Dept. Agr. Bul. 48.

Eckerson, Sophia H. 1917. "Microchemical studies in the progressive development of the wheat plant." Wash. Agr. Exp. Sta. Bul. 139.

Edgar, William C. 1912. "The story of a grain of wheat." D. Appleton, New York and London.

Einhof, H. 1805. "Chemische Analyse des Roggens (Secale cereale)." Neues Algem. J. d. chem. *5*, 131-153. Cited by Osborne (1907), and by Greaves (1911). Original not seen.

Ellis, J. H. 1919. "The stage of maturity of cutting wheat when affected with black stem rust." Agr. Gaz. Can., *6*, 971.

Eto, Itsuo. 1924. "Beiträge zur Kenntnis des Gliadins." J. Biochem. (Japan) *3*, 373-392.

Evans, G. 1909. "Annual variations in the character of Central Provinces wheats." Dept. Agr. Cent. Prov. and Berar (India), Bul. 3.

Failyer, G. H., and Willard, J. T. 1891. "Composition of Kaffir, corn, oats, and wheat at different stages of growth." Kans. Agr. Exp. Sta. Bul. 32.

Farrow, F. D., and Lowe, G. M. 1923. "The flow of starch paste through capillary tubes." J. Textile Inst., *14*, 414-440T.

Feilitzen, H. von. 1904. "Über den Einfluss des Saatgutes des Bodens und der Düngung auf die Beschaffenheit des Mehlkörpers des geernteten kornes bei Sommerweizen und Gerste." J. Landw., *52*, 401-412.

Fenyvessy, B. 1911. "Über den Einfluss einiger vegetabilischen Proteide auf den Weizenkleber." Z. Nahr. Genussm., *21*, 658-662.

Fernández, O., and Pizarroso, A. 1921. "El podor catalitico de las harinas." Anales soc. espan. fis. quin., *19*, 265-268.

Fitz, L. A. 1910. "Handling wheat from field to mill." U. S. Bur. Plant Ind. Circ. 68.

Fitz, L. A., Swanson, C. O., et al. 1916. "The milling and baking quality and chemical composition of wheat and flour as influenced by 1. Different methods of handling and storage. 2. Heat and moisture. 3. Germination." Kansas Agr. Exp. Sta. Tech. Bul. 1.

Fleurent, E. 1896a. "Sur la composition immédiate du gluten des céréales." Compt. rend., *123*, 327-330.

Fleurent, E. 1896b. "Sur une méthode chemique d'appreciation de la valeur boulangère des farines de blé." Compt. rend., *123:* 755-758.

Fleurent, E. 1896c. "Étude de la constitution immediate du gluten des différentes cereales; influence de cette constitution." 2nd Congress Internat. chim. Appl. II. 64-70.

Fleurent, E. 1901. "Étude d'un densimètre destiné à la détermination de la valeur boulangère des farines de blé." Compt. rend., *132*, 1421-1423.

Fleurent, E. 1904. "Recherches sur l'action exercée par differente agents physiques et chimiques sur le gluten des farines de blé; conditions du dosage de cet élément." Bul. soc. chim. Paris, 3 ser. *33*, 81-103.

Fleurent, E. 1905a. "Sur le dosage rationnel du gluten dans les farines de blé. Compt. rend., *140*, 99-101.

Fleurent, E. 1905b. "Détermination de la valeur boulangère des farines de blé au moyen du gliadimètre." Ann. chim. anal. chim. appl. *8*, 6-9.

Fleurent, E. 1906. "Sur le blanchiment des farines de blé." Compt. rend., *142*, 180-182.

Ford, J. S., and Guthrie, J. M. 1908. "The amylolytic and proteolytic ferments of wheaten flours, and their relation to 'baking value.' " J. Soc. Chem. Ind., *27*: 389-393.

Fornet, Artur. 1922a. "Rohfasertypen. Eine neue Methode zur Bestimmung und Identifizierung von Mehlprodukten in Gebäcken und anderen Zubereitungen." Z. ges. Getreidew., *14*, 58-61.

Fornet, Artur. 1922b. "Rohfasertypen. Eine neue Methode zur Bestimmung und Identifizierung von Mahlprodukten in Gebäcken, Speisen, usw." Chem. Z., *46*, 969-970.

Frank, W. L. 1922a. "Texture classification of wheat." J. Am. Assoc. Cereal Chemists, *7*, 174-180.

Frank, W. L. 1922b. "Chemical test for heat damage in wheat." J. Am. Assoc. Cereal Chemists, *7*, 218.

Frank, W. L. 1923. "Short weight of flour and feed." J. Am. Assoc. Cereal Chemists, *8*, 19-47.

Frank, W. L., and Campbell, R. L. 1922. "Moisture content vs. net weight." Natl. Miller, *27*, 40-41, 70.

Freeman, G. F. 1918. "Producing bread making wheats for warm climates." J. Heredity, *9*, 211-226.

Fryer, J. R. 1921. "A comparison of some physical properties of immaturely frosted and non-frosted seeds of wheat and oats." Proc. West. Can. Socy. Agron. 2nd Ann. Meeting, 46-56.

Gardner, F. D. 1910. "Milling and baking tests." Penn. Agr. Exp. Sta. Bul. 97.

Gelissen, H. C. J. H. 1924. "Treating flour with highly oxygenated compounds." U. S. Patent 1,483,546, dated Feb. 12th, 1924.

Gericke, W. F. 1920. "On the protein content of wheat." Science, *52*, 446.

Gericke, W. F. 1921. "Influence of temperature on the relations between salt proportions and the early growth of wheat." Am. J. Bot., *2*, 59-62.

Gericke, W. F. 1922a. "Certain relations between the protein content of wheat and the length of the growing period of the head-bearing stalk." Soil Sci., *13*, 135-138.

Gericke, W. F. 1922b. "Certain relations between root development and tillering in wheat; significance in the production of high protein wheat." Am. J. Bot., *9*, 366-369.

Gericke, W. F. 1922c. "Differences effected in the protein content of grain by applications of nitrogen made at different growing periods of the plant." Soil Sci., *14*, 103-109.

Gerum, J. 1920. "Die Pentosane als Grundlage zur Ermittlung des Ausmahlungsgrades der Mehle." Z. Nahr. Genussm., *39*, 65-69.

Gerum, J., and Metzger. 1923 "Zur Kenntniss des Weizenklebers." Z. Nahr. Genussm., *46*, 74-86.

Girard, A. 1895. "Appreciation de la valeur boulangère des farines; dosage des debris d'envellope et de germe susceptibles de diminuer la qualité du pain." Compt. rend., *121*, 858.

Girard, A., and Fleurent, E. 1903. "Le froment et sa mouture." Gauthier-Villars, Paris.

Goebel, Lee H. 1923. "Artificial maturing and bleaching with agene." Modern Miller, Jan. 20, 1923, 24.

Gordon, M. 1922. "The development of endosperm in cereals." Roy. Soc. Victoria Proc. *34,* 105-116.

Gore, H. C. 1925. "The effect of hydrogen ion concentration on the estimation of diastatic power by the polarimetric method." J. Am. Chem. Soc., *47,* 281-283.

Gortner, R. A. 1920. "Is wild vetch seed a safe feed." Breeders Gaz., May 6th, 1920, 1230-1232.

Gortner, R. A. 1924. "Viscosity as a measure of gluten quality." Cereal Chem., *1:* 75-81.

Gortner, R. A., and Doherty, E. H. 1918. "Hydration capacity of gluten from 'strong' and 'weak' flours." J. Agr. Res., *13:* 389-419.

Greaves, J. E. 1911. "Some factors influencing the quantitative determination of gliadin." Univ. Calif. Pub. Physiol., *4,* No. 6, 31-74.

Greaves, J. E., and Carter, E. G. 1923. "The influence of irrigation on the composition of grains and the relationship to nutrition." J. Biol. Chem., *58,* 531-541.

Griess, Peter. 1879. "Bemerkungen zu der Abhandlung der H. H. Weselsky und Benedikt über einige azoverbindungen." Ber. *12,* 426-428.

Gröh, J., and Friedl, G. 1914. "Beiträge zu den physikalisch-chemischen Eigenschaften der alkohollöslich Proteine des Weizens und Roggens." Biochem. Z., *66:* 154-164.

Grossfeld, J. 1920. "Einige Erfahrungen bei der Untersuchung und Beurteilung von Backwaren." Z. ges. Getreidew., *12,* 73-84.

Guess, H. A. 1900. "The gluten constituents of wheat and flour and their relation to bread-making qualities." J. Am. Chem. Soc., *22:* 263-268.

Günsberg, R. 1862. "Ueber die in Wasser löslichen Bestandtheile des Weizenklebers." J. prakt. Chem., *85,* 213-229.

Gurney, E. H., and Norris, G. 1901. "Further notes on the milling qualities of different varieties of wheat." Dept. Agr. N. S. Wales, Misc. Pub. 526.

Guthrie, F. B. 1899. "Notes on the milling qualities of the varieties of wheat most commonly grown in New South Wales." Dept. Agr. N. S. Wales, Misc. Pub. 307.

Guthrie, F. B. 1900. "Strong and weak flours." Dept. Agr. N. S. Wales, Misc. Pub. 427.

Guthrie, F. B. 1902. "The history of a grain of wheat." Dept. Agr. N. S. Wales, Misc. Pub. 529.

Guthrie, F. B., and Norris, G. W. 1902. "Note on the effect of manuring upon the milling properties of the grain." Agr. Gaz. N. S. Wales, *13, 727-729.*

Guthrie, F. B., and Norris, G. W. 1907. "Milling characteristics of Australasian wheats." Dept. Agr. N. S. Wales, Misc. Pub. 1,071.

Guthrie, F. B., and Norris, G. W. 1912a. "Wheat cut at different stages." Agr. Gaz. N. S. Wales, *23,* 1029-1034.

Guthrie, F. B., and Norris, G. W. 1912b. "Daily variation in moisture content of flours." Dept. Agr. N. S. Wales, Sci. Bul. 7, 18-20.

Guthrie, F. B., Norris, G. W., and Ward, J. G. 1921. "The influence of atmospheric variations on the weight of bagged wheat." Agr. Gaz. N. S. Wales, *32,* 200-202.

Haigh, L. D. 1912. "A study of the variations in chemical composition of the timothy and wheat plants during growth and ripening." Proc. 8th Int. Cong. App. Chem., *26,* 115-117.

Hale, Worth. 1910. "The bleaching of flour and the effect of nitrites on certain medicinal substances." U. S. Public Health and Marine Hosp. Serv., Hyg. Lab. Bul. 68.

Haley, Frank L. 1914. "Bleached flour." Biochem. Bul., *3,* 440-443.

Hall, A. D. 1904. "The question of quality in wheat." J. Bd. Agr. *11:* 321-337.

Hall, A. D. 1905. "Recent developments in agricultural science." Science, *22,* 449-464.

Halliburton, W. D. 1909. "The bleaching of flour." J. Hygiene, *9,* 170-180.

Halton, Philip. 1924. "The chemistry of the strength of wheat flour." J. Agr. Sci., *14,* 587-599.

Hammill, J. M. 1911. "On the bleaching of flour and the addition of so-called 'improvers' to flour." Rpts. Local Govt. Bd. (Great Brit.) Pub. Health and Med. Subjs., N. ser., *49*, 1-33.

Hankóczy, E. v. 1920. "Apparat für Kleberbewertung." Z. ges. Getreidew., *12*, 57-62.

Harcourt, R. 1901. "The comparative values of Ontario wheats for breadmaking purposes." Ontario (Canada) Agr. Coll. and Exp. Farm Bull. 115.

Harcourt, R. 1908. "Feeding value of frosted wheat." Ontario Agr. Coll. Report, 1908, 70-71.

Harcourt, R. 1909. "Wheat and flour testing." Ontario Agr. Coll. Report for 1908, Part VI, 79-87.

Harcourt, R. 1910. "Testing varieties of wheat." Ontario Agr. Coll. Report for 1909, 70-76.

Harcourt, R. 1911. "(1) Baking tests of varieties of wheat. (2) A study of the effect of bleaching on the quality of flour." Ontario Agr. Coll. Report for 1910, 82-87; 92-98.

Harcourt, R. 1912. "Baking tests of wheat grown on experimental plots." Ontario Agr. Coll. Report for 1911, 75-81.

Harcourt, R. 1915. "A study of the influence of bleaching on the baking quality of flour." Canadian Miller and Cerealist, *7*, 133-136.

Harcourt, R., and Purdy, M. A. 1910. "Flour and breadmaking." Ontario Agr. Coll. Bul. 180.

Hardy, W. B. 1910. "An analysis of the factors contributing to strength in wheaten flours." Sup. 4, J. Bd. Agr. (Eng.) *17*: 52-56.

Hare, R. F. 1904. "Effect of soil moisture on the amount of protein in wheat" 15th Rept. New Mex. Agr. Exp. Sta., 43-46.

Harlan, Harry V. 1920. "Daily development of kernels of Hannschen barley from flowering to maturity at Aberdeen, Idaho." J. Agr. Res., *19*, 393-429.

Harlan, H. V., and Anthony, S. B. 1920. "Development of barley kernels in normal and clipped spikes and the limitations of awnless and hooded varieties." J. Agr. Res., *19*, 431-472.

Harlan, H. V., and Pope, M. N. 1923. "Water content of barley kernels during growth and maturation." J. Agr. Res., *23*, 333-360.

Harper, D. H. 1889. "The chemistry of wheat under different conditions." Minn. Agr. Exp. Sta. Bul. *7*, 65-84.

Harris, F. S. 1914. "Effects of variations in moisture content on certain properties of a soil and on the growth of wheat." Cornell (N. Y.) Agr. Exp. Sta. Bul. 352.

Harris, F. S., and Thomas, G. 1914. "The change in weight of grain in arid regions during storage." Utah Agr. Exp. Sta. Bul. 130.

Harvey, H. W., and Wood, T. B. 1911. "A method of determining the baking strength of single ears of wheat." Rept. Brit. Assoc. Adv. Sci., 1911, 597-598.

Hayes, H. K. 1923. "Inheritance of kernel and spike characters in crosses between varieties of Triticum vulgare." Studies in the Biol. Sci., Univ. of Minn., No. 4, 163-182.

Hayes, H. K., Bailey, C. H., Arny, A. C., and Olson, P. J. 1917. "The color classification of wheat." J. Am. Soc. Agron., *9*, 281-284.

Hayes, Willet M., and Boss, Andrew. 1899. "Wheat. Varieties, breeding, cultivation." Minn. Agr. Exp. Sta. Bul. 62.

Headden, W. P. 1915a. "Yellowberry in wheat." Colo. Agr. Exp. Sta. Bul. 205.

Headden, W. P. 1915b. "A study of Colorado wheat. Part I." Colo. Agr. Exp. Sta. Bul. 208.

Headden, W. P. 1916a. "A study of Colorado wheat. Part II." Colo. Agr. Exp. Sta. Bul. 217.

Headden, W. P. 1916b. "A study of Colorado wheat. Part III." Colo. Agr. Exp. Sta. Bul. 219.

Headden, W. P. 1918a. "The properties of Colorado wheat." Colo. Agr. Exp. Sta. Bul. 237.

Headden, W. P. 1918b. "A study of Colorado wheat. Part IV." Colo. Agr. Exp. Sta. Bul. 244.

Headden, W. P. 1918c. "A study of Colorado Wheat." Colo. Agr. Exp. Sta. Bul. 247.

Heinrich, Remhold. 1867. "Untersuchungen über den Stoffwechsel während der vegetation der Weizenpflanze." Ann. Landw. Preussen, *50*, 314-333.

Heinrich, Remhold. 1871. "Mineralbestandtheile dem Samenkorne der Weizenpflanze während der Entwicklung vom Fruchtknoten bis zur Überreife." Ann. Landw. Preussen, *57*, 31-49.

Hendel, Julius, and Bailey, C. H. 1924. "The quality of gluten of flour streams as determined by the viscosity of water suspensions." Cereal Chem., *1*, 320-324.

Henderson, L. J., Cohn, E. J., Cathcart, P. H., Wachman, J. D., and Fenn, W. O. 1919. "A study of the action of acid and alkali on gluten." J. Gen. Physiol., *1:* 459-472.

Henderson, L. J., Fenn, W. O., and Cohn, E. J. 1919. "Influence of electrolytes upon the viscosity of dough." J. Gen. Physiol., *1:* 387-397.

Henry, J. 1903. "Marche de l'absorption de l'azote par les céréales." Bul. Agr. (Brussels), *19*, 154-156.

Herman, H. S., and Hall, W. 1921. "Variation in moisture content of flour during storage." J. Am. Assoc. Cereal Chemists, *6*, 10.

Hickman, J. F. 1892. "Field experiments with wheat shrinkage in the granary." Ohio Agr. Exp. Sta. Bul. *42*, 86-87.

Hoagland, R. 1911. "The determination of gliadin or alcohol-soluble protein in wheat flour." J. Ind. Eng. Chem., *3*, 838-842.

Hoffman, J. F. 1908. "Über Backfähigheit und Bleichen der Mehle." Wochenschr. Brau., *25*, 108-110.

Hoffman, J. F. 1918. "Über die Selbsterhitzung der Getreide und andere Nahrstoffen." Dingler's Polytechn. J., *333*, 63-67.

Hoffman, W. F., and Gortner, R. A. 1925. "Physico-chemical studies on proteins. I. The prolamines—their chemical composition in relation to acid and alkali binding." Colloid Symposium Monograph, pp. 209-368. Chemical Catalog Co., N. Y.

Howard, Albert, and Howard, Gabrielle L. C. 1908. "The milling and baking qualities of Indian wheats." Agr. Res. Inst. (Pusa, India) Bul. 14.

Howard, Albert, and Howard, Gabrielle L. C. 1911. "The milling and baking qualities of Indian wheat. No. 3. Some new Pusa hybrids tested in 1910." Agr. Res. Inst., (Pusa, India) Bul. 22.

Howard, Albert, Leake, H. M., and Howard, Gabrielle L. C. 1914. "The influence of environment on the milling and baking qualities of wheat in India. No. 3. The experiments of 1911-12." Mem. Dept. Agr. in India, Bot. series *6*, No. 8, 233-266.

Humphries, A. E. 1905. "The improvement of English wheats." Nat'l Assoc. Brit. and Irish Millers, Liverpool, 1905.

Humphries, A. E., and Biffin, R. H. 1907. "The improvement of English wheat." J. Agr. Sci., *2*, 1-17.

Humphries, A. E., and Simpson, A. G. 1909. "Gas-making capacity as a factor in the estimation of strength in wheaten flour." Seventh Int. Cong. Appl. Chem. Sect. VIa, 27-38.

Hurd, Annie May. 1923. "Hydrogen ion concentration and varietal resistance of wheat to stemrust and other diseases." J. Agr. Res., *23*, 373-386.

Ilosvay, M. L. 1889. "Sur les réactions des acides azoteux et azotique." Bul. soc. chim. Paris, 3rd ser., *2*, 347-351.

Jacobs, B. R. 1915. "Report to the Amer. Soc. Mlg. and Baking Tech. on collaborative studies." Not published.

Jacobs, B. R., and Rask, O. S. 1920. "Laboratory control of wheat flour milling." J. Ind. Eng. Chem., *12*, 899-903.

Jago, Wm. 1915. "Effect of admixture of corn starch with flour." In hearings on H. R. 16,675, and H. R. 21,453, Ways and Means Committee, House of Representatives, 63rd Cong., 27-30.

Jago, Wm., and Jago, W. C. 1911. "The technology of bread-making." Bakers Helper Company, Chicago, Ills.

Jensen, G. H. 1918. "Studies on the morphology of wheat." Wash. Agr. Exp. Sta. Bul. 150.

Jensen, J. L. 1899. "Sammenhaengen mellem Klima og Kornarternes Kaernestorrelse." Tiddskr. Landbr. Planteavl, *5*, 138-147.

Jessen-Hansen, H. 1911. "Études sur la farine de froment. 1. Influence de la concentration en ions hydrogène sur la valeur boulangère de la farine." Compt. rend. trav. lab. Carlsberg, *10*, 170-206.

Johnson, Arnold H. 1923. "The relation of gas retention by wheat flour doughs to flour strength and fermentation." A thesis filed in the library of the University of Minnesota, Minneapolis.

Johnson, A. H., and Bailey, C. H. 1924. "A physico-chemical study of

cracker dough fermentation." Cereal Chem., *1*, 327-410.
Johnson, A. H., and Bailey, C. H. 1925. "Gluten of flour and gas retention of wheat flour doughs." Cereal Chem., *2*, 95-106.
Jones, D. Breese, and Gersdorff, C. E. F. 1923. "Proteins of wheat bran. I. Isolation and elementary analysis of a globulin, albumin and prolamine." J. Biol. Chem., *58*, 117-131.
Jones, J. S., and Colver, C. W. 1916. "Dry-farmed and irrigated wheats." Idaho Agr. Exp. Sta. Bul. 88.
Jones, J. S., and Colver, C. W. 1918. "Performance records of some eastern wheats in Idaho." Idaho Agr. Exp. Sta. Bul. 103.
Jones, J. S., Colver, C. W., and Fishburn, H. P. 1918. "Irrigation and the protein content of wheat." Idaho Agr. Exp. Sta. Bul. 109.
Jones, J. S., Fishburn, H. P., and Colver, C. W. 1911. "A report on the milling properties of Idaho wheats." Idaho Agr. Exp. Sta. Bul. 72.

Kalning, H., and Schleimer, A. 1913. "Die chemische Zusammensetzung des Weizens und seiner Mahlprodukte." Z. ges. Getreidew., *5*, 199-207.
Kansas State Board of Agriculture. 1920. "Wheat in Kansas." Report for the quarter ending September, 1920.
Karchevski. 1901. "The influence of the fluctuation of temperature on the respiration of seeds and embryos of wheat." (Abstract) Exp. Sta. Rec., *14*, 839, 1903. (Original article in Mem. Warsaw Univ. 9, 114 p. 1901. Not seen.)
Kedzie, R. C. 1882. "The ripening of wheat." Rept. Mich. Bd. Agr. 1881-1882, 233-238.
Kedzie, R. C. 1893. "Composition of wheat at different periods of ripening." Mich. Agr. Exp. Sta. Bul. 101, 1-12.
Keenan, George L. 1923. "Significance of wheat hairs in microscopical examination of flour." U. S. Dept. Agr. Bul. 1130.
Keenan, George L., and Lyons, Mary A. 1920. "The microscopical examination of flour." U. S. Dept. Agr. Bul. 839, revised August, 1922.
Kent-Jones, D. W. 1924. "Modern Cereal Chemistry." Northern Publishing Co., Liverpool, Eng.

Kepner, B. H. 1914. "Physical factors which influence the percentage of wet and dry gluten in wheaten flour." J. Ind. Eng. Chem., *6*, 481.
Kharchenko, V. A. 1903. "Nitrogen in wheat grains and their size in dependence on precipitation and temperature." Ann. Inst. Agron. Moscow, *9*, 304-311. Abstract in Exp. Sta. Rec., *16*, 260. Original not seen.
Kick, Frederick. 1888. "Flour manufacture." Translated by Powles, London.
Kjeldahl, J. 1896. "Untersuchungen über das optische Verhalten einiger vegetabilischer Einweisskörper." Bied. Centr. agr. Chem. *25*, 197-199.
König, J. 1903. "Chemie der Menslichen Nahrungs—und Genussmittel." Bd. I, 413-461. Julius Springer, Berlin.
König, J., and Rintelen, P. 1904. "Beziehungen zwischen dem Klebergehalt und der Backfähigkeit eines Weizenmehles." Z. Nahr. Genussm., *8*, 721-728.
Koning, C. J., and Moorj, W. C., Jr. 1914. "Bijdrage tot het ouderzoek van brood." Chem. Weekblad, *11*, 1064-1066.
Kosutány, T. 1903. "Uber Weizen and Weizenmehle." J. Landw., *51*, 139-161.
Kosutány, T. 1907. "Der ungarische Weizen and das ungarische Mehl." Budapest.
Kosutány, T. 1909. "Die Mehlanalyse mit Rücksicht auf die Backfähigkeit des Mehles." Proc. 7th Int. Cong. App. Chem., Sect. VIa, 56-68.
Kozmin, Peter A. 1917. "Flour milling." Translated from the Russian by M. Falkner and Theodor Fjelstrup. George Rutledge and Sons, Ltd., London.
Kraft, John. 1913. "Patent flour and its origin." Operative Miller, *18*, 450.
Kress, C. B. 1924. "Gluten quality." Cereal Chem., *1*: 247-250.
Kuhl, Hugo. 1911. "Flour and its deterioration when in store." Die Muehle. Translation in Operative Miller, *16*, 691.

Ladd, E. F. Undated. "Suggested standards and definitions."
Ladd, E. F. 1907. "Bleaching of flour." Proc. Assoc. State and Nat'l Food and Dairy Depts. for 1907, 196-209.

Ladd, E. F. 1908a. "Chemically treated flour." No. Dak. Agr. Exp. Sta., Spec. Bul. 9.

Ladd, E. F. 1908b. "1. Milling and baking test. 2. Tests of flour sold in North Dakota." No. Dak. Agr. Exp. Sta. Bul. 82, 717-793.

Ladd, E. F. 1912. "Protein content of hard red spring and durum wheats from the same farm." Twenty-third Annual report of the No. Dak. Agr. Exp. Sta., Part III, 343.

Ladd, E. F. 1915. "Red durum or durum No. 5." No. Dak. Agr. Exp. Sta. Spec. Bul. Food Dept., *3*, 349-360.

Ladd, E. F., and Bailey, C. H. 1911. "Wheat investigations. Milling, baking and chemical tests." No. Dak. Agr. Exp. Sta. Bul. 93.

Ladd, E. F., and Bassett, H. B. 1909. "Bleaching of flour." J. Biol. Chem., *6*, 75-86.

Ladd, E. F., and Stallings, R. E. 1906. "Bleaching of flour." No. Dak. Agr. Exp. Sta. Bul. 72, 219-235.

Ladd, E. F., and White, H. L. 1909. "Effect of bleached flour extracts on rabbits." No. Dak. Agr. Exp. Sta., Rpt. for 1909, 145-153.

Lawellin, S. J. 1924. "Bleaching with nitrogen trichloride." Natl. Miller, *29*, 53, 86-88.

Lawes, J. B., and Gilbert, J. H. 1857. "Some points on the composition of wheat." London.

Lawes, J. B., and Gilbert, J. H. 1884. "On the composition of the ash of wheat and wheat straw." J. Chem. Socy., *45*, 305-407.

Leavitt, S., and LeClerc, J. A. 1909. "Change in the composition of unground cereals during storage." J. Ind. Eng. Chem., *5*, 299-302.

LeClerc, J. A. 1906. "The effect of climatic conditions on the composition of durum wheat." Yearbook U. S. Dept. Agr., 1906, 199-212.

LeClerc, J. A. 1920. "Report of the referee on cereal products." J. Assoc. Off. Agr. Chemists, *4*, 180-183.

LeClerc, J. A., Bailey, L. H., and Wessling, H. L. 1918. "Milling and baking tests of einkorn, emmer, spelt and Polish wheat." J. Am. Soc. Agron., *10*, 215-217.

LeClerc, J. A., and Breazeale, J. F. 1908. "Plant food removed from ripening plants by rain or dew."
U. S. Dept. Agr. Yearbook, 1908, 389-402.

LeClerc, J. A., and Breazeale, J. F. 1911. "The translocation of plant food and elaboration of organic material in wheat seedlings." U. S. Dept. Agr., Bur. Chem. Bul. 138.

LeClerc, J. A., and Leavitt, S. 1909. "The influence of environment on the composition of wheat." Proc. 7th Int. Cong. App. Chem., Sect. VII, 136-147.

LeClerc, J. A., and Leavitt, S. 1910. "Tri-local experiments on the influence of environment on the composition of wheat." U. S. Dept. Agr., Bur. Chem., Bul. 128.

LeClerc, J. G., Wessling, H. L., Bailey, L. H., and Gordon, W. O. 1919. "Composition and baking value of different particles of flour." Operative Miller, *24*, 257-258.

LeClerc, J. A., and Yoder, P. A. 1914. "Environmental influences on the physical and chemical characteristics of wheat." J. Agr. Res., *1*, 275-291.

Leith, B. D. 1919. "The milling and baking qualities of Wisconsin grown wheats." Wis. Agr. Exp. Sta. Bul. 43.

Lewis, J. H., Wells, H. G., Hoffmann, W. F., and Gortner, R. A. 1924. "An immunological and chemical study of the alcohol-soluble proteins of cereals." Proc. Soc. Exp. Biol. Med. *22*, 185-187.

Liebermann, L. 1901. "Apparat und Verfahren zur Bestimmung der Qualität des Weizenklebers." Z. Nahr. Genussm. *4*, 1009-1016.

Liebermann, L. V., and Andriska, V. 1911. "Ein neues Verfahren zur Bestimmung des Feinheitsgrades der Weizenmehle." Z. Nahr. Genussm. *22*, 291-294.

Liebig, H. J. von. 1909. "Über den Zuckergehalt der feinen Weizenmehle, der Weizenmehlteige unter der vergorener Mehlteige sowie über die diastatische Kraft der Weizenmehle." Landw. Jahrb. *38*, 251-271.

Liebig, J. 1841. "Ueber die stickstoffhaltigen Nahrungsmittel des Pflanzenreiches." Ann. Chem. Pharm. *39*, 129-160.

Liebscher, G. 1887. "Der Verlauf der Nahrstoffaufnahme und seine Bedeutung für die Dungerlehre." J. Landw. *35*, 335-518.

Liechti, P. 1909. "Die Prüfung von Mehlen auf Grund ihres gehalten an Katalase." Chem. Ztg. *23*, 1057.

Lindet, L., and Ammann, L. 1907. "Sur le pouvoir rotatoire des proteines extraites de farines de céréales par l'alcool aqueux." Bul. Soc. Chem. *1*, 968-974.

Lipman, C. B., and Taylor, J. K. 1922. "Proof of the power of the wheat plant to fix atmospheric nitrogen." Science, *56*, 605-606.

Lipman, C. B., and Taylor, J. K. 1924. "Do green plants have the power of fixing elementary nitrogen from the atmosphere?" J. Franklin Inst. *198*, 475-506.

Löew, O. 1901. "Catalase, a new enzyme of general occurrence." U. S. Dept. Agr., Rept. 68.

Love, H. H., and Craig, W. T. 1919. "The synthetic production of wild wheat forms." J. Heredity, *10*, 51-64.

Lucanus, B. 1862. "Ueber das Reifen und Nachreifen des Getreides." Landw. Versuchsstat., *4*, 147-166.

Lüers, H. 1919. "Beiträge zur Kolloidchemie des Brotes. III. Kolloidchemische Studien am Roggen und Weizenkleber mit besonderer Berücksichtigung des Kleber- und Backfähigkeitsprobleme." Kolloid-Z., *25:* 177-196.

Lüers, H. 1920. "Studien über die Reifung der Cerealien." Biochem. Z., *104*, 30-81.

Lüers, H., and Landauer, M. 1922. "Der Isoelektrische Punkt Des Pflanzlichen Albumins 'Leucosin.'" Z. Elektrochem. *28*, 341-347.

Lüers, H., and Ostwald, W. 1919. "Beiträge zur Kolloidchemie des Brotes. II. Zur Viskosimetrie der Mehle." Kolloid-Z. 25: 82-90; 116-136.

Lüers, H., and Ostwald, W. 1920. "Beiträge zur Kolloidchemie des Brotes. V. Die kolloide Quellung des Weizenklebers." Kolloid-Z. *26:* 66-67.

Lüers, H., and Schwarz, M. 1925. "Über die Beziehung der Viscosität zur Backfähigkeit der Mehle." Z. Nahr. Genussm. *49*, 75-89.

Lyon, T. L. 1905. "Improving the quality of wheat." U. S. Dept. Agr. Bur. Pl. Ind. Bul. 78.

Lyon, T. L. 1910. "The relation of wheat to climate and soil." Proc. Am. Soc. Agron., *1*, 108-125.

Lyon, T. L., and Keyser, A. 1905. "Winter wheat." Nebr. Agr. Exp. Sta. Bul. 89.

McCall, A. G., and Richards, P. E. 1918. "Mineral food requirements of the wheat plant at different stages of its growth." J. Am. Soc. Agron., *10*, 127-134.

McClendon, J. F., and Hathaway, J. C. 1924. "Inverse relation between iodin in food and drink and goiter, simple and exophthalmic." J. Am. Med. Assoc., *82*, 1668-1672.

McDowell, R. H. 1895. "Wheat-cutting at different dates." Nevada Agr. Exp. Sta. Bul. 30.

Macfarlane, T. 1905. "On the determination of the constituents of gluten." Trans. Roy. Socy. Canada, ser. 2, *2*, sect. 3, 17-24.

McGill, A. 1910. "Flour. Nitrite reacting nitrogen in." Laby. Inland Rev. Dept. (Canada) Bul. 206.

McGinnis, F. W., and Taylor, G. S. 1923. "The effect of respiration upon the proteins percentage of wheat, oats and barley." J. Agr. Res. *24*, 1041-1048.

McHargue, J. S. 1924. "The significance of the occurrence of copper, iron, manganese and zinc in forage crops and cereals." Paper presented before the Am. Chem. Soc., Ithaca, N. Y.

MacIntire, W. H. 1911. "The influence of fertilizers upon the composition of wheat." Penn. State Coll. Report, 1910-1911, 173-193.

McLaren, L. H. 1923. "Varieties of durum wheat for semolina millers." J. Am. Assoc. Cereal Chemists, *8*, 62-66.

Manchester, H. H. 1922. "A pictorial history of milling." Northwestern Miller, *129*, No. 11, 1143, March 15th, 1922.

Mangels, C. E. 1923. "Recent durum wheat investigations at the North Dakota Experiment Station." J. Am. Assoc. Cereal Chemists, *8*, 110-117.

Mangels, C. E. 1924. "Effect of storage on baking quality of common and durum wheats." Cereal Chem., *1*, 168-178.

Mangels, C. E., and Sanderson, T. 1925. "The correlation of the protein content of hard red spring wheat with physical characteristics and baking quality." Cereal Chem., *2*, 107-112.

Mann, A., and Harlan, H. V. 1915. "Morphology of the barley kernel with reference to its enzym-secreting areas." U. S. Dept. Agr. Bul. 183.

Marion. 1909. "Des modifications subrès par les farines avec l'âge." Seventh Int. Cong. Appl. Chem., Sect. VIa, 45-47.

Marion, F. 1920. "Action de l'eau oxygénée sur les farines." Compt. rend., *171*, 804-806.

Marotka, D., and Kaminka, R. 1922. "La decomposizione dell' acqua ossigenata come mezzo di determinazione del grado di abburattamento della farine." Giorn. chim. ind. applicata (Italy), *4*, 249-251.

Martin, F. J. 1920a. "The distribution of enzymes and proteins in the endosperm of the wheat berry." J. Soc. Chem. Ind., *39*, 327T.

Martin, F. J. 1920b. "Properties affecting strength in wheaten flour." J. Soc. Chem. Ind., *39*: 246-251T.

Maschhaupt, J. G. 1922. "De samenstelling onzer landbougewassen in opeenvolgende groeiperioden." Verslag. Land. Onderzoek. Riijkslandbouwproefsta. No. 27, 125-132.

Mathewson, W. E. 1906. "The optical rotation of gliadin in certain organic solvents." J. Am. Chem. Soc., *28*, 1482-1485.

Maurizio, A. 1920. "Die Backfähigheit des Weizens und ihre Bestimmung." Land. Jahrb. *31*: 179-234.

Mayer, W. 1857. "Untersuchungen der vorzuglichsten Cerealian aus den Provinzen Bayerns zunächst auf ihren Gehalt an Phosphorsäure und Stickstoff." Ergibnisse landw. und agr.-chemie Versuche, Heft *1*, 1-45.

Melikov, P. 1900. "Investigation of the wheat of southern Russia." Zhur. Opuitn. Agron. *1*, 256-267. Abstract in Exp. Sta. Rec., *13*, 451. Original not seen.

Merl, Th., and Daimer, J. 1921. "Studien über Mehlkatalase." Z. Nahr. Genussm. *42*, 273-290.

Metzger, J. E. 1922. "Garlic and other factors influencing grades of wheat." Maryland Agr. Exp. Sta. Bul. 246.

Miller, E. H. 1912. "A new method for the detection and estimation of small quantities of nitrous acid." Analyst, *37*, 345.

Miller, Edgar S. 1921. "Wheat tempering." Operative Miller, *26*, 202-204.

Miller, M. 1909. "Beiträge zur chemischen Kenntnis der Weizenmehle." Z. ges. Getreidew., *1*, 194-200; 214-222; 238-244.

Miller, M. 1912. "Betrachtungen über den Humphries-Prozess und seinen Wert für die Getriedemüllerei." S. B. B. Zeitung, Heft. 4. Cited by K. Mohs, Z. Nahr. Genussm., *46*, 364. Original not seen.

Millers Gazette. 1911. "The Humphries' process of treating flour." Millers Gaz., *34*, 546-548.

Miller, R. C. 1915. "Milling and baking tests of wheat containing admixtures of rye, corn cockle, kinghead and vetch." U. S. Dept. Agr. Bul. 328.

Millon, E. 1854. "Ueber die Zusammensetzung des Weizens." J. prakt. Chem. *61*, 344-351.

Mohs, K. 1921. "Ueber das Humphries-Verfahren." Z. ges. Getreidew., *12*, 89-103; 113-120.

Mohs, K. 1922. "Neue Erkenntnisse auf dem Gebiete der Müllerei und Bäckerei." Dresden and Leipzig.

Mohs, K. 1924. "The size of pores in baked bread." Cereal Chem., *1*, 149-151.

Monier-Williams, G. W. 1911. "On the chemical changes produced in flour by bleaching." Repts. Local Govt. Bd. (Great Brit.) Pub. Health and Med. Subjs., n. ser., No. 49, 34-65.

Monier-Williams, G. W. 1912. "Report to the local government board on the nature of the coloring matter of flour and its relation to processes of natural and artificial bleaching." Repts. Local Govt. Bd. (Great Brit.) Pub. Health and Med. Subjs., n. ser., No. 73, 10 pp.

Moore, B., and Wilson, J. T. 1914. "The effects of nitrogen peroxide on the constituents of flour in relation to the commercial practice of bleaching flour with that reagent. J. Hyg., *13*, 438-466.

Morgan, J. O. 1911. "The effect of soil moisture on the availability of plant nutrients in the soil." Proc. Am. Soc. Agron., *3*, 191-249.

National Association of British and Irish Millers. 1911. "Report of the Home Grown Wheat Committee for the cereal years 1909-1910." London.

National Association of British and Irish Millers. 1913. "Report of the Home Grown Wheat Committee for the seasons 1910-1911 and 1911-1912." London.

Neidig, R. E., and Snyder, R. S. 1922. "The effect of available nitrogen on the protein content and yield of wheat." Idaho Agr. Exp. Sta. Res. Bul. I.

Neumann, M. P. 1909. "Lagerungsversuch mit Weizenmehl." Z. ges. Getreidew., *1*, 281-283.

Neumann, M. P. 1911. "Über den Einfluss der Lagerung and Trocknung auf die Beschaffenheit and Backfähigkeit des Weizenmehles." Z. ges. Getreidew., *3*, 83-92.

Neumann, M. P. 1924. "Ueber den Einfluss der Düngung auf die Backfähigkeit des Getreides." Z. Pflanzenernährung u. Düngung, *3*, 9-16.

Neumann, M. P., and Mohs, K. 1909. "Studien über die Teiggarung Beziehungen zwischen Gärungsumfang und Starkeabbau." Z. ges. Getreidew., *1*, 89-96.

Newton, Robert. 1922. "A comparative study of winter wheat varieties with special reference to winter-killing." J. Agr. Sci., *12*, 1-19.

Newton, Robert. 1923. "The nature and practical measurement of frost resistance in winter wheat." Univ. of Alberta (Canada), Coll. Agr. Research Bul. *1*.

Norton, F. A. 1906. "Crude gluten." J. Am. Chem. Soc., *28*, 8-25.

Norzi, Grazia. 1915. "Di un metodo atto a svelare le alterazioni iniziale del farine." Giorn. farm. chim. *64*, 533-538.

Oliver, C. E. 1913. "The Miller and Milling Engineer." Indianapolis.

Olsen, A. G., and Fine, M. S. 1924. "Influence of temperature on optimum hydrogen ion concentration for the diastatic activity of malt." Cereal Chem., *1*, 215-221.

Olson, G. A. 1912a. "The effect of modifying the gluten surrounding of flour." Proc. 8th Int. Cong. Appl. Chem., *18*, 283-300.

Olson, G. A. 1912b. "Composition of dry gluten and its relation to the protein content of flour." J. Ind. Eng. Chem., *4*, 206-209.

Olson, G. A. 1913. "The quantitative estimation of gliadin in flour and gluten." J. Ind. Eng. Chem., *5*, 917-922.

Olson, G. A. 1914. "The quantitative estimation of the salt-soluble proteins in wheat flour." J. Ind. Eng. Chem., *6*, 211-214.

Olson, G. A. 1917a. "Gluten development in the wheat kernel." Wash. Agr. Exp. Sta. Bul. 142.

Olson, G. A. 1917b. "Wheat and flour investigations." Wash. Agr. Exp. Sta. Bul. 144.

Olson, G. A. 1923. "A study of the factors affecting the nitrogen content of wheat and of the changes that occur during the development of wheat." J. Agr. Res., *24*, 939-953.

Osborne, T. B. 1907. "The proteins of the wheat kernel." Carnegie Institution of Washington, Publication No. 84, Washington, D. C.

Osborne, T. B. 1912. "The vegetable proteins." Longmans, Green and Company, N. Y.

Osborne, T. B., and Harris, I. F. 1902. "Die Nucleinsäure des Weizenembryos." Z. physiol. Chem., *36*, 85-133.

Osborne, T. B., and Harris, I. F. 1903a. "Nitrogen in protein bodies." J. Am. Chem. Soc., *25*, 323-353.

Osborne, T. B., and Harris, I. F. 1903b. "The specific rotation of some vegetable proteins." J. Am. Chem. Soc., *25*, 842-848.

Osborne, T. B., Van Slyke, D. D., Leavenworth, C. S., and Vinograd, M. 1915. "Some products of hydrolysis of gliadin, lactalbumin, and the protein of the rice kernel." J. Biol. Chem., *22*, 259-280.

Ostwald, W. 1919. "Beiträge zur Kolloidchemie des Brotes. I. Ueber Kolloidchemie Probleme bei der Brotbereitung." Kolloid-Z. 25: 26-45.

Palmer, L. S. 1924. "The carotinoids and related pigments. The chromolipins." Chem. Cat. Co., New York.

Passerini, N. 1894. "Sulla maturazione del frumento." Staz. sper. agr. ital., *26*, 138-150.

Peterson, Anna C. 1922. "A study of the buffer values of wheat flour extracts by means of the hydrogen electrode." A thesis filed in the library of the University of Minnesota, Minneapolis.

Pettibone, C. J. V., and Kennedy, Cornelia. 1916. "Translocation of seed protein reserves in the growing corn seedling." J. Biol. Chem., *26*, 519-525.

Phillips, C. Louise. 1921. "Heating and spoiling of grain." Abstracts and references. U. S. Bur. of Markets and Crop Estimates.

Phillips, C. Louise. 1922. "Foreign material in grain and grain screenings." Abstracts and references.

U. S. Dept. Agr. Bur. Agr. Economics. Ploti Agricultural Experiment Station. 1901. "Report of the chemical laboratory for the year 1901." Sept. Rap. An. Sta. Expt. Agron. Ploti, 1901. Abstract in Exp. Sta. Rec. *14*, 340. Original not seen.

Prandi, O., and Perracini, F. 1917. "Sull' analisi delle farine per la determinazione del titolo di resa." Staz. sper. agr. ital., *50*, 391-393.

Preul, F. 1908. "Untersuchungen über den Einfluss verschieden hohen Wassergehaltes des Bodens in den einzelnen Vegetationsstadien bei verschiedenem Bodenreichtum auf die Entwickelung der Sommerweizenpflanze." J. Landw. *56*, 229-271.

Prianishinkov, D. 1900. "The influence of the moisture of the soil on the growth of plants." Zhur. Opuitn. Agr. *1*, 13. Abstract in Exp. Sta. Rec. *13*, 631. Original not seen.

Rabak, Frank. 1920. "The effect of mold upon the oil in corn." J. Ind. Eng. Chem., *12*, 46-48.

Rammstedt, O. 1909. "Die verschiedenen Stickstoffsubstanzen des Weizenmehles in Beziehung gebraucht zu dem Volumen des Gebackes." Z. ges. Getreidew. *1*, 286-291.

Rammstedt, O. 1910. "Kritische Betrachtungen über die Feinheitsbestimmungsmethoden der Weizen- und Roggenmehl." Z. Öffentl. Chem., *16*, 231-243.

Rask, O. S. 1922. "Determination of chlorine in bleached and natural flour." J. Assoc. Official Agr. Chemists, *6*, 68-71.

Rask, O. S., and Alsberg, C. L. 1924. "A viscosimetric study of wheat starches." Cereal Chem., *1*: 7-25.

Raynaud, Brunerie, and Paturel, G. 1910. "De l'influence des engrais chimiques sur la composition des graines de cereales." Prog. Agr. et Vit. *53*, 777-780.

Reimund, F. 1916. "Mehlatmung und Lagerschwund." Neue Bäcker u. Konditor-Ztg., *11*, 158-160.

Richardson, Clifford. 1883. "An investigation of the composition of American wheat and corn." U. S. Dept. Agr. Div. of Chem. Bul. 1.

Richardson, Clifford. 1884. "An investigation of the composition of American wheat and corn." U. S. Dept. Agr. Bur. Chem. Bul. 4.

Ritthausen, H. 1872. "Die Eiweisskörper der Getreidearten, Hülsenfruchte und Oelsamen." Max Cohen und Sohn. Bonn.

Ritthausen, H., and Pott, R. 1873. "Untersuchungen über das Einfluss einer an stickstoff und phosphorsäure reichen Düngung auf die Zusammensetzung der Pflanze und der Samen von Sommerweizen." Landw. Vers. Stat. *16*, 384-399.

Roberts, H. F. 1910. "A quantitative method for the determination of hardness in wheat." Kans. Agr. Exp. Sta. Bul. 167.

Roberts, H. F. 1919. "Yellow berry in hard winter wheat." J. Agr. Res., *18*, 155-169.

Roberts, H. F., and Freeman, G. F. 1908. "The yellow-berry problem in Kansas hard winter wheats." Kans. Agr. Exp. Sta. Bul. 156.

Robertson, T. B., and Greaves, J. E. 1911. "On the refractive indices of solutions of certain proteins. V. Gliadin." J. Biol. Chem., *9*, 181-184.

Rockwood, E. W. 1910. "The effect of bleaching upon the digestibility of wheat flour." J. Biol. Chem., *8*, 327-340.

Rousseaux, E., and Sirot. 1913. "Les matières azotées solubles comme facteur d'appréciation des farines." Compt. rend., *156*, 723-725.

Rousseaux, E., and Sirot, M. 1918. "Les matières azotées solubles comme indice de la valeur boulangère des farines." Compt. rend., *166*, 190-192.

Rousseaux, E., and Sirot, M. 1920. "Les matières azotées et l'acide phosphorique dans le maturation et la germination du blé." Compt. rend., *171*, 578-580.

Rumsey, L. A. 1922. "Diastatic enzymes of wheat flour and their relation to flour strength." Amer. Inst. Baking Bul. 8.

Salamon, A. G. 1908. "Discussion of paper by Baker and Hulton." J. Soc. Chem. Ind., *27*, 375.

Salvini, G., and Silvestri, G. 1912. "Su di un nuovo saggio practico per apprezzare la qualita delle farine." Giorn. farm. chim. *61*, 459-460.

Sanderson, T. 1913a. "A study of the effect of sprouted wheat on the milling and baking quality of an average No. 2 northern wheat." No.

BIBLIOGRAPHY

Dak. Agr. Exp. Sta., Spec. Food Bul. *2*, 205-210.
Sanderson, T. 1913b. "Wheat storage." No. Dak. Agr. Exp. Sta., Spec. Food Bul. *2*, 350-352.
Sanderson, T. 1914a. "A study of the variation in a fifty-pound sack of flour during storage." No. Dak. Agr. Exp. Sta., Spec. Food Bul. *3*, 14-31.
Sanderson, T. 1914b. "A further study of the variation in weight of a fifty-pound sack of flour during storage." No. Dak. Agr. Exp. Sta., Spec. Food Bul. *3*, 250-259.
Sanderson, T. 1918. "Flour from wheat of different test weights." No. Dak. Agr. Exp. Sta., Spec. Food Bul. *5*, 40-73.
Sanderson, T. 1920. "Value of red durum or D 5 wheat." No. Dak. Agr. Exp. Sta., Spec. Food Bul. *5*, 507-517.
Saunders, C. E. 1907. "Quality in wheat." Cent. Exp. Farm (Canada) Bul. 57, Part I.
Saunders, C. E. 1909. "The inheritance of 'strength' in wheat." J. Agr. Sci., *3*, 218-222.
Saunders, C. E. 1910. "Effect of storage on wheat and flour." Cent. Exp. Farm (Canada) Rept. 1909, 205-206.
Saunders, C. E. 1911. "Effect of storage on wheat and flour." Cent. Exp. Farm (Canada) Rept. 1910, 168.
Saunders, C. E. 1921. "The effect of premature harvesting on the wheat kernel." Scientific Agr. (Canada), Feb., 1921, 74-77.
Saunders, C. E., Nichols, R. W., and Cowan, P. R. 1921. "Researches in regard to wheat, flour and bread." Cent. Exp. Farm (Canada) Bul. 97.
Sax, Karl. 1921. "Chromosome relationships in wheat." Science, *54*, 413-415.
Schindler, F. 1893. "Der Weizen." Berlin.
Schleimer, A. 1911. "Studien über das Gliadin, den alkohollöslichen Anteil des Weizenklebers." Z. ges. Getreidew. *3*, 138-140.
Schneidewind, W. 1909. "Die bisherigen Ergebnisse der Seitens der Versuchsstation Halle a. S. ausgeführten Backversuche." Z. ges. Getreidew. *1*, 7-13.
Schoen, M. 1923. "Preserving flour and meal." British Patent No. 203,661, July 15th, 1923.
Scholler, F. 1922. "Berechnung des Ausmahlungsgrades der Mehle." Z. Nahr. Genussm. *44*, 348-351.
Schribaux, E. 1901. "Sur la création de blés riches en gluten." J. Agr. Prat. *1*, 274-277.
Schulze, B. 1904. "Studien über die Entwickelung der Roggen und Weizenpflanze." Landw. Jahrb., *33*, 405-441.
Seelhorst, C. von. 1900. "Neuer Beitrag zur Frage des Einflusses des Wassergehalts des Bodens auf die Entwicklung der Pflanzes." J. Landw. *48*, 165-177.
Shanahan, J. D., Leighty, C. E., and Boerner, E. G. 1910. "American export corn (Maize) in Europe." U. S. Dept. Agr. Bur. Plant Indus. Circ. 55.
Sharp, Paul F. 1924. "Wheat and flour studies II. Aging I. The change in hydrogen ion concentration of wheat and mill products with age." Cereal Chem., *1*, 117-132.
Sharp, P. F. 1925. "Wheat and flour studies III. The amino nitrogen content of the immature wheat kernel and the effect of freezing." Cereal Chemistry, *2*, 12-38.
Sharp, P. F., and Gortner, R. A. 1923. "Viscosity as a measure of hydration capacity of wheat flour and its relation to baking strength." Minn. Agr. Exp. Sta. Tech. Bul. 19.
Sharp, P. F., Gortner, R. A., and Johnson, A. H. 1923. "The physicochemical properties of strong and weak flours VII. The physical state of the gluten as influencing the loaf volume." J. Phys. Chem., *27*, 942-947.
Shaw, G. W. 1913. "Studies upon influences affecting the protein content of wheat." Univ. Calif. Pub's. Agr. Sci., *1*, No. 5, 63-126.
Shaw, G. W., and Gaumnitz, A. 1911. "California white wheats." Calif. Agr. Exp. Sta. Bul. 212.
Shaw, G. W., and Walters, E. H. 1911. "A progress report upon soil and climatic factors influencing the composition of wheat." Calif. Agr. Exp. Sta. Bul. 216.
Shaw, R. H. 1906. "A proposed method for examining bleached flour." J. Am. Chem. Soc., *28*, 687-688.
Shepard, J. H. 1902. "Macaroni wheat in South Dakota. The chemical and milling properties of macaroni wheat." S. Dak. Agr. Exp. Sta. Bul. 77, 39-42.

Shepard, J. H. 1903. "Macaroni wheat. Its milling and chemical characteristics." So. Dak. Agr. Exp. Sta. Bul. 82.

Shepard, J. H. 1905. "Macaroni wheat. Its milling and chemical characteristics and its adaptation for making bread and macaroni." So. Dak. Agr. Exp. Sta. Bul. 92.

Shepard, J. H. 1906. "Macaroni or durum wheats." So. Dak. Agr. Exp. Sta. Bul. 99.

Shepard, J. H. 1908. "Nitrous acid as an antiseptic." Amer. Food J., Sept. 15th, 1908, 11-14.

Sheringa, K. 1916. "Het bleeken van meel met behulp van stikstofdioxyde." Chem. Weekblad. *13*, 840-849.

Sherwood, R. C. 1925. "The control of diastatic activity in wheat flour." Thesis filed in the library of the University of Minnesota, Minneapolis.

Shollenberger, J. H. 1919. "Moisture in wheat and mill products." U. S. Dept. Agr. Bul. 788.

Shollenberger, J. H. 1921a. "The influence of relative humidity and moisture content of wheat on milling yields and moisture content of flour." U. S. Dept. Agr. Bul. 1013.

Shollenberger, J. H. 1921b. "Influence of the size of flour particles on the baking quality." National Miller, Dec. 1921, pp. 29-30.

Shollenberger, J. H. 1923a. "Relation of test weight to milling quality of wheat. Test weight per bushel." Compiled by E. G. Boerner and C. Louise Phillips. U. S. Bur. Agr. Economics (pp. 25-28).

Shollenberger, J. H. 1923b. "Influence of relative humidity on flour milling results." U. S. Dept. Agr., Press Release of April 1923.

Shollenberger, J. H. 1925. "Relation of flour yield to test weight per bushel." Amer. Miller, *53*, 293.

Shollenberger, J. H., and Clark, J. A. 1924. "Milling and baking experiments with American wheat varieties." U. S. Dept. Agr. Bul. 1183.

Shutt, F. T. 1905. "The effect of rust on the straw and grain of wheat." J. Am. Chem. Soc., *27*, 366-369.

Shutt, F. T. 1907. "Quality in wheat. Part II. The relationship of composition to bread-making value." Cent. Exp. Farm (Canada) Bul. 57, 37-51.

Shutt, F. T. 1908. "Wheat. The composition of the grain as influenced by environment." Cent. Exp. Farm (Canada) Report, 1908, 135-140.

Shutt, F. T. 1909a. "The composition of the grain as influenced by the soil moisture content." Cent. Exp. Farm (Canada) Report, 1909, 140-144.

Shutt, F. T. 1909b. "Influence of age on wheat and flour." Cent. Exp. Farm (Canada) Report, 1909, 144-147.

Shutt, F. T. 1910. "Flour—the relationship of composition to bread-making value." Seventh Int. Cong. App. Chem., Sect. VII, 108-135.

Shutt, F. T. 1911a. "Wheat—the composition of the grain as influenced by the soil moisture content." Cent. Exp. Farm (Canada) Report, 1910, 193-194.

Shutt, F. T. 1911b. "Bleached flour." Cent. Exp. Farm (Canada), Report for 1910, 196-203.

Shutt, F. T. 1911c. "Influence of age on wheat and flour." Cent. Exp. Farm (Canada) Report, 1911, 168-171.

Shutt, F. T. 1922. "The development of the wheat kernel." Interim report of the Dominion Chemist, for the year ending Mar. 31, 1922, 77-78.

Smith, C. D. 1901. "Shrinkage of farm products." Mich. Agr. Exp. Sta. Bul. 191.

Smith, Geo. T. 1913. "Origin of patent flour." Operative Miller, *18*, 605.

Snyder, H. 1893. "The draft of the wheat plant upon the soil in the different stages of its growth." Minn. Agr. Exp. Sta. Bul. 29, 152-160.

Snyder, H. 1899a. "The proteids of wheat flour." Minn. Agr. Exp. Sta. Bul. 63, 519-533.

Snyder, H. 1901. "Studies on bread and bread-making." U. S. Dept. Agr., Office Exp. Sta's. Bul. 101.

Snyder, H. 1904a. "Wheat and flour investigations." Minn. Agr. Exp. Sta. Bul. 85.

Snyder, H. 1904b. "The determination of gliadin in wheat flour by means of the polariscope." J. Am. Chem. Soc., *26*, 263-266.

Snyder, H. 1905a. "1. Heavy and light weight seeds (pp. 214-218). 2. Starchy and glutenous grains (pp. 219-225). 3. Rusted Wheat (pp. 228-231)." Minn. Agr. Exp. Sta. Bul. 90.

Snyder, H. 1905b. "Testing wheat

flour for commercial purposes." J. Am. Chem. Soc., *27*, 1068-1074.

Snyder, H. 1906. "Report on the separation of vegetable proteids." U. S. Dept. Agr. Bur. Chem. Bul. *105*, 88-90.

Snyder, H. 1907. "Soil investigations. 2. Influence of fertilizers upon the composition and quality of wheat." Minn. Agr. Exp. Sta. Bul. 102.

Snyder, H. 1908a. "Flour bleaching." Minn. Agr. Exp. Sta. Bul. 111.

Snyder, H. 1908b. "Influence of fertilizers upon the composition of wheat." J. Am. Chem. Soc., *30*, 604-608.

Snyder, H. 1923a. "Wheat flour: its weight and moisture content." Pub. by Miller's Natl. Federation, Chicago, 2nd ed.

Snyder, H. 1923b. "The nomenclature of wheat flours." Pub. by Miller's Natl. Federation, Chicago.

Soc. Anon. des Rizeries Francaises. 1924. "Flour." British Patent No. 211,505, Feb. 16.

Sorenson, S. P. L. 1924. "Breadmaking." Bakers' Weekly, *43*, 65-69.

Soule, A. M., and Vanatter, P. O. 1903. "Influence of climate and soil on the composition and milling qualities of winter wheat." Tenn. Agr. Exp. Sta. Bul. *16*, No. 4, 51-88.

Spragg, F. A. 1912. "Wheat improvement." Mich. Agr. Exp. Sta. Bul. 268.

Spragg, F. A., and Clark, A. J. 1916. "Red Rock wheat." Mich. Agr. Exp. Sta. Circ. 31.

Steensma, F. A. 1916. "Is het gebruik van meel, gebleekt stikstofperoxyde, schadelijk voor gezonheid." Chem. Weekblad, *13*, 849-854.

Stein, Hans. 1904. "Beiträge zur Kenntniss der Weizenmehl." Z. Nahr. Genussm. *7*, 730-742.

Stewart, R., and Greaves, J. E. 1908. "The milling quality of wheat." Utah Agr. Exp. Sta. Bul. 103.

Stewart, R., and Hirst, C. T. 1913. "The chemical, milling and baking values of Utah wheats." Utah Agr. Exp. Sta. Bul. 125.

Stoa, T. E. 1921. "Varietal trials with spring wheat." No. Dak. Agr. Exp. Sta. Bul. 149.

Stockham, W. L. 1912a. "The influence of the amount of rainfall on the yield and quality of wheat." No. Dak. Food Commissioner's 23rd Report, 305-324.

Stockham, W. L. 1912b. "The relation of the amount of moisture in the soil at the time of maturity and the protein content of hard spring wheat." No. Dak. Food Commissioner's 23rd Report, 329-331.

Stockham, W. L. 1912c. "A study of the mean temperature during the growing season on the quality and grade of wheat." No. Dak. Food Commissioner's 23rd Report, 404-431.

Stockham, W. L. 1914. "Studies of wheat quality under North Dakota conditions." No. Dak. Agr. Exp. Sta., Spec. Food Bul., *3*, 129-140.

Stockham, W. L. 1917. "The capacity of wheat and mill products for moisture." No. Dak. Agr. Exp. Sta. Bul. *120*, 97-131.

Stockham, W. L. 1920. "Some factors related to the quality of wheat and the strength of flour." No. Dak. Agr. Exp. Sta. Bul. 139.

Stoecklin, L. 1917. "Pain et farines toxiques, caractérisation et dosage des Sapotoxines." Ann. fals. *10*, 561-572.

Stork, C. T. 1922. "The Novadel process." J. Am. Assoc. Cereal Chemists, *7*, 185-186.

Sutton, G. L. 1921. "The absorption of moisture by wheat grain and its relation to the humidity of the atmosphere." J. and Proc. Roy. Socy. West. Aust. *6*, 75-87.

Swanson, C. O. 1912. "Acidity in wheat flour." J. Ind. Eng. Chem., *4*, 274-278.

Swanson, C. O. 1924. "Factors which influence the quantity of protein in wheat." Cereal Chem., *1*, 279-288.

Swanson, C. O., and Calvin, J. W. 1913. "A preliminary study on the conditions which affect the activity of the amylolytic enzymes in wheat flour." J. Am. Chem. Soc., *35*, 1635-1643.

Swanson, C. O., and Tague, E. L. 1916. "A study of certain conditions which affect the activity of proteolytic enzymes in wheat flour." J. Am. Chem. Soc., *38*, 1098-1109.

Swanson, C. O., Willard, J. T., and Fitz, L. A. 1915. "Kansas flours." Kans. Agr. Exp. Sta. Bul. 202.

Taddei. 1820. "(Abstract of work)." Thompson's Ann. Phil., *15*, 390.

Tague, E. L. 1920. "Changes taking place in the tempering of wheat." J. Agr. Res., *20*, 271-275.

Tague, E. L. 1925a. "The iso-electric

points of gliadin and glutenin." J. Am. Chem. Soc., *47*, 418-422.

Tague, E. L. 1925b. "The solubility of gliadin." Cereal Chem., *2*, 117-127.

Teller, G. L. 1896. "Concerning wheat and its products." Ark. Agr. Exp. Sta. Bul. 42, Part 2.

Teller, G. L. 1898. "A report of progress of investigations in the chemistry of wheat." Ark. Agr. Exp. Sta. Bul. 53, 53-81.

Teller, G. L. 1909. "Differences in flour grades and the cause." Operative Miller, *14*, 301-302.

Teller, G. L. 1912. "The carbohydrates of wheat and wheat products and changes in same during development of the grain." Proc. 8th Int. Cong. App. Chem., Sect. VIa, *13*, 273-278.

Thatcher, R. W. 1908. "Some experiments in breeding high nitrogen wheat." Proc. Am. Soc. Agron., *1*, 126-131.

Thatcher, R. W. 1910. "Factors which determine the composition of wheat." Proc. Am. Soc. Agron., *1*, 131-135.

Thatcher, R. W. 1911. "The composition and milling quality of Washington wheats." Wash. Agr. Exp Sta. Bul. 100, Part II, 25-44.

Thatcher, R. W. 1913a. "The progressive development of the wheat kernel." J. Am. Soc. Agron., *5*, 203-213.

Thatcher, R. W. 1913b. "The chemical composition of wheat." Wash. Agr. Exp. Sta. Bul. 111.

Thatcher, R. W. 1915. "The progressive development of the wheat kernel—II." J. Am. Soc. Agron., *7*, 273-282.

Thatcher, R. W., Olson, G. A., and Hadlock, W. L. 1911. "Wheat and flour investigations." Wash. Agr. Exp. Sta. Bul. 100.

Thom, C., and Hunter, A. C. 1924. "Hygienic fundamentals of food handling." New York.

Thomas, Adrian, and Dox, A. W. 1925. "Untersuchungen über die Natriumsalze der Nucleinsäure aus Weizenkeimen." Z. physiol. chem. *142*, 1-13.

Thomas, L. M. 1916. "The origin, characteristics and quality of Humpback wheat." U. S. Dept. Agr. Bul. 478.

Thomas, L. M. 1917a. "Characteristics and quality of Montana-grown wheat." U. S. Dept. Agr. Bul. 522.

Thomas, L. M. 1917b. "A comparison of several classes of American wheats and a consideration of some factors influencing quality." U. S. Dept. Agr. Bul. 557.

Thomson, R. T. 1914. "Notes on flour. 1. Acidity of flour. 2. Natural and artificial bleaching of flour. 3. Sulfates and lime in flour." Analyst, *39*, 519-529.

Tollens, B. 1902. "The ash constituents of plants." Exp. Sta. Rec., *13*, 305-317.

Traphagen, F. W. 1902. "Hardening of soft wheat." Mont. Agr. Exp. Sta. Report, 1902. Note p. 59.

Tschermak, E. 1914. "Die Verwertung der Bastartierung für phylogenetische Fragen in der Getreidegruppe." Z. Pflanzenzucht. *2*, 303-304.

U. S. Bureau of Markets and Crop Estimates. 1921. "Garlic in wheat." Weekly News Letter, U. S. Dept. Agr., Bur. Markets and Crop Estimates, Oct. 19, 1921.

U. S. Dept. of Agriculture. 1910. "Adulterated and misbranded wheat flour." Notice of Judgment No. 382, Food and Drugs Act.

U. S. Dept. of Agriculture. 1911. "Adulteration and misbranding of bleached flour." Notice of Judgment No. 722, Food and Drugs Act.

U. S. Department of Agriculture, Bureau of Chemistry. 1920. "Service and regulatory announcements 26. Bleached Flour" (page 21).

Upson, F. W., and Calvin, J. W. 1915. "On the colloidal swelling of wheat gluten." J. Am. Chem. Soc., *37*, 1295-1304.

Upson, F. W., and Calvin, J. W. 1916. "The colloidal swelling of wheat gluten in relation to milling and baking." Nebr. Agr. Exp. Sta. Res. Bul. 8.

Utt, C. A. A. 1914. "Some characteristics of chlorine bleached flour." J. Ind. Eng. Chem., *6*, 908-909.

Van Slyke, D. D. 1911. "The analysis of proteins by the determination of the chemical groups characteristic of the different amino acids." J. Biol. Chem., *10*, 15-55.

Van Slyke, D. D. 1922. "On the measurement of buffer values and on the relationship of buffer value to the dissociation constant of the buffer and the concentration and reaction of the buffer solution." J. Biol. Chem., *52*, 525-570.

Vedrödi, V. 1893. "Untersuchungen von Mehlsorten nebst einer neuen Methode zur Bestimmung der Feinheit der Mehle." Z. angew. Chem. Jahrgang 1893, 691-696.

Vickery, H. B. 1922. "The rate of hydrolysis of wheat gliadin." J. Biol. Chem., *53*, 495-511.

Vickery, H. B. 1923. "A product of mild acid hydrolysis of wheat gliadin." J. Biol. Chem., *56*, 415-428.

Vignon, L., and Conturier, F. 1901. "Sur certaines causes de variation de la richesse en gluten des blés." Compt. rend., *132*, 791-794.

Vinassa, E. 1895. "Über Mikroscopische Mehluntersuchung." Z. Nahr. Untersuch. Hyg. und Waarenkunde, *9*, 53-54.

Vuaflart, L. 1908. "Sur les caractères des vielles farines." Ann. chim. anal. appl. *13*, 437-438.

Waldron, L. R., Stoa, T. E., and Mangels, C. E. 1922. "Kota wheat." No. Dak. Agr. Exp. Sta. Circular 19.

Warington, Robert. 1881. "Note on the appearance of nitrous acid during the evaporation of water." J. Chem. Socy. (London), *39*, 229-234.

Weaver, H. E. 1921. "The relation of ash and gluten in wheat flour." J. Am. Assoc. Cereal Chemists, *6*, No. 2, 11-13.

Weaver, H. E. 1922. "Flour bleaching." Amer. Miller, *50*, 743-744.

Weaver, H. E. 1925. "Hydrogen ions and their application to mill control." Proc. Convention Am. Assoc. Cereal Chemists, to be published in Cereal Chemistry.

Weaver, H. E., and Wood, J. C. 1920. "The proteoclastic enzymes." J. Am. Assoc. Cereal Chemists, *5*, 6-11.

Weaver, J. E. 1916. "The effect of certain rusts upon the transpiration of their hosts." Minnesota Botanical Studies *4*, part 4, 379-406.

Wehmer, C. 1911. "Die Pflanzenstoffe." Jena.

Weil, L. 1909. "Die Erkennung gebleichter Mehl." Chem. Ztg. *33*, 29.

Weiss, Freeman. 1924. "The effect of rust infection upon the water requirement of wheat." J. Agr. Res., *27*, 107-118.

Wells, H. G., and Hedenburg, O. F. 1916. "The toxicity of carotin." J. Biol. Chem., *27*, 213-216.

Wells, H. G., and Osborne, T. B. 1911. "The biological reactions of the vegetable proteins. I. Anaphylaxis." J. Infect. Dis., *8*, 66-124.

Wells, H. G., and Osborne, T. B. 1913. "Is the specificity of the anaphylaxis reaction dependent on the chemical constitution of the proteins or on their biological relation? The biological reactions of the vegetable proteins." J. Infect. Dis., *12*, 341-358.

Wender, Neumann. 1905. "Feinheitsbestimmung der Mehle." Z. Nahr. Genussm. *10*, 747-756.

Wender, N., and Lewin, D. 1904. "Die katalytischen Eigenschaften des Getreides und der Mehle." Öster. Chem. Ztg. *7*, 173-175.

Wesener, J. A., and Teller, G. L. 1907. "Bleaching of flour as now practiced by millers." Amer. Food J., *2*, 9-15.

Wesener, J. A., and Teller, G. L. 1911. "The aging of flour and its effect on digestion." J. Ind. Eng. Chem., *3*, 912-919.

Whitcomb, W. O., Day, W. F., and Blish, M. J. 1921. "Milling and baking studies with wheat." Montana Agr. Exp. Sta. Bul. 147.

White, H. L. 1909. "The acidity of water extracts of flour." No. Dak. Agr. Exp. Sta., 20th Rept., 65-67.

White, H. L. 1912. "Flour storage investigations." No. Dak. Food Comm. Rept., 1911, 141-143.

Whymper, R. 1909. "Microscopical study of changes occurring in starch granules during germination of wheat." Proc. 7th Int. Cong. App. Chem., Sect. VIa, 7-13.

Whymper, R. 1920. "Colloid problems in bread-making." Third report on colloid chemistry, Brit. Assoc. Adv. Sci., 61-74.

Widtsoe, J. A. 1902. "Irrigation experiments in 1901." Utah·Agr. Exp. Sta. Bul. 80 (Note p. 148).

Widtsoe, J. A., and Stewart, Robert. 1920. "The chemical composition of crops as influenced by different quantities of irrigation water." Utah Agr. Exp. Sta. Bul. 120.

Wiedmann, Fr. 1921. "Bestimmung der Kleibestandteile im Mehl." Z. Nahr. Genussm., *41*, 236-237.

Wiley, H. W. 1901. "Influence of environment on the chemical composition of plants." Yearbook U. S. Dept. Agr., 1901, 219-318. (Note p. 305.)

Wiley, H. W. 1921. "Deposition filed

with the Directors of the New York Produce Exchange in a hearing on complaint Case No. 3396."

Wiley, H. W., et al. 1898. "Foods and food adulterants: cereals and cereal products." U. S. Dept. Agr., Bur. Chem. Bul. 13, Part 9.

Wilfarth, H., Römer, H., and Wimmer, G. 1905. "Über die Nahrstoffaufnahme der Pflanzen in verschiedenen Zeiten ihres Wachstums." Landw. Versuchsstat. 63, 1-71.

Wilfarth, H., and Wimmer, G. 1903. "Die Wirkungen des Stickstoff-Phosphorsäure-und Kalimangels auf die Pflanzen." J. Landw. 51, 129-138.

Wilhoit, A. D. 1915. "Annual report of the Chief Inspector of Grain to the Minn. Ry. & Warehouse Comm." (p. 57.)

Wilhoit, A. D. 1916. "Annual report of the Chief Inspector of Grain to the Minn. Ry. & Warehouse Comm." (pp. 25-28.)

Willard, J. T. 1911. "Changes in weight of stored flour and butter." Bul. Kans. Bd. Health, 7, 9-14.

Willard, J. T., and Swanson, C. O. 1911. "Milling tests of wheat and baking tests of flour." Kans. Agr. Exp. Sta. Bul. 177.

Willard, J. T., and Swanson, C. O. 1913. "The baking qualities of flour as influenced by certain chemical substances, milling by-products and germination of the wheat." Kans. Agr. Exp. Sta. Bul. 190.

Willstätter, R., and Escher, H. H. 1910. "Ueber den Farbstoff der Tomate." Z. physiol. Chem. 64, 47-61.

Willstätter, R., and Mieg, Walter. 1907. "Ueber die gelben Begleiter des Chlorophylls." Ann. 355, 1-28.

Willstätter, R., and Stoll, A. 1913. "Untersuchungen über Chlorophyll. Methoden und Ergibnisse." Berlin.

Winton, A. L. 1911. "Color of flour and a method for the determination of the 'gasoline color value.'" U. S. Dept. Agr. Bur. Chem. Bul. 137, 144-148.

Winton, A. L., and Hansen, A. W. 1912. "The effects of time and temperature on the amounts of acidity and nitrous nitrogen extracted from flour." U. S. Dept. Agr., Bur. Chem. Bul. 152, 114-116.

Winton, A. L., and Shanley, E. J. 1908. "Simple tests for detecting bleaching in flour." U. S. Dept. Agr., Bur. Chem. Bul. 122, 216-217.

Wohltmann, F. 1906. "Die Einwirkung der Witterung auf die Zusammensetzung der Weizenkörner." Biedermann's Centrbl. Agr. Chem., 35, 41-44.

Wood, T. B. 1906-08. "A new chemical test for 'strength' in wheat flour." Proc. Cambridge Phil. Soc. (England), 14, 115-118.

Wood, T. B. 1907. "The chemistry of the strength of wheat flour. I. The size of the loaf." J. Agr. Sci., 2, 139-160. "II. The shape of the loaf." Ibid., 267-277.

Wood, T. B. 1912. "Apparatus for the comparison of the opacities of liquids." W. J. Pye and Co., Cambridge, Eng.

Wood, T. B., and Hardy, W. B. 1909. "Electrolytes and colloids: the physical state of gluten." Proc. Roy. Socy. (Eng.), B, 81, 38-43.

Woodman, H. E. 1922. "The chemistry of the strength of wheat flour." J. Agr. Sci., 12, 231-243.

Woodman, H. E., and Engledow, F. L. 1924. "A chemical study of the development of the wheat grain." J. Agr. Sci., 14, 563-586.

Woods, Chas. D., and Merrill, L. H. 1903. "Wheats and flours of Aroostook County." Maine Agr. Exp. Sta. Bul. 97.

Woolcott, F. W. 1921. "The effect of atmospheric conditions on milling." Operative Miller, 26, 218.

Working, E. B. 1924. "Lipoids, a factor influencing gluten quality." Cereal Chem., 1, 153-158.

Zade, Dr. 1914. "Serologische Studien an Leguminosen und Gramineen." Z. Pflanzenzucht. 2, 101-151.

Zak, J. 1924. "Estimation of deterioration of grain products and of flour by chemical analysis." Chem. Listy. 18, 72-86. Abstract in Chem. Abst. 18, 1721. Original not seen.

Zinn, J. 1920. "Wheat investigations. I. Pure lines." Maine Agr. Exp. Sta. Bul. 285.

Zinn, J. 1923. "Correlations between various characters of wheat and flour as determined from published data from chemical, milling and baking tests of a number of American wheats." J. Agr. Res., 23, 529-548.

AUTHOR INDEX

Acree, 210.
Adler, 238.
Adorjan, J., 39.
Agr. Gazette, N. S. Wales, 103.
Alcock, A. W., 115.
Allen, R. M., 173.
Alsberg, C. L., 241, 256, 285, 286, 287.
Alsop, J. N., 200.
Alway, F. J., 172, 204, 207, 209, 213, 215, 216, 234.
Ames, J. W., 74, 75, 76.
Ammann, L., 249.
Amos, P. A., 122.
Andriska, V., 146.
Anderson, R. J., 288.
d'Andre, H., 271.
Andrews, J., 199, 200.
Andrews, S., 199, 200.
Anthony, S. B., 43.
Arny, A. C., 105.
Armstrong, E. F., 234.
Arpin, M., 29, 179, 213, 214, 223, 255.
Ashton, J., 14.
Association of Operative Millers, 131, 173, 175, 219.
Australian Advisory Council of Science and Industry, 118.
Avery, 199, 210.
Azzi, G., 41.

Bailey, C. H., 24, 25, 28, 61, 89, 91, 93, 94, 101, 102, 105, 106, 110, 111, 112, 113, 114, 118, 129, 130, 131, 142, 149, 150, 151, 156, 162, 164, 165, 166, 167, 168, 179, 180, 181, 183, 184, 185, 220, 221, 222, 223, 228, 229, 230, 237, 240, 243, 245, 258, 259, 260, 261, 262, 263, 266, 281, 282, 283, 284.
Bailey, L. H., 17, 181, 285.
Baker, J. C., 224, 225.
Baker, J. L., 189, 193, 233.
Ball, C. D., 291, 292.
Ball, C. R., 17, 27.
Balland, A., 63, 182, 188, 211, 212, 255, 258, 267.
Bassett, H. B., 209, 210, 212.
Baston, G. H., 117.
Bates, E. N., 110.
Beans, 199.
Beccari, 242.
Bedford, S. A., 50.
Bell, H. G., 183, 187.
Belval, H., 41, 48.

Benard, 255.
Bertrand, G., 160, 194.
Beyer, C., 118.
Biffin, R. H., 33, 34, 35, 50.
Birchard, F. J., 88, 115, 116, 117, 118.
Bodenstein, M., 201.
Boerner, E. G., 118.
Bogdan, 72.
Boland, 270.
Boltz, G. E., 75, 76.
Bornand, M., 163.
Boss, A., 22, 33.
Bowen, J. C., 131, 132.
Brahm, C., 212.
Brandeis, E., 188.
Breazeale, J. F., 38, 39, 40.
Bremer, W., 258, 265.
Brenchley, W. E., 42, 43.
Blish, M. J., 26, 87, 88, 190, 219, 243, 245, 246, 248, 249, 252, 264, 265, 267, 272, 276, 278.
Brewer, W. H., 118, 119.
Briggs, C. H., 63, 64, 229.
Briggs, L. J., 50, 51, 52.
Brunerie, 74.
Buchanan, J. H., 234.
Buchwald, J., 116, 133, 144, 213, 215.
Buck, C. F., 219.
Burlakow, G., 110, 158.

Calendoli, E., 147.
Calvin, J. W., 235, 269.
Campbell, R. L., 179.
Carter, E. G., 69.
Cathcart, P. H., 269.
Cerkez, S., 156.
Choate, Helen, 37, 38.
Chopin, M., 270, 281.
Clark, A. J., 22.
Clark, J. A., 17, 20, 21, 24, 25, 27, 28, 30, 31, 57, 151, 197, 261.
Cobb, N. L., 192.
Clark, V. L., 172.
Cobb, N. A., 140, 141, 142, 143, 192.
Cohn, E. J., 250, 252, 269, 280.
Coleman, D. A., 104, 120, 197, 198.
Colin, H., 41, 48.
Collatz, F. A., 164, 165, 221, 237, 239.
Colver, C. W., 26, 68, 69, 83.
Conturier, F., 72.
Corbould, M. K., 22, 100.
Cowan, P. R., 77, 109, 182, 190, 191, 287.

315

AUTHOR INDEX

Coward, K. H., 198.
Cox, J. H., 104, 118.
Craig, W. T., 17.
Cross, R. J., 246, 248, 267.

Daimer, J., 163.
Dakin, H. D., 249.
Davidson, J., 78.
Day, W. F., 26, 88, 190.
Dedriek, B. W., 123, 125, 126, 129, 173, 202.
Deherain, P. P., 43, 58.
Dempwolf, O., 148, 151.
Dendy, A., 112.
Dienst, K., 117.
Dill, D. B., 255, 256, 257.
Dingwall, A., 249, 251, 273, 277, 278.
Dobrescu, I. M., 63.
Doherty, E. H., 269, 284.
Dox, A. W., 254.
Dunlap, F. L., 216, 219, 220, 221.
Dupont, C., 43, 58.
Duvel, L., 118.
Duvel, J. W. T., 104, 118.

Eckerson, S. H., 42.
Edgar, W. C., 15.
Einhof, H., 242.
Elkington, H. D., 112.
Ellis, J. H., 91.
Engledow, F. L., 46, 47.
Escher, H. H., 196.
Eto, I., 245, 272, 282.
Evans, Oliver, 14.
Evans, G., 66.

Failyer, G. H., 50.
Farrer, Wm., 22.
Farrow, F. D., 279.
Fegan, Elmer, 224.
Feilitzen, H. V., 33.
Fellows, H. C., 120.
Fenn, W. O., 269, 280.
Fenyvessy, B., 261, 264.
Fernandez, O., 163.
Fine, M. S., 238.
Fishburn, H. P., 26, 68.
Fitz, L. A., 97, 108, 109, 137, 138, 139, 169, 178, 182, 189, 190, 207, 230.
Fleurent, E., 122, 123, 210, 216, 232, 245, 250, 255, 264.
Florell, V. H., 27.
Ford, J. S., 233, 234, 266.
Fornet, A., 147.
Foster, 270.
Frank, W. L., 100, 107, 179.
Freeman, G. F., 33, 34.
Frichot, 199.
Friedl, G., 273.
Fryer, J. R., 89.

Gardner, F. D., 22.

Gaumnitz, A., 27.
Gelissen, H. C. J. H., 227.
Gericke, W. F., 39, 63, 79.
Gersdorff, C. E. F., 142, 253.
Gerum, J., 152, 153, 154, 256.
Gilbert, J. H., 39, 57, 70, 71.
Girard, A., 143, 232, 264.
Girardin, J., 255.
Goebel, L. H., 225.
Gordon, M., 42.
Gordon, W. O., 285.
Gore, H. C., 237.
Gortner, R. A., 102, 142, 168, 216, 239, 245, 246, 248, 250, 252, 261, 265, 267, 269, 274, 275, 276, 279, 282, 284.
Greaves, J. E., 28, 69, 243, 244, 250.
Griess, P., 215.
Gröh, J., 273.
Grossfeld, J., 152, 155.
Guess, H. A., 258, 264.
Günsberg, R., 245.
Gurjar, A. M., 89, 111, 112, 113, 116.
Gurney, E. H., 22, 58.
Guthrie, J. M., 233, 234.
Guthrie, F. B., 22, 50, 73, 119, 122, 178, 266.

Hackel, 16.
Hadlock, W. L., 27.
Haigh, L. D., 40.
Hale, W., 203, 213, 215.
Haley, F. L., 214.
Hall, A. D., 43, 58, 258.
Hall, W., 179.
Halliburton, W. D., 214.
Halton, P., 250, 251.
Hamill, J. M., 169, 205, 215.
Hankóczy, E., 270.
Hansen, A. W., 213.
Harcourt, R., 23, 87, 95, 96, 157, 190, 213.
Hardy, W. B., 268.
Hare, R. F., 59.
Harlan, H. V., 37, 43, 46, 49.
Harper, D. H., 87.
Harris, F. S., 48, 67, 68, 119.
Harris, I. F., 253, 254.
Harris, I. F., 249.
Hartzell, Stella, 234.
Harvey, H. W., 271.
Hathaway, J. C., 153.
Hayes, H. K., 34, 35, 105.
Hays, W. M., 22, 23.
Headden, W. P., 68, 76, 77, 90, 106.
Hedenburg, O. F., 223.
Heinrich, R., 42.
Hendel, J., 91, 142, 168.
Henderson, L. J., 269, 280, 281.
Henry, J., 39.
Herman, R. S., 219.
Herman, H. S., 179.
Hickman, J. F., 119.

AUTHOR INDEX

Hilgard, 118.
Hirst, C. T., 28, 67.
Hoagland, R., 243.
Hoffman, J. F., 117, 213.
Hoffman, W. F., 245, 246, 248, 250, 267.
Howard Wheat and Flour Testing Laboratory, 92.
Howard, Albert, 29.
Howard, Gabrielle, 29.
Hulett, 201, 205.
Hulton, H. F. E., 189, 193, 233.
Humphries, A. E., 33, 34, 50, 228, 231, 234, 258.
Hunter, A. C., 188.
Hurd, A. M., 90.

Ilosvay, M. L., 215.

Jacobs, B. R., 135, 136, 153, 154, 255.
Jago, Wm., 148, 262.
Jago, W. C., 148.
James, 270.
Jensen, G. H., 42.
Jensen, J. L., 57.
Jessen Hansen, H., 192, 282.
Johnson, A. H., 159, 166, 183, 184, 185, 220, 223, 240, 262, 263, 266, 283, 284.
Jones, D. B., 142, 253.
Jones, J. S., 26, 68, 69, 83.
Jordan, 72.

Kalning, H., 157, 158.
Kansas State Board of Agriculture, 92, 108.
Karchevski, 110, 158.
Kedzie, R. C., 42.
Keenan, G. L., 145, 146, 195.
Kennedy, C., 38.
Kent-Jones, D. W., 123, 168, 193, 199, 228, 229.
Kepner, B. H., 255.
Keyser, A., 105.
Kharchenko, V. A., 58.
Kick, F., 122.
Kjeldahl, J., 243, 249.
König, J., 56, 264.
Koning, C. J., 153.
Kosntany, T., 246, 264, 270.
Kozmin, P. A., 14, 123.
Kraft, J., 173.
Kress, C. B., 270.
Kuhl, H., 187.

Ladd, E. F., 24, 31, 130, 171, 172, 209, 210, 211, 212, 213, 214, 216, 261.
Landauer, M., 253.
Lawes, J. B., 39, 57, 70, 71.
Leake, H. M., 29.
Leavenworth, C. S., 246.
Leavitt, S., 59, 60.
Le Clerc, J. A., 17, 38, 39, 40, 59, 60, 66, 78, 81, 82, 110, 255, 285.

Leighty, C. E., 27, 118.
Leith, B. D., 23, 106.
Le Vesconte, A. M., 262, 281, 282, 283, 284.
Lewin, D., 160.
Lewis, J. H., 267.
Liebermann, L., 146, 270.
Liebig, H. J., 232, 234.
Liebig, J., 242.
Liebermann, L., 270.
Liebscher, G., 39.
Liechti, P., 161.
Lindet, L., 249.
Lipman, C. B., 40.
Löew, O., 160.
Love, H. H., 17.
Lowe, G. M., 279.
Lucanus, B., 42.
Lüers, H., 49, 168, 169, 253, 269, 272, 279.
Lunt, 204.
Lyons, M. A., 145, 146, 195.
Lyon, T. L., 33, 59, 105.

Macfarlane, T., 256.
MacIntire, W. H., 74.
McCall, A. G., 40.
McClendon, J. F., 153.
McDowell, R. H., 50.
McGill, A., 208.
McGinnis, F. W., 53.
McHargue, J. S., 152.
McLaren, P. H., 30.
Manchester, H. H., 13.
Mangels, C. E., 25, 31, 91, 107, 110, 151, 187, 260, 261.
Mann, A., 37.
Mann, G., 212, 213.
Marion, F., 162, 163, 183, 192.
Marshall, W. K., 194.
Martin, F. J., 158, 238, 264.
Martin, J. H., 17, 25.
Maschanpt, J. G., 40.
Mathewson, W. E., 249.
Maurizio, A., 233.
Mayer, W., 148.
Melikov, P., 57.
Merl, T., 163.
Merrill, L. H., 24.
Metzer, 152, 256.
Metzger, J. E., 103.
Meyer, 58.
Mieg, W., 196.
Miller, E. H., 216.
Miller, E. S., 129, 219.
Miller, M., 133, 161, 162.
Miller, R., 101, 196.
Millers' Gazette, 133.
Millers' National Federation, 174.
Millon, E., 254.
Mitchell, A. E., 215.
Mohs, K., 133, 230, 234, 269.

AUTHOR INDEX

Monier-Williams, G. W., 196, 198, 203, 206, 207, 209, 211, 214, 215.
Mooj, W. C., Jr., 153.
Moore, B., 210, 211, 212.
Morgan, J. O., 61, 77, 78.
Mutermilch, 160, 194.

Nabenhauer, F., 288.
National Association of British and Irish Millers, 29.
Naudain, G. G., 234.
Neidig, R. E., 77.
Neumann, M. P., 77, 133, 191, 213, 234.
Newton, R., 53.
Nichols, R. W., 77, 109, 182, 190, 191, 287.
Norris, G. W., 22, 50, 58, 73, 119, 178.
Norton, F. A., 256, 264.
Norzi, G., 188.

Oliver, C. E., 123.
Olsen, A. G., 238.
Olson, G. A., 27, 49, 50, 68, 98, 99, 243, 256, 257, 265.
Olson, P. J., 105.
Osborne, T. B., 242, 246, 247, 248, 249, 250, 252, 253, 254, 267.
Ostwald, Wo., 168, 169, 269, 278, 279, 284, 287.

Palmer, L. S., 196.
Passerini, N., 48.
Patterson, J. H., 145.
Paturel, G., 74.
Pecaud, M. T., 29, 179.
Pékar, 194.
Perracini, F., 153.
Perry, E. E., 241, 285, 286.
Peterson, A. C., 166, 167, 168, 213, 282.
Pettibone, C. J. V., 38.
Phillips, C. L., 103, 115, 117.
Pinckney, A. J., 249, 251, 272, 278.
Pinckney, R. M., 213.
Pizarroso, A., 163.
Ploti experiment station, 57.
Pope, M. N., 49.
Pott, R., 71.
Prandi, O., 153.
Preul, F., 66.
Prianishinkov, D., 64.
Purdy, M. A., 23.

Rabak, F., 188.
Rammstedt, O., 163, 264, 265.
Rask, O. J., 153, 154, 219, 287.
Raynaud, 74.
Regan, S. A., 104.
Reimund, F., 187.
Rejtö, 270.
Richards, P. E., 40.
Richardson, C., 134, 136, 177.
Rintelen, P., 264.

Ritthäusen, H., 71, 242.
Roberts, H. F., 33, 105, 106.
Robertson, T. B., 250.
Rockwood, E. W., 214.
Römer, H., 40.
Rousseaux, E., 49, 265.
Rumsey, L. A., 235, 236, 237, 238.
Rush, G. L., 110.

Sakamura, 16.
Salamon, A. G., 287.
Salmon, S. C., 24.
Salvini, G., 147.
Sanderson, T., 31, 91, 98, 107, 119, 178, 260, 261.
Sandstedt, R. M., 252, 265, 276.
Saunders, C. E., 29, 35, 51, 52, 77, 109, 182, 190, 191, 230, 287.
Sasse, 224.
Sax, K., 17.
Schindler, F., 57.
Schleimer, A., 157, 158, 244.
Schneidewind, W., 258.
Schoen, M., 188.
Scholler, F., 155.
Schribaux, E., 57.
Schulze, B., 53.
Schwarz, M., 279.
Stelhorst, C. Von, 65.
Shanahan, J. D., 118.
Shanley, E. J., 215.
Sharp, P. F., 50, 88, 89, 115, 142, 186, 190, 239, 245, 250, 252, 261, 265, 274, 275, 279, 282, 284.
Shaw, G. W., 27, 51, 62, 73, 80, 81.
Shaw, R. N., 216.
Shepard, J. H., 30, 213.
Sheringa, K., 206, 215.
Sherwood, R. C., 100, 102, 168, 187, 222, 237, 239.
Shollenberger, J. H., 20, 21, 28, 30, 57, 92, 93, 130, 132, 194, 261, 285.
Shutt, F. T., 29, 51, 52, 66, 67, 82, 83, 87, 90, 91, 110, 192, 199, 208, 209, 243, 258, 264, 265.
Silvestri, G., 147.
Simon, Henry, 133.
Sirot, M., 49, 265.
Smith, C. D., 119.
Smith, G. T., 173.
Smith, R. W., 25.
Snyder, H., 39, 73, 77, 90, 105, 170, 171, 174, 181, 193, 199, 201, 204, 207, 209, 212, 243, 249, 261, 264.
Soc. Anon. des Rizeries Francaises, 188.
Sorenson, S. P. L., 238.
Soule, A. M., 58, 73.
Spragg, F. A., 22.
Stallings, R. E., 211, 213.
Steensma, F. A., 215.
Stein, H., 255, 261.

AUTHOR INDEX

Stenius, J. A., 75, 76.
Stephens, D. E., 27.
Stewart, R., 28, 67, 69.
Stoa, T. E., 25, 151.
Stockham, W. L., 62, 63, 100, 109, 119, 160, 166, 178, 197, 235, 259, 261, 266, 287.
Stoecklin, L., 102.
Stoll, A., 196.
Stork, C. T., 226.
Sutton, G. L., 120.
Swain, R. E., 246, 248, 267.
Swanson, C. O., 24, 83, 97, 98, 137, 138, 139, 157, 169, 178, 182, 189, 190, 207, 230, 235, 266.

Taddei, 242.
Tague, E. L., 244.
Tague, E. L., 129, 246, 266, 272, 282.
Taylor, G. S., 53.
Taylor, J. K., 40.
Teller, G. L., 43, 95, 152, 157, 172, 207, 208, 210, 215, 243.
Thatcher, R. W., 27, 33, 43, 44, 45, 60, 61, 62, 81.
Thom, C., 181, 188.
Thomas, A., 254.
Thomas, G., 119.
Thomas, L. M., 18, 19, 20, 25, 26, 28, 92, 99, 258, 261.
Thomson, R. T., 208.
Tillman, 105.
Tollens, B., 73.
Traphagen, F. W., 64.
Treml, H., 215.
Tschermak, E., 16.

Upson, F. W., 269.
U. S. Food Administration, 93, 94.
U. S. Bureau of Markets and Crop Estimates, 103.
Upson, F. W., 269.
Utt, C. A. A., 219.

Vanatter, P. O., 58, 73.
Van Slyke, D. D., 246, 248.
Vedrödi, V., 149, 150.
Vickery, H. B., 247.
Vignon, L., 72.
Vinassa, E., 144.
Vinograd, M., 246.
Vuaflart, L., 183.

Wachmann, J. D., 269.
Waldron, L. R., 25, 151.
Walters, E. H., 80, 81.
Ward, J. G., 119.
Warington, R., 215.
Weaver, H. E., 140, 141, 213, 223, 224, 225, 266.
Weaver, J. E., 89.
Wehmer, C., 10.
Weigley, M., 283.
Weil, L., 215.
Weiss, F., 89.
Wells, H. G., 223, 267.
Wender, N., 160, 161.
Wesener, J. A., 207, 208, 210, 215, 216, 217.
Wessling, H. L., 285.
Whitcomb, W. O., 26, 88, 190.
White, H. L., 182, 214.
Whymper, R., 234, 287.
Widtsoe, J. A., 64, 65, 69.
Wiedmann, F., 147.
Wiley, H. W., 58, 181.
Wilfarth, H., 40.
Wilhoit, A. D., 88, 102.
Wilkins, S. D., 166.
Willard, J. T., 24, 50, 97, 98, 137, 138, 139, 157, 169, 178, 182, 189, 190, 207, 230.
Williams, 217.
Willstätter, R., 196, 197.
Wilson, J. T., 145, 210, 211, 212.
Wimmer, G., 40.
Winton, A. L., 182, 197, 198, 200, 213, 215.
Wohltmann, F., 58.
Wood, J. C., 219, 266.
Wood, T. B., 231, 232, 264, 267, 268, 271.
Woodman, H. E., 46, 47, 249, 250, 251, 272, 277.
Woods, C. D., 24.
Woolcott, F. W., 133.
Working, E. B., 287.

Yoder, P. A., 81, 82.

Zade, 16.
Zak, J., 183.
Zinn, J., 24, 254, 260, 264.

SUBJECT INDEX

Absorption of flour, and flour storage, 191.
Acidity of flour, correlation with ash content and color score, 149; changes on storage, 182; of bleached flour, 212 222.
Aging of flour, 177.
Agrostemma githago, 101, 102.
Aleurometer, 270.
Aleurone layer, 121.
Alkali soils, effect on wheat, 40, 72.
Alsop process of flour bleaching, 200, 202.
Aluminum salts, effect upon dough, 283.
Ambrosia trifida, 101.
Ammonia, anhydrous, in combination with chlorine in flour bleaching, 224.
Ammonium salts, as flour improvers, 224, 232; effect upon viscosity of dough, 280.
Ash content of flour, and flour grades, 148; composition of, 148; correlated with color and acidity, 149; correlated with specific conductivity of flour extracts, 165.
Attrition, effect upon flour properties, 284.
Australian wheat varieties, 22.

Benzoyl peroxide bleaching, 200, 226.
Bibliography, 293-314.
Bin-burned wheat, 100.
Bleaching flour, 127, 199.
Bokhara clover seed in wheat, 103.
Boland aleurometer, 270.
Bolting wheat, 124, 125.
Bran fragments, in different flour grades, 142; and flour color, 195.
Bran proteins, 142.
Breaking wheat in roller milling, 124.
Breeding wheat, 32.
Bromates, as flour improvers, 232; effect on dough viscosity, 280.
Bromine as a bleaching agent, 199.
Buffer value of flour grades, 166; of bleached flour, 223.

Calcium-magnesium ratio in flour grades, 152.
California wheats, 27.
Canadian wheats, 23, 29.

Carbohydrates of wheat plant, 41, 48.
Carotinoid pigments, of flour, 196, 197, 218.
Catalase activity, of germinated wheat, 38; of flour grades, 160.
Chlorine bleaching, generators, 217; effect on acidity of flour, 219; and chlorination of flour fat, 219; effect on baking strength, 221.
Chopin extensimeter, 262, 270, 281.
Chromosome numbers of wheat species, 17.
Class of flour, definition of, 176.
Classes of wheat, 18.
Climate and wheat composition, 57.
Color changes in stored flour, 182.
Color score correlated with ash content and acidity of flour, 149; factors influencing, 194.
Colloidal behavior of gluten, and dough, 267, 278.
Conditioning wheat, 123.
Conductivity of flour extracts, 164; as influenced by chlorine bleaching, 220.
Copper content of wheat products, 152.
Corn cockle, *Agrostemma githago*, 101, 195.
Criteria of flour grade, 142.
Crude gluten, correlation with protein content, 254; quantitative estimation, 255; composition of, 256, 257; colloidal behavior, 267, 268; testing properties of, 270.

Diastatic activity, of flour grades, 158; of stored flour, 189; of bleached flour, 214, 222; and flour strength, 232.
Digestibility of bleached flour, 213.
Dihydrosterol in wheat, 288.
Dimethylaniline hydrochloride as a reagent for detecting flour bleaching, 216.
Diphenylamine as a reagent in testing bleached flour, 216.
Drying wheat, 118.
Durum wheat, genetic classification, 16; sources of, 18; composition and baking properties, 19, 21, 24; comparison of varieties, 29, 31; hybrids with common wheat, 34, 35.

Einkorn, 16, 17.
Electrolytes, and gluten properties, 267, 268; and colloidal behavior of dough, 280.
Electrolytic generation, of nitrogen peroxide, 203; of chlorine, 217.
Electrolytic resistance of flour extracts, 164; as influenced by chlorine bleaching, 220.
Emmer, 16, 17.
Endosperm, structure, 121; percentage of wheat, 46, 122, 123; characteristics in yellow-berry wheat, 104; gluten content in different regions, 140, 142.
English flour nomenclature, 169.
Enzymic activity of flour grades, 158.
Ether extract (see fat, and lipoids)
Evaporation of moisture in milling, 130.
Extensimeter (see Chopin extensimeter)
Extraction, definition of, 176.

Farrer's wheat varieties, 22.
Fats, relation to flour grade, 154, 156; effect upon gluten quality and baking strength, 287; properties of, in wheat and flour, 291, 292.
Fertilizers, and wheat composition, 48, 69.
Fiber, in flour grades, 147.
Fleurent's test for bleached flour, 216.
Flour improvers, 232, 280, 282.
Flour mill streams, 134.
Flour strength, breeding for, 34; definition of, 228, 229; Enzyme phenomena in relation to, 228; correlation with gluten content, 257; and extensibility, 262; and gas retention, 262; and gliadin-glutenin ratio, 263; and colloidal behavior of dough, 278.
Flour yield, formula for computing, 154, 155.
Food administration milling requirements, 95.
French wheats, 29.
Frosted wheat, 86, 112.

Garlic in wheat, 103.
Gasoline color valve, of flour, 197, 198, 200; of durum semolinas, 30.
Gas-retention of dough, a method of determining, 283.
Genetic classification of wheat, 16.
German flour nomenclature, 169.
Germinated wheat, 38, 95, 112.
Gliadin, discovery by Einhof, 242; chemical characteristics, 243; isoelectric points, 245, 272; products of hydrolysis, 246, 247, 248, 267; specific rotation, 249; racemization, 249; molecular weight, 250; colloidal behavior, 272; refractive index dispersion, 273.
Gliadin-glutenin ratio, 263.
Gluten (see also crude gluten), distribution in endosperm, 140, 142; discovery of, by Beccari, 242; composition of, 256, 257; colloidal behavior, 267; testing devices, 269.
Glutenin, isoelectric point, 250; racemization, 251; rotation dispersion, 251; quantitative estimation, 252; molecular weight, 252; products of hydrolysis, 247, 248, 267; colloidal behavior, 274.
Glutenin-gliadin ratio and flour strength, 263.
Grades of flour, 169; definition of, 176.
Granulation, and flour color, 194; and baking strength, 285.
Griess-Ilosvay test for nitrites, 215.
Growing period and wheat composition, 57, 58, 63.
Growth of wheat, 37.
Gum guaiac as a test for benzoyl peroxide, 226.

Hairs, wheat, in flour, 144.
Hankóczy gluten tester, 270.
Hard spring wheat, sources of, 18; composition and baking properties, 19, 21; comparison of varieties, 22.
Hard winter wheats, sources of, 18; composition and baking properties, 19, 21, 26; comparison of varieties, 23, 26.
Harvey and Wood turbidity test of flour strength, 271.
Heat-damaged wheat, 100.
History of milling, 13.
Humidity, and hygroscopic moisture of flour, 180; and flour quality, 132; and milling yields, 133.
Humphries process, 133.
Hydrocyanic acid in wild vetch seed, 102.
Hydrogen ion concentration, changes in stored flour, 184, 187, 192; in bleached flour, 213, 220, 222; relation to diastatic activity, 236; optimum as a function of temperature, 238; relation to viscosity of glutenin suspensions, 275; effect upon dough, 276, 281.
Hydrogen ion concentration of wheat plants, 90.
Hydrogen peroxide as flour bleaching agent, 199.
Hygroscopic moisture, of wheat, 120; of flour, 180, 188.

Idaho wheats, 26.
Imbibitional capacity of glutenin, 274.

Immunological methods in differentiating proteins, 267.
Iodine value of flour fat, 210, 291, 292.
India (British East India) wheats, 29.
Iron, in flour, 84, 88.
Irrigation and wheat composition, 64.
Isoelectric point, of gliadin, 245, 282; of glutenin, 250, 282; of leucosin, 253.

Kanred wheat, 24.
Kernel development in wheat, 42.
Kingheads, *Ambrosia trifida,* 101, 196.
Kneading, effect upon dough properties, 284.
Kota wheat, 25.
Kress-James gluten tester, 270.

Ladd's scheme of flour classification, 172.
Leucosin, chemical characteristics, 252; isoelectric point, 253.
Levulosine in developing wheat kernel, 48.
Lime treatment of damaged wheat, 118.
Lipoids, relation to gluten properties and baking strength, 287.
Losses of moisture in milling, 130.

Magnesium-calcium ratio in flour grades, 152.
Maine wheats, 24.
Manganese content of wheat products, 152.
Marquis wheat, 25, 28.
Mealy wheat (see yellow-berry)
Michigan wheat varieties, 22.
Middlings, flour, 124.
Millers' National Federation, suggested standards for flour, 174.
Mixing of dough, effect upon properties, 284.
Moisture in developing wheat kernel, and protein synthesis, 45, 49, 50.
Moisture and heating of wheat, 111, 112, 115, 116.
Moisture, in roller milling, 128; changes in stored flour, 179.
Molds, and flour storage, 187, 188.
Montana wheats, 26, 239.

Nematode galls, 104.
Nitrite reaction in flour, 206, 207, 208; in flour fat, 210.
Nitrogen fertilizers and wheat composition, 71, 72, 73, 74, 75, 76, 77, 78, 79.
Nitrogen fixation by wheat plant, 40.
Nitrogen peroxide, generation with flaming arc, 200; generation chemically, 203; absorption of flour, 204, 205, 206; effect on color of flour, 200, 209; effect on flour fat, 210; effect on gluten, 211; effect on digestibility and toxicity of flour, 213; methods of detecting in flour, 215.
Nitrogen trichloride bleaching, 200, 224.
Nitrosyl chloride bleaching, 200, 216.
Nucleic acid of wheat, 254.

Ohio wheat varieties, 22.
Over-grinding, effect upon flour properties, 285.
Ozone, in flour bleaching, 199.
Ozonideperoxides as flour bleaching reagents, 227.

Pékar test, 194.
Pentosan content of flour grades, 153.
Pepsin, effect on flour strength, 266.
Per-acids, salts of, as flour improvers, 232.
Peraldehydes, as bleaching reagents, 227.
Peroxozonides in flour bleaching, 227.
Perozonides as bleaching reagents, 227.
Phosphates as flour improvers, 232, 283.
Phosphate fertilizer and wheat composition, 71, 72, 73, 74, 75, 76, 77.
Physico-chemical methods for determining flour grade, 164.
Phytin, and flour grades, 164.
Phytosterols of wheat, 288.
"Pile," as a measure of flour strength, 228.
Potash fertilizers and wheat composition, 73, 74, 75, 76, 77.
Proteins of wheat flour, 242.
Proteolytic activity, of flour grades, 160; in relation to strength, 265.
Purifier, middlings, 15, 126.

Quality constant "b," 275, 276.
Quern, 14.

Racemization, of gliadin, 249, 272; of glutenin, 251, 277.
Rainfall and wheat composition, 58, 61.
Reduction of middlings, 126.
Refractive index of flour extracts, 166.
Regtö dough tester, 270.
Respiration of ripening grain, 53.
Respiration, of stored wheat, 110; of flour, 187.
Roller milling, 121.
Roman milling, 13.
Rotation dispersion, of gliadin, 249, 272; of glutenin, 251, 277.
Rust infected wheat, 89.
Rye in wheat mixtures, 102, 103.

Saccharogenesis in flour doughs, 241.
Saunders' method of computing flour strength, 230.
Selection in wheat breeding, 34.

SUBJECT INDEX

Self-rising flour, detecting grade of, 146.
Semolinas, 124.
Serological tests in wheat classification, 16.
Shade, and wheat composition, 62.
Shaw's test for bleached flour, 216.
Sitosterol in wheat, 288.
Smutty wheat, 101, 104.
Soft red winter wheats, sources of, 18; composition and baking properties, 19, 21, 26; comparison of varieties, 22, 23.
Soil, and wheat composition, 48, 56, 69, 80, 82.
Soil moisture and wheat composition, 62, 65, 66, 67, 68, 78, 83.
Specific gravity of wheat, 106.
Spelt, 16, 17.
Sprouted wheat, 95, 112, 187, 239.
Standards for flour, proposed by Ladd, 172; proposed by Millers' National Federation, 174.
Starch, content as a measure of flour yield, 154, 155; size of granules in relation to flour strength, 234; viscosity of heated paste, 286, 287; percentage of, and flour strength, 261, 262; effect of over-grinding upon, 241, 285.
Stones, mill, 14.
Storage, of flour, 177, 206; of wheat, 108.
Strength of flour (see flour strength)
Structure of wheat kernel, 121.
Sulfur dioxide in flour bleaching, 199.
Sunlight, in flour bleaching, 199.
Sweating of wheat, 108.
Sweet clover seed in wheat, 103.

Temperature, and flour storage, 186, 190, 193.
Temperature and wheat respiration, 112.
Tempering wheat, 123.
Test weight (see weight per bushel)
Toxicity of bleached flour, 214, 223.
Tri-local experiments, and composition of wheat, 60, 61.
Tritico-nucleic acid, 254.

Triticum species, 16, 17.
Trypsin, effect on flour strength, 266.
Turbidity test of flour strength, 271.
Tyrosinase of flour grades, 160, 194.

Ultra violet spectra, of gliadin sols, 273; of glutenin sols, 277.
Utah wheats, 28.

Varieties of wheat, 22.
Vedrödi's classification of flour, 150.
Vetch, wild, *Vicia angustifolia*, 101, 102, 196.
Vicia angustifolia, 101, 102, 196.
Viscosity, of flour mill streams, 142, 168; of gliadin sols, 271; of glutenin suspensions, 274; of dough solutions, 279.

Water-soluble proteins and baking strength, 265.
Water wheels, 14.
Weight changes in stored flour, 177.
Weight of developing wheat kernels, 44, 46, 51, 52.
Weight per bushel and flour yield, 91.
Weil's modification of the Griess-Ilosvay test for nitrites in bleached flour, 215.
Wheat species, 16.
White wheats, sources of, 18; composition and baking properties, 19, 21, 26; comparison of varieties, 22, 27, 28.
Wild vetch, *Vicia angustifolia*, 101, 102, 196.
Wind mills, 14.
Wisconsin wheats, 23.
Woolcott air-conditioner in flour milling, 133.

Xanthophyll, in wheat, 198.

Yeast nutrition and gas production in doughs, 231.
Yellow berry, 33, 104.
Yield, definition of, 176.

Zinc content of wheat products, 152.